Springer Series in Reliability Engineering

Series Editor

Hoang Pham

For further volumes:
http://www.springer.com/series/6917

Maxim Finkelstein · Ji Hwan Cha

Stochastic Modeling for Reliability

Shocks, Burn-in and Heterogeneous Populations

Springer

Maxim Finkelstein
Department of Mathematical Statistics
University of the Free State
Bloemfontein
South Africa

and

Max Planck Institute for Demographic Research
Rostock
Germany

Ji Hwan Cha
Department of Statistics
Ewha Womans University
Seoul
Republic of Korea

ISSN 1614-7839
ISBN 978-1-4471-5027-5 ISBN 978-1-4471-5028-2 (eBook)
DOI 10.1007/978-1-4471-5028-2
Springer London Heidelberg New York Dordrecht

Library of Congress Control Number: 2013936794

Printed on acid-free paper

Springer is part of Springer Science+Business Media (www.springer.com)

To Olga and Veronica

—Maxim Finkelstein

To my loving mother Sul Ja Choi

—Ji Hwan Cha

Preface

This book is about reliability and reliability related stochastics. It focuses on shocks modeling, burn-in and heterogeneous populations. At the first sight, it looks that these three areas of research in stochastic modeling are not so close. However, it turns out that they can be naturally combined in the unified framework and some of the results of this kind have been already reported in our recent publications. Indeed, there is no pure homogeneity of items (industrial or biological) in real life. Therefore, it is only an assumption that makes the corresponding statistical analysis much easier. As most of the real life populations are heterogeneous, taking this property into account in reliability analysis of various problems is only increasing the adequacy of stochastic modeling. Furthermore, all objects are operating in a changing environment. One of the ways to model an impact of this environment is via the external shocks occurring in accordance with some stochastic point processes. We understand the term "shock" in a very broad sense as some instantaneous and potentially harmful event (e.g. electrical impulses of large magnitude, demands for energy in biological objects, insurance claims in finance etc.). Shock models are widely used in practical and theoretical reliability and in the other disciplines as well. Numerous shock models have been studied in the literature during the past 50 years. However, only a few of most recent publications deal with heterogeneous items subject to shock processes. Finally, we also focus on burn-in as a method of elimination of 'weak' items from heterogeneous populations. It is well-known that burn-in can be justified when the failure rate of items is initially decreasing (infant mortality). Heterogeneity of populations is one of the main causes for this remarkable shape of the failure rate. Burn-in is often performed in industry in the accelerated environment and this means that at certain instances shocks can play the role of this environment when the time of burn-in decreases.

Our presentation combines classical and recent results of other authors with our research over the past 5 years. The excellent encyclopedic books [4] and [5] give a broad picture of the modern mathematical reliability theory and also present useful sources of references. Along with the classical text [2], the excellent textbook [6], and a mathematically oriented reliability monograph [1], these books can be considered as complementary or further reading. The recent monograph of one of the authors [3] was also extensively used in this book, especially for the introductory Chap. 2 and for stochastic descriptions of heterogeneous populations.

We hope that our text will be useful for reliability researchers and practitioners and to graduate students in reliability or applied probability. It contains numerous stochastic models that can be of interest to applied mathematicians and statisticians.

This project started in a natural way. One of us was very much interested in his research in mathematical and applied aspects of burn-in, whereas the other published intensively on failure rate modeling for heterogeneous populations and various shocks models. Therefore, at a certain stage we decided to combine our efforts and consider burn-in via shocks and also burn-in for heterogeneous populations. Along with that some theoretical work on shocks modeling was initiated. When the critical mass of the obtained results in these directions reached a certain level, we decided to write them down in the form of the book. Of course, some introductory information had to be added along with classical, well-established results.

Maxim Finkelstein acknowledges the support of the University of the Free State, the National Research Foundation (South Africa) and the Max Planck Institute for Demographic Research (Germany).

Ji Hwan Cha's work was supported by the National Research Foundation of Korea (NRF), grant funded by the Korea government (MEST) (No. 2011-0017338). Ji Hwan Cha acknowledges the support of the Ewha Womans University (Republic of Korea).

We are also grateful to our colleagues, co-workers, and the students of Ji Hwan Cha (Hyunju Lee, Jihyun Kim, Haebyur Nam, and Eunjung Jang). Their support and discussions contributed a lot to this project. Finally, we are indebted to Grace Quinn, Anthony Doyle, and the Springer staff for their editorial work.

November 2012

Maxim Finkelstein
Ji Hwan Cha

References

1. Aven T, Jensen U (1999) Stochastic models in reliability. Springer, New York
2. Barlow RE, Proschan F (1975) Statistical theory of reliability and life testing. Holt, Renerhart and Winston, New York
3. Finkelstein M (2008) Failure rate modeling for reliability and risk. Springer, London
4. Lai CD, Xie M (2006) Stochastic ageing and dependence for reliability. Springer, New York
5. Marshall AW, Olkin I (2007) Life distributions. Springer, New York
6. Rausand M, Hoylandt A (2004) System reliability theory: models and statistical methods, 2nd edn. Wiley, New York

Contents

1 Introduction 1
1.1 Aim and Scope of the Book 1
1.2 Brief Overview 4
References 8

2 Basic Stochastics for Reliability Analysis 9
2.1 Failure Rate 9
2.2 Mean Remaining Lifetime 13
2.3 Monotonicity of the Failure Rate and the MRL Function 15
2.4 Point Processes 19
2.4.1 Characterization of Point Processes 19
2.4.2 Poisson Process 21
2.4.3 Renewal Process 23
2.5 Minimal Repair 25
2.6 General (Imperfect) Repair 30
2.6.1 Virtual Age 30
2.6.2 Models of General Repair 33
2.7 Multivariate Accelerated Life and Proportional Hazards Models 39
2.8 Simplest Stochastic Orders 45
References 47

3 Shocks and Degradation 51
3.1 Degradation as Stochastic Process 52
3.2 Shocks and Shot Noise Process 55
3.3 Asymptotic Properties 57
3.4 Extreme Shock Models 58
3.5 State-Dependent Probability of Termination 60
3.6 Termination with Recovery Time 63
3.7 Two Types of Shocks 66
3.8 Spatial Extreme Shock Model 69
3.9 Shock-Based Theory of Biological Aging 72
References 76

4 Advanced Theory for Poisson Shock Models 79
4.1 The Terminating Shock Process with Independent Wear Increments 80
4.1.1 General Setting 80
4.1.2 Exponentially Distributed Boundary 82
4.1.3 Deterministic Boundary. 85
4.2 History-Dependent Termination Probability 88
4.3 Shot Noise Process for the Failure Rate 94
4.3.1 Shot Noise Process Without Critical Shocks 94
4.3.2 Shot Noise Process with Critical Shocks and Deterioration 96
4.4 Extreme Shock Model with Delayed Termination. 105
4.5 Cumulative Shock Model with Initiated Wear Processes 109
4.6 'Curable' Shock Processes. 113
4.7 Stress–Strength Model with Delay and Cure 117
4.8 Survival of Systems with Protection Subject to Two Types of External Attacks 119
4.9 Geometric Process of Shocks. 126
4.10 Information-Based Thinning of Shock Processes 133
4.10.1 General Setting 133
4.10.2 Formal Description of the Information-Dependent Thinning 136
4.10.3 Stress–Strength Type Classification Model 137
References 140

5 Heterogeneous Populations 143
5.1 Failure Rate of Mixture of Two Distributions 145
5.2 Continuous Mixtures. 148
5.3 Examples. 151
5.3.1 Weibull and Gompertz Distributions. 151
5.3.2 Reliability Theory of Aging. 152
5.4 Mixture Failure Rate for Large t 154
5.5 Mortality Plateaus. 159
5.6 Inverse Problem 160
5.7 The Failure Rate Dynamics in Heterogeneous Populations . . . 162
5.8 Stochastic Intensity for Minimal Repairs in Heterogeneous Populations 170
5.9 Preventive Maintenance in Heterogeneous Populations 176
5.10 Population Mortality at Advanced Ages (Demographic Application) 183
5.10.1 Fixed and Evolving (Changing) Heterogeneity 184
5.10.2 Fixed Heterogeneity 186
5.10.3 Vitality Models and Lifetime Distributions 189
5.11 On the Rate of Aging in Heterogeneous Populations. 194
References 198

6 The Basics of Burn-in . . . 201
6.1 Population Distribution for Burn-in . . . 203
6.2 Optimal Burn-in for Performance Criteria . . . 206
6.3 Optimal Burn-in for Minimizing Costs . . . 217
6.4 Models for Accelerated Burn-in Procedures . . . 221
6.4.1 Failure Rate Model for Accelerated Burn-in Procedure . . . 221
6.4.2 Optimal Burn-in Time . . . 223
6.4.3 Proportional Hazards and Additive Hazards Models . . . 230
6.4.4 Relationships Between the Models . . . 234
References . . . 235

7 Burn-in for Repairable Systems . . . 237
7.1 Burn-in and Maintenance Policies: Initial Models . . . 237
7.1.1 Model 1 . . . 238
7.1.2 Model 2 . . . 239
7.1.3 Model 3 . . . 241
7.2 Burn-in Procedures for General Failure Model . . . 243
7.2.1 Constant Probability Model . . . 244
7.2.2 Time-Dependent Probability Model . . . 251
7.3 Accelerated Burn-in and Maintenance Policy . . . 255
7.3.1 Model 1 . . . 255
7.3.2 Model 2 . . . 257
7.3.3 Model 3 . . . 259
References . . . 260

8 Burn-in for Heterogeneous Populations . . . 261
8.1 Discrete Mixtures . . . 262
8.1.1 Ordered Subpopulations and the Effect of Burn-in . . . 262
8.1.2 Optimal Burn-in Time for Performance Quality Measures . . . 264
8.2 Continuous Mixtures . . . 276
8.2.1 The Effect of Burn-in . . . 276
8.2.2 Optimal Burn-in Time for Performance Quality Measures . . . 279
8.2.3 Examples . . . 283
8.3 Burn-in for Minimizing Risks . . . 286
8.3.1 Burn-in for Avoiding Large Risks: Discrete Mixture . . . 286
8.3.2 Burn-in for Avoiding Large Risks: Continuous Mixture . . . 292
8.3.3 Optimal Burn-in Based on Conservative Measures . . . 294

8.4 Burn-in for Repairable Items 298
8.4.1 Basic Setup 298
8.4.2 Optimal Burn-in for Minimizing Weighted Risks 300
8.4.3 Optimal Burn-in for Minimizing Expected Number of Repairs 304
References 311

9 Shocks as Burn-in 313
9.1 Discrete Mixtures 314
9.1.1 General Setting 314
9.1.2 Optimal Severity for Population Quality Measures . . . 316
9.1.3 Optimal Severity for Minimizing Expected Costs 321
9.2 Continuous Mixtures 325
9.2.1 The Impact of Shocks on Mixed Populations 325
9.2.2 The Impact of Shocks on an Item 329
9.2.3 Shock's Severity 330
9.2.4 The Cost of Burn-in and Optimal Problem 333
9.3 Burn-in for Minimizing Risks 338
9.3.1 Discrete Mixtures 338
9.3.2 Continuous Mixtures 344
9.3.3 Optimal Shock Burn-in Based on Conservative Measures 346
9.4 Burn-in for Systems in Environment with Shocks 349
9.4.1 Strength–Stress Shock Model 350
9.4.2 Optimal Level of Shock's Severity 352
9.4.3 Burn-in Procedure Combining Shock and Conventional Burn-in 356
References 361

10 Stochastic Models for Environmental Stress Screening 363
10.1 Stress–Strength Type ESS Model 364
10.1.1 Stochastic Model for ESS 364
10.1.2 Optimal Severity 371
10.2 ESS Model with Wear Increments 374
10.2.1 Stochastic Model 374
10.2.2 Optimal Severity 380
References 384

Index 385

Chapter 1
Introduction

1.1 Aim and Scope of the Book

As the title suggests, the book is devoted to stochastic models for reliability. This very wide topic is naturally 'censored' by the current research interests of the authors in the field which are: *shock models, burn-in* and *stochastic modeling in heterogeneous populations*. At first sight, it seems that these three areas of research are rather 'independent'. However, it turns out that they can be naturally combined in the unified framework and some of the results of this kind have already been reported in our recent publications. As most of the real-life populations are heterogeneous, taking this property into account in reliability analysis of various problems is only increasing the adequacy of the corresponding modeling. Furthermore, all objects are operating in a changing environment. One of the ways to model an impact of this environment is *via* the external shocks occurring in accordance with some point process (e.g., the Poisson process or the renewal process). By a 'shock' we understand an 'instantaneous', potentially harmful event. Depending on its magnitude, a shock can destroy an operating system (failure), leave it unchanged (as good as old), or, e.g., increase its wear (deterioration) on some increment. Numerous shock models were developed and reported in the reliability-related literature during the past 50 years. However, only a few papers (mostly of the authors) deal with shocks in heterogeneous populations and with shocks as a method of burn-in.

Burn-in is a method of 'elimination' of initial failures in field usage. To burn-in a component or a system means to subject it to a period of simulated use prior to the actual operation. Due to the high failure rate at the early stages of a component's life, burn-in has been widely accepted as an effective method of screening out early failures before systems are actually used in field operation. Under the assumption of decreasing or bathtub-shaped failure rate functions, various problems of determining optimal burn-in have been intensively studied in the literature. In the conventional burn-in, the main parameter of the burn-in procedure is its duration. However, in order to shorten the length of this procedure, burn-in is often performed in an accelerated environment. This indicates that high environmental

M. Finkelstein and J. H. Cha, *Stochastic Modeling for Reliability*,
Springer Series in Reliability Engineering, DOI: 10.1007/978-1-4471-5028-2_1,

stress can be more effective in eliminating weak items from a population. In this case, obviously, the larger values of stress should correspond to the shorter duration of burn-in. By letting the stress to increase, we can end up (as some limit) with very short (negligible) durations, in other words, with *shocks*.

One of the essential features of conventional burn-in is that it is performed for the items with decreasing (at least, initially) failure rate. Indeed, by burning-in items for some time we eventually decrease the failure rate for future usage. One of the main causes that 'force' the failure rate to decrease is heterogeneity of populations of items: the weakest subpopulations are dying out first. When a population consists of ordered (in some suitable stochastic sense) subpopulations, the population failure rate is usually initially decreasing. It can have the bathtub or a more complex shape as well. It turns out that under certain assumptions, burn-in for populations of heterogeneous items can be justified even in the case when the population failure rate is increasing. This counter intuitive finding among others shows the importance of taking into account heterogeneity of the manufactured items.

We consider the positive (non-negative) random variables, which are called *lifetimes*. The time to failure of an engineering component or a system is a lifetime, as is the time to death of an organism. The number of casualties after an accident and the wear accumulated by a degrading system are also positive random variables. Although we deal here mostly with engineering applications, the reliability-based approach to lifetime modeling for organisms is one of the important topics for several meaningful examples and applications in the book. Obviously, the human organism is not a machine, but nothing prevents us from using stochastic reasoning developed in reliability theory for life span modeling of organisms.

An important tool and characteristic for reliability analysis in our book is the failure rate function that describes the lifetime. It is well known that the failure rate function can be interpreted as the probability (risk) of failure in an infinitesimal unit interval of time. Owing to this interpretation and some other properties, its importance in reliability, survival analysis, risk analysis, and other disciplines is hard to overestimate. For example, the increasing failure rate of an object is an indication of its deterioration or aging of some kind, which is an important property in various applications. Many engineering (especially mechanical) items are characterized by the processes of "wear and tear" and, therefore, their lifetimes are described by an increasing failure rate. The failure (mortality) rate of humans at adult ages is also increasing. The empirical Gompertz law of human mortality defines the exponentially increasing mortality rate. On the other hand, the constant failure rate is usually an indication of a non-aging property, whereas a decreasing failure rate can describe, e.g., a period of "infant mortality" when early failures, bugs, etc., are eliminated or corrected. This, as was mentioned, is also very important for justification of burn-in, which is usually performed with items characterized by the decreasing or bathtub failure rate. Therefore, the shape of the failure rate plays an important role in reliability analysis. When the lifetime distribution function $F(t)$ is absolutely continuous, the failure rate $\lambda(t)$ can be defined as $F'(t)/(1-F(t))$. In this case, there exists a simple, well-known exponential representation for $F(t)$ (Sect. 2.1). It defines an important characterization of the distribution function *via* the failure rate $\lambda(t)$. Moreover, the failure rate contains information about the chances of failure of an operating object in the next sufficiently small interval of time. Therefore, the shape of $\lambda(t)$ is often much more informative in the described sense than, for example, the shapes of the distribution function or of the probability density function. On the other hand, the mean remaining lifetime contains information about the remaining life span and in combination with the failure rate creates a useful tool for reliability analysis.

In this text, we consider several generalizations of the 'classical' notion of the failure rate $\lambda(t)$. One of them is the *random failure rate*. Engineering and biological objects usually operate in a random environment. This random environment can be described by a stochastic process $\{Z_t, t \geq 0\}$ (e.g., a point process of shocks) or by a random variable Z as a special case. Therefore, the failure rate, which corresponds to a lifetime T, can also be considered as a stochastic processes $\lambda(t, Z_t)$ or $\lambda(t, Z)$. These functions should be understood conditionally on realizations $\lambda(t|z(u),\ 0 \leq u \leq t)$ and $\lambda(t|Z = z)$, respectively. Similar considerations are valid for the corresponding distribution functions $F(t, Z_t)$ and $F(t, Z)$.

Another important generalization of the conventional failure rate $\lambda(t)$ deals with repairable systems and considers the failure rate of a repairable component as an intensity process (stochastic intensity) $\{\lambda_t, t \geq 0\}$. The 'randomness' of the failure rate in this case is due to random times of repair. Assume for simplicity that the repair action is perfect and instantaneous. This means that after each repair a component is 'as good as new'. Let the governing failure rate for this component be $\lambda(t)$. Then the intensity process at time t for this simplest case of perfect repair is defined as

$$\lambda_t = \lambda(t - T_-),$$

where T_- denotes the random time of the last repair (renewal) before t. Therefore, the probability of a failure in $[t, t + dt)$ is $\lambda(t - T_-)dt$, which should also be understood conditionally on realizations of T_-. This and a more general notion of stochastic intensity for general orderly point processes will be intensively exploited throughout the book.

Our presentation combines classical and recent results of other authors with our research findings of recent years. We discuss the subject mostly using necessary tools and approaches and do not intend to present a self-sufficient textbook on reliability theory. The choice of topics is driven by the research interests of the authors. The excellent encyclopedic books by Lai and Xie [6] and Marshall and Olkin [7] give a broad picture of modern mathematical reliability theory and also present the up-to-date reference sources. Along with the classical text by Barlow and Proschan [2], an excellent textbook by Rausand and Hoylandt [8] and a mathematically oriented reliability monograph by Aven and Jensen [1], these books can be considered the first-choice complementary or further general reading. On the other hand, a useful introduction to burn-in can be found in Jensen and

Petersen [5], whereas numerous relevant facts and results on stochastics for heterogeneous populations are covered in Finkelstein [4].

The book is mostly targeted at researchers and 'quantitative engineers'. The first two chapters, however, can be used by undergraduate students as a supplement to a basic course in reliability. This means that the reader should be familiar with the basics of reliability theory. The other parts can form a basis for graduate courses on shocks modeling, burn-in, and on mixture failure rate modeling for students in probability, statistics, and engineering.

Note that all necessary acronyms and nomenclatures are defined below in the appropriate parts of the text, when the corresponding symbol or abbreviation is used for the first time. For convenience, where appropriate, these explanations are often repeated later in the text as well. This means that each section is self-sufficient in terms of notation.

1.2 Brief Overview

Chapter 2 is devoted to reliability basics and can be viewed as a brief introduction to some reliability notions and results that are extensively used in the rest of the book. We pay considerable attention to the crucial reliability notions of the failure (hazard) rate and the remaining (residual) life functions. The shapes of the failure rate and of the mean remaining life function are especially important for the presentation of chapters devoted to burn-in and heterogeneous populations. On the other hand, sections devoted to basic properties of stochastic point processes are helpful for the presentation of Chaps. 3 and 4 that deal with the theory and applications of shock models. Note that, in this chapter, we mostly consider only those facts, definitions, and properties that are necessary for further presentation and do not aim at a general introduction to reliability theory.

Chapter 3 deals mostly with basic shock models and their simplest applications. Along with discussing some general approaches and results, we present the necessary material for describing our recent results on shocks modeling in Chap. 4. As in the other chapters of this book, we do not intend to perform a comprehensive literature review of this topic, but rather concentrate on notions and results that are vital for further presentation. We understand the term "shock" in a very broad sense as some instantaneous, potentially harmful event (e.g., electrical impulses of large magnitude, demands for energy in biological objects, insurance claims in finance, etc.). It is important to analyze the consequences of shocks for a system (object) that can be *basically* two-fold. First, under certain assumptions, we can consider shocks that can either 'kill' a system, or be successfully survived without any impact on its future performance (as good as old). The corresponding models are usually called the *extreme shock models*, whereas the setting when each shock results in an additive damage (wear) to a system is often described in terms of the *cumulative shock models*. In the latter case, the failure occurs when the cumulative effect of shocks reaches some deterministic or random level and, therefore, this setting is useful for modeling degradation (wear). We first briefly discuss several simplest stochastic models of wear that are helpful in describing basic cumulative shock models. In the rest of the chapter, we mostly consider the basic facts about the extreme and cumulative shock models and also describe several meaningful modifications and applications of the extreme shock modeling.

In Chap. 4, we extend and generalize approaches and results of the previous chapter to various reliability-related settings of a more complex nature. We relax some assumptions of the traditional models except the one that defines the underlying shock process as the nonhomogeneous Poisson process (NHPP). Only in the last section of this chapter, we suggest an alternative to the Poisson process to be called *the geometric point process*. It is remarkable that although the members of the class of geometric processes do not possess the property of independent increments, some shock models for this class can be effectively described without specifying the corresponding dependence structure. The chapter is rather technical in nature, however, the formulation of results is reasonably simple and is illustrated by meaningful examples. In extreme shock models, only an impact of the current, possibly fatal shock is usually taken into account, whereas in cumulative shock models, an impact of the preceding shocks is accumulated as well. In this chapter, we also combine extreme shock models with specific cumulative shock models and derive probabilities of interest, e.g., the probability that the process will not be terminated during a 'mission time'. We also consider some meaningful interpretations and examples.

Chapter 5 deals with heterogeneity in stochastic modeling. Homogeneity of objects is the unique property that is very rare in nature and in industry. Therefore, one can hardly find homogeneous populations in real life, however, most of reliability modeling deals with a homogeneous case. Due to instability of production processes, environmental and other factors, most populations of manufactured items in real life are heterogeneous. Similar considerations are obviously true for biological items (organisms). Neglecting heterogeneity can lead to serious errors in reliability analysis of items and, as a consequence, to crucial economic losses. Stochastic analysis of heterogeneous populations presents a significant challenge to developing mathematical descriptions of the corresponding reliability indices. Mixtures of distributions usually present an effective mathematical tool for modeling heterogeneity, especially when we are interested in the failure rate, which is the conditional characteristic. In heterogeneous populations, the analysis of the shape of the mixture (population) failure rate starts to be even more meaningful. It is well known, e.g., that mixtures of decreasing failure rate (DFR) distributions are always DFR. On the other hand, mixtures of increasing failure rate (IFR) distributions can decrease, at least, in some intervals of time. Note that the IFR distributions are often used to model lifetimes governed by the aging processes. Therefore, the operation of mixing can dramatically change the pattern of population aging, e.g., from positive aging (IFR) to negative aging (DFR). These properties are very important when considering burn-in for heterogeneous populations of manufactured items. In this chapter, we first present a brief survey of results relevant for our further discussion in this and the subsequent chapters.

In the rest of the chapter, some new applications of the mixture failure rate modeling are discussed and basic facts to be used in the subsequent chapters are presented.

In Chap. 6, we introduce the concept of burn-in and review the 'initial research' in this area. Burn-in is a method of elimination of initial failures (infant mortality) in items before they are shipped to customers or put into field operation. It is important to obtain an optimal duration of burn-in, because, if this procedure is too short, then the items with shorter lifetimes will still remain in the population. On the other hand, if the procedure is too long, then it decreases the life spans of items with 'normal' lifetimes and also results in additional costs. By investigating the relationship between the population failure rate and the corresponding performance quality measures, we illustrate how the burn-in procedure can be justified for items with initially decreasing failure rates. First, we review some important 'classical' papers that consider minimization of various cost functions for the given criteria of optimization. Burn-in is generally considered to be expensive and, therefore, the length of burn-in is usually limited. Furthermore, for today's highly reliable products, many latent failures or weak components require a long time to detect or identify. Thus, as stated in Block and Savits [3], for decreasing the length of this procedure, burn-in is often performed in an accelerated environment. Therefore, in the last part of this chapter, we introduce several stochastic models for accelerated burn-in.

Chapter 7 mostly deals with burn-in for repairable items. When a non-repairable item fails during burn-in, and this case was considered in the previous chapter, it is just scraped and discarded. However, when dealing with expensive products or complex devices, the complete product will not be typically discarded because of failure during burn-in, but rather a repair will be performed. Following an influential survey by Block and Savits [3], there has been intensive research on burn-in for repairable systems. The main directions of recent studies include the following: (i) various reliability models which jointly deal with burn-in and maintenance; (ii) burn-in procedures for general failure models; (iii) stochastic models for accelerated burn-in. In this chapter, recent developments on burn-in methodology will be reviewed mainly focusing on the burn-in procedures for minimally repairable systems. The general repair models for burn-in can constitute an interesting and challenging topic for further studies.

Chapter 8 is devoted to burn-in for heterogeneous populations of items. In Chaps. 6 and 7, burn-in procedures for homogeneous populations have been discussed. Burn-in can be usually justified when the failure rate of a population is decreasing or bathtub-shaped. It is well known that heterogeneity of populations is often the reason for the initial decrease in the failure rate. In this chapter, the optimal burn-in procedure is investigated without assuming that the population failure rate is bathtub-shaped. We consider first the mixed population composed of two ordered subpopulations—the subpopulation of the strong items (items with 'normal' lifetimes) and that of the weak items (items with shorter lifetimes). Then the continuous mixture model is also discussed in detail. Our goal is to describe optimization of various characteristics of the performance quality of items after burn-in. It is well known that when the failure rate of a component is increasing there is no need to perform the burn-in procedure and only when it is decreasing or non-monotonic there is a possibility for burn-in. We show that this reasoning is usually valid only for homogeneous populations. However, when we deal with heterogeneous populations the situation can be dramatically different and burn-in can be justified even for increasing failure rates. Furthermore, for heterogeneous populations, there exist the risks of selecting items with poor reliability characteristics (i.e., with large failure rates), which is undesirable in practice. Therefore, to account for this situation, we also develop the special burn-in procedure that minimizes these specific risks.

In Chap. 9 we apply the stochastic theory of shocks described in the previous parts of this book to burn-in modeling. In conventional burn-in, the main parameter of the burn-in procedure is its duration. However, in order to shorten the length of this procedure, burn-in is often performed in an accelerated environment. This indicates that a large environmental stress can be effective in eliminating weak items from a population. In this case, obviously, the larger values of stress should correspond to the shorter duration of burn-in. By letting the stress to increase, we can end up (as some limit) with very short (negligible) durations, in other words, with *shocks*. Then the stress level can be considered as a controllable parameter for the corresponding optimization, which in a loose sense is an analog of the burn-in duration in accelerated burn-in. This general reasoning suggests that 'electrical', 'thermal', and 'mechanical' shocks can be used for burn-in in heterogeneous populations of items. Therefore, in this chapter, we consider shocks (i.e., 'instantaneous' stresses of large level) as a method of burn-in and develop the corresponding optimization model. As in the previous chapters, we also assume that our population is the mixture of stochastically ordered subpopulations. As before, we consider both discrete and continuous mixture models. Under this and some other natural assumptions, we discuss the problem of determining the optimal severity level of a stress. We also develop a burn-in model for items that operate in the environment with shocks. For this we assume that there are two competing risk causes of failure—the 'usual' one (in accordance with aging processes in a system) and environmental shocks. A new type of burn-in via the controlled (laboratory) test shocks is considered and the problem of obtaining the optimal level (severity) of these shocks is investigated as well.

Chapter 10 describes Environmental Stress Screening (ESS) as another (although related to burn-in) method of eliminating weak items. There are different ways of improving reliability characteristics of manufactured items. The most common methodology adopted in the industry, as described in the previous chapters, is burn-in, which is a method of 'elimination' of initial failures (infant mortality). Usually, to burn-in a component or a system means to subject it to a fixed time period of simulated use prior to actual operation. Thus, the 'sufficient condition' for employing the *traditional* burn-in is the initially decreasing failure rate. It should be noted, however, that not all populations of engineering items that contain 'weaker' items to be eliminated exhibit this shape of the failure rate. For example, the 'weakness' of some manufactured items can result from *latent defects* that can create *additional* failure modes. The failure rate in this case is not necessarily decreasing and, therefore, traditional burn-in should not be applied. However, by applying the short-time excessive stress, the weaker items in the population with increasing failure rate can be eliminated by the ESS and, therefore, the reliability characteristics of the population of items that have successfully passed the ESS test can be improved. This is a crucial distinction of the ESS from burn-in. Another important distinction of the considered model from burn-in is that the ESS can also create new defects in items that were previously defect-free. In this chapter, we develop stochastic models for the ESS, analyze its effect on the population characteristics of the screened items, and describe related optimization problems. We assume that, due to substandard materials of faulty manufacturing process, some of the manufactured items are susceptible to additional cause of failure (failure mode), i.e., shocks (such as electrical or mechanical shocks). We define the ESS as a procedure of applying a shock of the controlled magnitude, i.e, a short-time excessive stress.

References

1. Aven T, Jensen U (1999) Stochastic models in reliability. Springer, New York
2. Barlow RE, Proschan F (1975). Statistical theory of reliability and life testing. Holt, Renerhart & Winston, New York
3. Block HW, Savits TH (1997) Burn-in. Stat Sci 12:1–19
4. Finkelstein M (2008) Failure rate modelling for reliability and risk. Springer, London
5. Jensen F, Petersen NE (1982) Burn-in. John Wiley, New York
6. Lai CD, Xie M (2006). Stochastic ageing and dependence for reliability. Springer, New York
7. Marshall AW, Olkin I (2007). Life distributions. Springer, New York
8. Rausand M, Hoylandt A (2004) System reliability theory: models and statistical methods, 2nd edn. Wiley, New York

Chapter 2
Basic Stochastics for Reliability Analysis

In this introductory chapter, we partially follow, revise, and expand the relevant portions of Chaps. 2 and 4 of Finkelstein [25] and also add other material that should be helpful when reading the rest of this book. Therefore, we will often refer to this chapter in the subsequent parts of the text. It covers the notions and some basic properties of the failure rate, the mean residual lifetime, stochastic point processes, minimal and general repair, multivariate accelerated and proportional hazards models and, finally, the simplest stochastic orders.

2.1 Failure Rate

Throughout this book we will use the term "failure rate" which is equivalent to the widely used synonym "hazard rate". The choice of the term is just the matter of taste and habit for us. The importance of this notion to reliability analysis is hard to overestimate. The failure rate defines the probability that an operating object will fail in the next sufficiently small unit interval of time and, therefore, plays an exceptional role in reliability engineering, survival analysis, and other disciplines that mostly deal with positive (nonnegative) random variables. They are often called *lifetimes*. As a random variable, a lifetime is completely characterized by its distribution function. A realization of a lifetime is usually manifested by a failure, death or some other 'end event'. Therefore, information on the probability of failure of an operating item in the next (usually sufficiently small) unit interval of time is really important in reliability analysis. If the failure rate function is increasing, then our object is usually degrading in some suitable stochastic sense. For example, it is well-known that the failure (mortality) rate of adult humans increases exponentially with time, whereas the failure rate of many mechanically wearing devices is often increasing as a power function (Weibull law). Thus, understanding and analyzing the shape of the failure rate is an essential part of reliability and survival analysis.

M. Finkelstein and J. H. Cha, *Stochastic Modeling for Reliability*, Springer Series in Reliability Engineering, DOI: 10.1007/978-1-4471-5028-2_2,

Let $T \geq 0$ be a continuous lifetime random variable with a cumulative distribution function (Cdf)

$$F(t) = \begin{cases} P(T \leq t), & t \geq 0, \\ 0, & t < 0. \end{cases}$$

Unless stated specifically (e.g., in Chap. 4), we will implicitly assume that this distribution is 'proper', i.e., $F^{-1}(1) = \infty$, and that $F(0) = 0$. The support of $F(t)$ will usually be $[0, \infty)$, although other intervals of $\Re_{+} = [0, \infty)$ will also be used especially when considering the limiting behavior of mixture failure rates in Chap. 5. We can view T as some time to failure (death) of a technical device (organism), but other interpretations and parameterizations are possible as well. Inter-arrival times in a sequence of ordered events or the amount of monotonically accumulated damage on the failure of a mechanical item are also relevant examples of 'lifetimes'.

Denote the expectation of the lifetime variable $E[T]$ by m and assume that it is finite, i.e., $m < \infty$. Assume also that $F(t)$ is absolutely continuous and, therefore, the probability density function (pdf) $f(t) = F'(t)$ exists (almost everywhere). In accordance with the definition of $E[T]$ and integrating by parts:

$$m = \lim_{t \to \infty} \int_0^t x f(x) \mathrm{d}x$$

$$= \lim_{t \to \infty} \left[-t\bar{F}(t) + \int_0^t \bar{F}(x) \mathrm{d}x \right],$$

where

$$\bar{F}(t) = 1 - F(t) = P(T > t).$$

Assuming that $0 < \int_0^\infty \bar{F}(x) \mathrm{d}x < \infty$, it is easy to conclude that

$$m = \int_0^\infty \bar{F}(x) \mathrm{d}x, \tag{2.1}$$

which is a well-known fact for lifetime distributions. Thus, the area under the survival curve defines the mean of T.

Let an item with a lifetime T and the Cdf $F(t)$ start operating at $t = 0$ and let it be operable (alive) at time $t = x$. The remaining (residual) lifetime is of significant interest in reliability and survival analysis. Denote the corresponding random variable by T_x. Its Cdf $F_x(t)$ is obtained using the law of conditional probability (on condition that an item is operable at $t = x$), i.e.,

$$F_x(t) = P(T_x \le t) = \frac{P(x < T \le x + t)}{P(T > x)} = \frac{F(x+t) - F(x)}{\overline{F}(x)}. \tag{2.2}$$

Therefore, the corresponding conditional survival probability is given by

$$\overline{F}_x(t) = P(T_x > t) = \frac{\overline{F}(x + t)}{\overline{F}(x)}. \tag{2.3}$$

We have everything in place now for defining the failure rate, which is crucial for reliability analysis and other disciplines that deal with lifetimes. Consider an interval of time $(t, t + \Delta t]$. We are interested in the probability of failure in this interval given that it did not occur before in $[0, t]$. This probability can be interpreted as the risk of failure (or of some other harmful event) in $(t, t + \Delta t]$ given the stated condition. Thus

$$P(t < T \le t + \Delta t | T > t) = \frac{P(t < T \le t + \Delta t)}{P(T > t)} = \frac{F(t + \Delta t) - F(t)}{\overline{F}(t)}.$$

As the pdf $f(t)$ exists, the failure rate is defined as the following limit

$$\lambda(t) = \lim_{\Delta t \to 0} \frac{P(t < T \le t + \Delta t | T > t)}{\Delta t} = \lim_{\Delta t \to 0} \frac{F(t + \Delta t) - F(t)}{\overline{F}(t)\Delta t} = \frac{f(t)}{\overline{F}(t)}. \tag{2.4}$$

Therefore, when $\Delta(t)$ is sufficiently small,

$$P(t < T \le t + \Delta t | T > t) \approx \lambda(t)\Delta t,$$

which gives a very popular and important interpretation of $\lambda(t)\Delta t$ as an *approximate* conditional probability of a failure in $(t, t + \Delta t]$. Note that, the similar product for the density function, $f(t)\Delta t$ defines the corresponding approximate *unconditional* probability of a failure in $(t, t + \Delta t]$. It is very likely that, owing to this interpretation, failure rate plays a pivotal role in reliability analysis, survival analysis and other fields. In actuarial and demographic disciplines, it is usually called the *force of mortality* or the *mortality rate*.

Definition 2.1 The failure rate $\lambda(t)$, which corresponds to the absolutely continuous Cdf $F(t)$, is defined by Eq. (2.4) and is approximately equal to the probability of a failure in a small unit interval of time $(t, t + \Delta t]$ given that no failure has occurred in $[0, t]$.

As $f(t) = F'(t)$, we can view Eq. (2.4) as the first-order differential equation (with respect to $F(t)$ with the initial condition $F(0) = 0$. Integration of this equation results in the *main exponential formula of reliability and survival analysis:*

$$F(t) = 1 - \exp\left(-\int_0^t \lambda(u)\mathrm{d}u\right). \tag{2.5}$$

It is clear now that for the proper distribution,

$$\lim_{t\to\infty}\int_0^t \lambda(u)\mathrm{d}u = \infty,$$

which is the necessary and sufficient condition for an arbitrary positive function to be a failure rate for some proper distribution. The finite limit corresponds to improper distributions that will be considered in Chap. 4 with respect to the cure models (see the relevant definitions in Sect. 4.7).

The importance of Eq. (2.5) is hard to overestimate as it presents a simple characterization of $F(t)$ via the failure rate. Therefore, along with the Cdf $F(t)$ and the pdf $f(t)$, the failure rate $\lambda(t)$ uniquely describes a lifetime T. At many instances, however, especially for lifetimes, this characterization is more convenient, which is often due to the meaningful probabilistic interpretation of the probability $\lambda(t)\Delta t$ and the simplicity of Eq. (2.5).

The failure rate can also be defined for the discrete distributions. Let our random variable T have support $\mathrm{N}^+ = \{1, 2, \ldots\}$. Then the analogue of the density for continuous distributions is the following probability

$$f(n) = P(T = n),\ n = 1, 2, \ldots$$

and the corresponding survival function is

$$\bar{F}(n) = P(T > n) = \sum_{i=n+1}^{\infty} f(i),\ n = 1, 2, \ldots.$$

Similar to (2.4), the discrete failure rate is defined as the following quotient

$$\lambda(n) = \frac{f(n)}{\bar{F}(n-1)} = \frac{\bar{F}(n-1) - \bar{F}(n)}{\bar{F}(n-1)},$$

which is now the (exact) conditional probability of failure at time n given that the failure did not happen before. Therefore, in contrast to $\lambda(t)$, the failure rate of discrete distributions is less or equal to 1. On the other hand, similar to $\lambda(t)$, the necessary and sufficient condition for a sequence $\lambda(n), n \geq 1$ to be a failure rate is

$$\sum_{i=1}^{\infty} \lambda(i) = \infty.$$

Various properties of discrete failure rates can be found, e.g., in Lai and Xie [37]. However, in this book, we will mostly consider the absolutely continuous lifetime distributions.

2.2 Mean Remaining Lifetime

Along with the failure rate, the mean remaining lifetime is also the main reliability characteristic. It turns out (see Eq. 2.10) that, similar to Eq. (2.5), the mean remaining lifetime function also uniquely defines the corresponding Cdf. How much longer will an item of age t survive? This question is vital for reliability analysis, survival analysis, actuarial applications and other disciplines. The distribution of this remaining time is defined by Eq. (2.2), where for the sake of notation, the variable x has been interchanged with the variable t.

Assume that $E[T] \equiv m < \infty$. Denote the mean remaining lifetime (MRL) function by $E[T_t] \equiv m(t)$, $m(0) = m$. It defines the mean lifetime left for an item of age t and plays a crucial role in reliability analysis, survival analysis, demography and other disciplines. In demography, for example, this important population characteristic is called the "life expectancy at time t" and in risk analysis the term "mean excess time" is often used.

Whereas the failure rate function at t provides information on a random variable T about a small interval after t, the MRL function at t considers information about the whole remaining interval (t, ∞) [27]. Therefore, these two characteristics complement each other, and reliability analysis of, e.g., engineering systems is often carried out with respect to both of them. It will be shown in this section that, similar to the failure rate, the MRL function also uniquely defines the Cdf of T and that the corresponding exponential representation is also valid. In accordance with Eqs. (2.1) and (2.3),

$$\begin{aligned} m(t) = E[T_t] &= E[T - t | T > t] \\ &= \int_0^\infty \bar{F}_t(u)\mathrm{d}u \\ &= \frac{\int_t^\infty \bar{F}(u)\mathrm{d}u}{\bar{F}(t)}. \end{aligned} \tag{2.6}$$

Definition 2.2 The MRL function $m(t) = E[T_t]$, $m(0) \equiv m < \infty$, is defined by Eq. (2.6), obtained by integrating the survival function of the remaining lifetime T_t.

In accordance with Eq. (2.3) and exponential representation (2.5), the survival function for T_t can be written as

$$\overline{F}_t(x) = P(T_t > x) = \exp\left\{-\int\limits_t^{t+x} \lambda(u)\mathrm{d}u\right\}, \tag{2.7}$$

which also means that the failure rate that corresponds to the distribution $F_t(x)$ is

$$\lambda_t(x) = \lambda(t + x). \tag{2.8}$$

The first simple observation based on Eq. (2.7) tells us that if the failure rate is increasing (decreasing) in $[0, \infty)$, then (for each fixed $x > 0$) the function $\overline{F}_t(x)$ is decreasing (increasing) in t. Therefore, the MRL function $m(t) = \int_0^\infty \overline{F}_t(x)\mathrm{d}x$ is decreasing (increasing). The inverse is generally not true, i.e., a decreasing $m(t)$ does not necessarily lead to an increasing $\lambda(t)$.

An interesting relationship can be obtained between the MRL and the reciprocal of the failure rate [7]:

$$\begin{aligned} m(t) &= \int\limits_0^\infty \overline{F}_t(u)\mathrm{d}u \\ &= \int\limits_0^\infty \lambda(t+u)\overline{F}(t+u)/\lambda(t+u)\overline{F}(t)\mathrm{d}u \\ &= E\left[\frac{1}{\lambda(T)}|T > t\right]. \end{aligned}$$

Specifically, for $t = 0$,

$$m(0) = E\left[\frac{1}{\lambda(T)}\right],$$

which means that the mean time to failure is the expectation of the reciprocal of the failure rate (in the defined sense). For the exponential distribution with the constant failure rate λ, obviously, $m = 1/\lambda$. Thus, the foregoing relationship for $m(t)$ shows the origin of departures from this simple equality.

Assume that $m(t)$ is differentiable. Differentiation in (2.6) yields

$$\begin{aligned} m'(t) &= \frac{\lambda(t)\int_t^\infty \overline{F}(u)\mathrm{d}u - \overline{F}(t)}{\overline{F}(t)} \\ &= \lambda(t)m(t) - 1. \end{aligned} \tag{2.9}$$

From Eq. (2.9) the following relationship between the failure rate and the MRL function is obtained:

$$\lambda(t) = \frac{m'(t) + 1}{m(t)}.$$

This simple but meaningful equation plays an important role in analyzing the shapes of the MRL and failure rate functions.

The following useful exponential representation for $F(t)$ via the MRL function [compare with (2.5)] also describes the relationship between the MRL function and the reciprocal of the failure rate [40]

$$\overline{F}(t) = \frac{m}{m(t)} \exp\left\{ - \int_0^t \frac{1}{m(u)} \mathrm{d}u \right\}. \quad (2.10)$$

Equation (2.10) can be used for 'constructing' distribution functions when $m(t)$ is specified. Zahedi [48] shows that in this case, differentiable functions $m(t)$ should satisfy the following conditions:

- $m(t) > 0, \ t \in [0, \infty)$;
- $m(0) < \infty$;
- $m'(t) > -1, \ t \in (0, \infty)$;
- $\int_0^\infty \frac{1}{m(u)} \mathrm{d}u = \infty$.

The first condition is obvious. The second means that we are considering distributions with the finite first moment. The third condition is obtained from Eq. (2.8) and states that $\lambda(t)m(t)$ is strictly positive for $t > 0$. Note that, $m(0)\lambda(0) = 0$ when $\lambda(0) = 0$. The last condition states that $F(t)$ is a proper distribution as $\lim_{t \to \infty} \overline{F}(t) = 0$ in this case.

2.3 Monotonicity of the Failure Rate and the MRL Function

Monotonicity properties of the failure rate and the MRL functions are important in different applications. As the failure rate defines the conditional probability of failure in $(t, t + \mathrm{d}t]$, the shape of this function can describe the aging properties of the corresponding distributions, which are crucial for modeling at many instances.

Survival and failure data are frequently modeled by monotone failure rates. This may be inappropriate when, e.g., the course of a disease is such that the mortality reaches a peak after some finite interval of time and then declines [28]. In such case, the failure rate has an upside-down bathtub (UBT) shape and the data should be analyzed with the help of, e.g., lognormal or inverse Gaussian distributions. On the other hand, many engineering devices possess a period of 'infant mortality' when the failure rate declines in an initial time interval, reaches a minimum, and then increases. In such a case, the failure rate has a bathtub (BT) shape and can be modeled, e.g., by mixtures of distributions (see Chap. 5).

If $\lambda(t)$ increases (decreases) in time, then we say that the corresponding distribution belongs to the increasing (decreasing) failure rate [IFR (DFR)] class. These are the simplest nonparametric classes of aging distributions. Unless stated specifically, as usual, by increasing (decreasing) we understand nondecreasing (nonincreasing). On the other hand, as already mentioned, the increasing (decreasing) failure rate results in the decreasing (increasing) MRL function (DMRL and IMRL classes, respectively).

It is well-known that the lognormal and the inverse Gaussian distributions have a UBT failure rate. We will see in Chap. 5 that many mixing models with an increasing baseline failure rate result in the UBT shape of the mixture (observed) failure rate. For example, mixing in a family of increasing (as a power function) failure rates (the Weibull law) 'produces' the UBT shape of the observed failure rate. From this point of view, the BT shape is 'less natural' and often results as a combination of different standard distributions defined for different time intervals. For example, infant mortality in $[0, t_0]$ is usually described by some DFR distribution in this interval, whereas the wear out in (t_0, ∞) is modeled by an IFR distribution. However, mixing of specific distributions can also result in the BT shape of the failure rate as, e.g., in Navarro and Hernandez [43].

It turns out that the function

$$g(t) = -\frac{f'(t)}{f(t)} \tag{2.11}$$

appears to be extremely helpful in the study of the shape of the failure rate $\lambda(t) = f(t)/\overline{F}(t)$. This function contains useful information about $\lambda(t)$ and is much simpler because it does not involve $\overline{F}(t)$. In particular, the shape of $g(t)$ often defines the shape of $\lambda(t)$ [28].

The rationale behind this statement becomes apparent when $\lim_{t\to\infty} f(t) = 0$. Indeed, by using L'Hopital's rule: $\lim_{t\to\infty} \lambda(t) = \lim_{t\to\infty} f(t)/\overline{F}(t) = \lim_{t\to\infty} -f'(t)/f(t)$.

The following theorem is a 'more modern' variation of the famous result by Glaser [31].

Theorem 2.1 [38]. *Let the density $f(t)$ of a lifetime random variable be strictly positive and differentiable on $(0, \infty)$, such that $\lim_{t\to\infty} f(t) = 0$. Then*

(i) *If $g(t)$ is increasing, then the failure rate $\lambda(t)$ is also increasing.*
(ii) *If $g(t)$ is decreasing, then $\lambda(t)$ is also decreasing.*
(iii) *If there exists t_1 for which $g(t)$ is decreasing in $t \le t_1$ and increasing in $t \ge t_1$, then there exists t_2 ($0 \le t_2 \le t_1$), such that $\lambda(t)$ is decreasing in $t \le t_2$ and increasing in $t \ge t_2$.*
(iv) *If there exists t_1 for which $g(t)$ is increasing in $t \le t_1$ and decreasing in $t \ge t_1$, then there exists t_2 ($0 \le t_2 \le t_1$), such that $\lambda(t)$ is increasing in $t \le t_2$ and decreasing in $t \ge t_2$.*

This important theorem states that monotonicity properties of $\lambda(t)$ are defined by those of $g(t)$, and because $g(t)$ is often much simpler than $\lambda(t)$, its analysis is more convenient. The simplest meaningful example is the standard normal distribution. Although it is not a lifetime distribution, the application of Theorem 2.1 is very impressive in this case. Indeed, the failure rate of the normal distribution does not have an explicit expression, whereas the function $\eta(t)$, as can easily be verified, is very simple:

$$g(t) = (t - \mu)/\sigma^2,$$

where μ and σ are the corresponding mean and the standard deviation, respectively. Therefore, as $g(t)$ is increasing, the failure rate is also increasing, which is a well-known fact for the normal distribution. Note that Gupta and Warren [30] generalized Glaser's theorem to the case where $\lambda(t)$ has two or more turning points.

Example 2.1 Failure Rate of the Lognormal Distribution.

A random variable $T \geq 0$ follows the lognormal distribution if $Y = \ln T$ is normally distributed. Therefore, we assume that Y is $N(\alpha, \sigma^2)$, where α and σ^2 are the mean and the variance of Y, respectively. The Cdf in this case is given by

$$F(t) = \Phi\left\{\frac{\ln t - \alpha}{\sigma}\right\}, \ t \geq 0,$$

where, as usual, $\Phi(\cdot)$ denotes the standard normal distribution function. The pdf is given by

$$f(t) = \frac{\exp\left\{-\frac{(\ln t - \alpha)^2}{2\sigma^2}\right\}}{(t\sqrt{2\pi}\sigma)},$$

and it can be shown [37] that the failure rate is

$$\lambda(t) = \frac{1}{t\sqrt{2\pi}\sigma} \frac{\exp\left\{-\frac{(\ln a t)^2}{2\sigma^2}\right\}}{1 - \Phi\left\{\frac{\ln a t}{\sigma}\right\}}, \ a \equiv \exp\{-\alpha\}.$$

The function $g(t)$ for the lognormal distribution is

$$g(t) = -\frac{f'(t)}{f(t)} = \frac{1}{\sigma^2 t}(\sigma^2 + \ln t - \alpha).$$

It can be shown that $g(t) \in \text{UBT}$ [37] and taking into account that

$$\lim_{t \to 0} \lambda(t) = 0, \ \lim_{t \to \infty} \lambda(t) = 0$$

it can be concluded that $\lambda(t) \in \text{UBT}$ as well.

then $m(t)$is decreasing;

- *If $m'(0) > 0$, then $m(t) \in UBT$.*

Let $\lambda(t)$ be a differentiable UBT failure rate in $[0, \infty)$.

- *If $m'(0) \geq 0$, then $m(t)$ is increasing;*
- *If $m'(0) < 0$, then $m(t) \in BT$.*

Corollary 2.1 *Let $\lambda(0) = 0$. If $\lambda(t)$ is a differentiable UBT failure rate, then $m(t)$ has a bathtub shape.*

Example 2.2 [29] Consider a lifetime distribution with $\lambda(t) \in$ BT, $t \in [0, \infty)$ of the following specific form:

$$\lambda(t) = \frac{(1 + 2.3t^2) - 4.6t}{1 + 2.3t^2}.$$

It can easily be obtained using Eq. (2.6) that the corresponding MRL is

$$m(t) = \frac{1}{1 + 2.3t^2},$$

which is a decreasing function. Obviously, the condition $\lambda(0) \leq 1/m(0)$ is satisfied.

2.4 Point Processes

Applied probabilistic analysis of point processes and, specifically, of shock processes is one of the main topics of this book. Various shock models are considered in most of the subsequent chapters. Therefore, in this introductory chapter, we discuss relevant properties of the point processes that are used throughout our book.

2.4.1 Characterization of Point Processes

The randomly occurring time points (instantaneous events) can be described by a stochastic point (counting) process $\{N(t),\ t \geq 0\}$ with a state space $\{0, 1, 2, \ldots\}$. For any $s, t \geq 0$ with $s < t$, the increment

$$N(s, t) \equiv N(t) - N(s)$$

is equal to the number of points that occur in $[s, t)$ and $N(s) \leq N(t)$ for $s \leq t$.

Assume that our process is orderly, which means that there are no multiple occurrences, i.e., the probability of the occurrence of more than one event in a small interval of length Δt is $o(\Delta t)$. Assuming the limits exist, the rate of this process $\lambda_r(t)$ is defined as

$$\begin{aligned}\lambda_r(t) &= \lim_{\Delta t \to 0} \frac{P(N(t, t + \Delta t) = 1)}{\Delta t} \\ &= \lim_{\Delta t \to 0} \frac{E[N(t, t + \Delta t)]}{\Delta t}\end{aligned}$$

We use a subscript r here, which stands for "rate", to avoid confusion with the notation for the 'ordinary' failure rate of an item $\lambda(t)$. However, in the forthcoming chapters, where it does not lead to confusion, the corresponding notation will be $\lambda(t)$ or $v(t)$. Thus, $\lambda_r(t)\mathrm{d}t$ can be interpreted as an approximate probability of an event occurrence in $[t + \mathrm{d}t)$. The mean number of events in $[0, t)$ is given by the cumulative rate

$$E[N(0, t)] \equiv \Lambda_r(t) = \int_0^t \lambda_r(u)\mathrm{d}u.$$

The rate $\lambda_r(t)$ does not completely define the point process and, therefore, a more detailed description should be used for this type of characterization. The heuristic definition of the corresponding stochastic process that is sufficient for our presentation (see [2, 3] for mathematical details) is as follows.

Definition 2.3 An intensity process (stochastic intensity) $\lambda_t, t \geq 0$ of an orderly point process $N(t),\ t \geq 0$ is defined as the following limit:

$$\begin{aligned}\lambda_t &= \lim_{\Delta t \to 0} \frac{P(N(t, t + \Delta t) = 1 | \mathrm{H}_{t-})}{\Delta t} \\ &= \lim_{\Delta t \to 0} \frac{E[N(t, t + \Delta t) | H_{t-}]}{\Delta t},\end{aligned} \tag{2.12}$$

where $\mathrm{H}_{t-} = \{N(s) : 0 \leq s < t\}$ is an internal filtration (history) of the point process in $[0, t)$, i.e., the set of all point events in $[0, t)$.

This definition can be written in a compact form via the following conditional expectation:

$$\lambda_t \mathrm{d}t = E[dN(t) | \mathrm{H}_{t-}]. \tag{2.13}$$

Thus the deterministic rate $\lambda_r(t)$ 'turns into' the corresponding stochastic process. More precisely: the rate of the orderly point process $\lambda_r(t)$ can be viewed as the expectation of the intensity process $\lambda_t, t \geq 0$ over the entire space of possible histories, i.e., $\lambda_r(t) = E[\lambda_t]$. Note that the term "complete intensity function" for $\lambda_r(t)$ is also sometime used in the literature (e.g., Cox and Isham [15]).

The intensity process completely defines (characterizes) the corresponding point process. We will consider several meaningful examples of $\lambda_t,\ t \geq 0$ in the subsections to follow.

Relation (2.13) can be also written as

$$E[dN(t) - \lambda_t \mathrm{d}t | \mathrm{H}_{t-}] = 0. \tag{2.14}$$

Thus, if we define the process

$$M(t) = N(t) - \int_0^t \lambda_s \mathrm{d}s,$$

Eq. (2.14) can be rewritten as

$$E[dM(t) | \mathrm{H}_{t-}] = 0,$$

which is the intuitive definition of the martingale (see, e.g., Aalen et al. [1]). Thus, the intuitive definition of the intensity process (2.13) is equivalent to asserting that the counting process minus the cumulative intensity process,

$$\Lambda_t = \int_0^t \lambda_s \, \mathrm{d}s$$

is a martingale.

2.4.2 Poisson Process

The simplest point process is where the points occur 'totally randomly'. The following definition is formulated in terms of conditional characteristics and is equivalent to the standard definitions of the Poisson process [44].

Definition 2.4 The nonhomogeneous Poisson process (NHPP) is an orderly point process such that its intensity process is equal to the rate, i.e.,

$$\lambda_t = \lambda(t | \mathrm{H}_{t-}) = \lambda_r(t). \tag{2.15}$$

Obviously, the property of independent increments holds automatically for this process. When $\lambda_r(t) \equiv \lambda_r$, the process is called the homogeneous Poisson process, or just the Poisson process. The number of events in any interval of length d is given by

$$\Pr[N(d) = n] = \exp\{-\Lambda_r(d)\} \frac{(\Lambda_r(d))^n}{n!}, \tag{2.16}$$

where $\Lambda_r(t) = \int_0^t \lambda_r(u)\mathrm{d}u$ is the cumulative rate. The distribution of time since $t = x$ up to the next event, in accordance with Eq. (2.3), is

$$F(t|x) = 1 - \exp\left\{ -\int_x^{x+t} \lambda_r(u)\mathrm{d}u \right\}. \tag{2.17}$$

Therefore, the time to the first event for a Poisson process that starts at $t = 0$ is described by the Cdf with the failure rate $\lambda_r(t)$.

Let the arrival times in the NHPP with rate $\lambda_r(t)$ be denoted by S_i, $i = 1, 2, \ldots$, $S_0 = 0$. The following remarkable property will be used extensively in Chap. 4. Consider the time-transformed process with arrival times

$$\tilde{S}_0 = 0, \quad \tilde{S}_i = \Lambda_r(S_i) \equiv \int_0^{S_i} \lambda_r(u)\mathrm{d}u.$$

It can be shown that the process defined by $\tilde{S}_i, i = 1, 2, \ldots$ is a homogeneous Poisson process with the rate equal to 1, i.e., $\tilde{\lambda}_r(t) = 1$. This can be described formally by the following theorem:

Theorem 2.5 [18]. *Let $\Lambda_r(t)$, $t \geq 0$ be a positive-valued, continuous, non-decreasing function. Then the random variables S_i, $i = 1, 2, \ldots$, $S_0 = 0$ are the arrival times corresponding to a nonhomogeneous Poisson process with the cumulative rate $\Lambda_r(t)$ if and only if $\Lambda_r(S_i)$ are the arrival times corresponding to a homogeneous Poisson process with rate* 1.

The importance of this result in reliability applications is hard to overestimate. While considering various shock models, we will use this theorem in combination with the following result:

Theorem 2.6 [16] Let S_i, $i = 1, 2, \ldots$, $S_0 = 0$ *be the arrival times of a nonhomogeneous Poisson process with a continuous cumulative rate function* $\Lambda_r(t)$ *Then, conditional on the number of events* $N(t_0) = n$, *the arrival times* S_i, $i = 1, 2, \ldots$ *are distributed as order statistics from a sample with distribution function* $F(t) = \Lambda(t)/\Lambda(t_0)$ for $t \in [0, t_0]$.

Finally, we will briefly describe the operation of thinning of the Poisson process, which will be also studied in Chap. 4 in a much more general setting. Assume that a function $\lambda_r(t)$ is bounded by the rate of the homogeneous Poisson process, i.e, $\lambda_r(t) \leq \lambda_r < \infty$. Suppose now that each event from the process with rate λ_r is counted with probability $\lambda_r(t)/\lambda$, then the resulting thinned process of counted events is the nonhomogeneous Poisson process with rate $\lambda_r(t)$ [44]. This operation can be generalized to the case when the initial nonhomogeneous Poisson process with rate $\lambda_r(t)$ is thinned with the time-dependent probability $p(t)$, which results in the thinned process with rate $p(t)\lambda_r(t)$(see also the Brown-Proschan model of Sect. 2.5).

2.4.3 Renewal Process

As the renewal process is the main tool and the basis for probabilistic analysis of repairable items, we will consider this process in more detail.

Let $\{X_i\}_{i\geq 1}$ denote a sequence of i.i.d. lifetime random variables with common Cdf $F(t)$. Therefore, X_i, $i\geq 1$ are the copies of some generic X. Let the corresponding arrival times be defined as

$$S_0 = 0, \quad S_n = \sum_1^n X_i,$$

where X_i can also be interpreted as the interarrival times or cycles, i.e., times between successive renewals. Obviously, this setting corresponds to perfect, instantaneous repair. Define the corresponding point process as

$$N(t) = \sup\{n : S_n \leq t\} = \sum_1^\infty I(S_n \leq t),$$

where, as usual, the indicator is equal to 1 if $S_n \leq t$ and is equal to 0 otherwise.

Definition 2.5 The described counting process $\{N(t),\ t\geq 0\}$ and the point process $S_n, n = 0, 1, 2, \ldots$ are both called *renewal processes*.

The rate of the process defined by Eq. (2.12) is called the *renewal density function* in this specific case. Denote this function by $h(t)$. Similar to the general setting, the corresponding cumulative function defines the mean number of events (renewals) in $[0, t)$, i.e.,

$$H(t) = E[N(t)] = \int_0^t h(u)\mathrm{d}u.$$

The function $H(t)$ is called the *renewal function* and is the main object of study in renewal theory. This function also plays an important role in different applications, as, e.g., it defines the mean number of repairs or overhauls of equipment in $[0, t)$. Applying the operation of expectation to $N(t)$ results in the following relationship for $H(t)$:

$$H(t) = \sum_1^\infty F^{(n)}(t), \tag{2.18}$$

where $F^{(n)}(t)$ denotes the n-fold convolution of $F(t)$ with itself. Assume that $F(t)$ is absolutely continuous and, therefore, the density $f(t)$ exists. Denote by

Glaser's approach was generalized by Block et al. [12] by considering the ratio of two functions $G(t) = N(t)/D(t)$, where the functions on the right-hand side are continuously differentiable and $D(t)$ is positive and strictly monotone. Similar to (2.11), we define the function $g(t)$ as

$$g(t) = \frac{N'(t)}{D'(t)}.$$

These authors show that the monotonicity properties of $G(t)$ are 'close' to those of $g(t)$. Consider, for example, the MRL function

$$m(t) = \frac{\int_t^\infty \overline{F}(u)\mathrm{d}u}{\overline{F}(t)}.$$

We can use it as $G(t)$. It is remarkable that $g(t)$ in this case is simply the reciprocal of the failure rate, i.e.,

$$g(t) = \frac{\overline{F}(t)}{f(t)} = \frac{1}{\lambda(t)}.$$

Therefore, the functions $m(t)$ and $1/\lambda(t)$ can be close in some suitable sense, as already stated before.

Glaser's theorem defines sufficient conditions for BT (UBT) shapes of the failure rate. The next theorems (see [25] for the proofs) establish important relationships between the shapes of $\lambda(t)$ and $m(t)$. The first one is obvious and, in fact, has already been mentioned before.

Theorem 2.2 *If $\lambda(t)$ is increasing then $m(t)$ is decreasing.*

Thus, a monotone failure rate always corresponds to a monotone MRL function. The inverse is true only under additional conditions.

Theorem 2.3 *Let the MRL function $m(t)$ be twice differentiable and the failure rate $\lambda(t)$ be differentiable in $(0, \infty)$. If $m(t)$ is decreasing (increasing) and is a convex (concave) function, then $\lambda(t)$ is increasing (decreasing).*

Theorem 2.3 gives the sufficient conditions for the monotonicity of the failure rate in terms of the monotonicity of $m(t)$. The following theorem generalizes the foregoing results to a non-monotone case [25, 29, 41]. It states that the BT (UBT) failure rate under certain assumptions can correspond to a monotone MRL function (compare with Theorem 2.3, which gives a simpler correspondence rule).

Theorem 2.4 *Let $\lambda(t)$ be a differentiable BT failure rate in $[0, \infty)$.*

- *If*

$$m'(0) = \lambda(0)m(0) - 1 \leq 0,$$

$$\lambda_t = \sum_{n \geq 0} \lambda(t - S_n) I(S_n \leq t < S_{n+1}), \quad t \geq 0, \tag{2.24}$$

where $\text{H}_{t-} = 0 \leq S_1 < S_2 < \ldots < S_{N(t)}$ is the history of the renewal process in $[0, t)$ Thus, at each fixed t, the intensity process can also be compactly written as $\lambda(t - S_{N(t)})$, where $S_{N(t)}$ is the random time of the last renewal. This means that the whole history of the process in this case reduces only to the time since the last renewal. In fact, this simplification makes the process mathematically tractable.

In contrast to the Poisson process, when the underlying Cdf $F(t)$ is nonexponential, the renewal process does not possess the Markov property and, therefore, its increments are not independent. However, the Markov property is preserved only at renewal times, as the process restarts after each renewal.

Asymptotic behavior of renewal processes is also usually of interest in different applications. A well-known result [44] states the intuitively expected asymptotic properties for the renewal function and the renewal density function as $t \to \infty$, i.e.,

$$H(t) = \frac{t}{m}[1 + o(1)], \quad h(t) = \frac{1}{m}[1 + o(1)], \tag{2.25}$$

where we assume that $E[X] = m < \infty$ exists. Thus, in contrast to the Poisson process with the rate defined by an 'arbitrary' function $\lambda_r(t)$, the rate of the renewal process tends to a constant as $t \to \infty$.

2.5 Minimal Repair

The renewal points of the renewal process can be interpreted as instants of perfect repair of a repairable system. But in reality the repair is usually not perfect. Therefore, researches came up with different models of imperfect repair. The first in this row was the, so-called, minimal repair. The concept of minimal repair is crucial for analyzing the performance and maintenance policies of repairable systems. It will be also of prime interest for burn-in and heterogeneity modeling of the forthcoming chapters of this book. It is the simplest and the best understood type of imperfect repair in applications. Minimal repair was introduced by Barlow and Hunter [8] and was later studied and applied in numerous publications devoted to modeling of repair and maintenance of various systems. It was also independently used in bio-demographic studies [47].

The term *minimal repair* is meaningful. In contrast to an overhaul (perfect repair), it usually describes a minor maintenance or repair operation. The mathematical definition is as follows.

Definition 2.6 The survival function of an item (with the Cdf $F(t)$ and the failure rate $\lambda(t)$) that had failed and was instantaneously minimally repaired at age x is

$$\frac{\overline{F}(x+t)}{\overline{F}(x)} = \exp\left\{ -\int\limits_{x}^{x+t} \lambda(u)\mathrm{d}u \right\}. \tag{2.26}$$

In accordance with Eq. (2.3), this is exactly the survival function of the remaining lifetime of an item of age x. Therefore, the failure rate just after the minimal repair is $\lambda(x)$, i.e., the same as it was prior the repair. This means that minimal repair does not change anything in the future stochastic behavior of an item, as if a failure did not occur. It is often described as the repair that returns an item to the state it had been in prior to the failure. Sometimes this state is called *as bad as old*. The term *state* should be clarified. In fact, the state in this case depends only on the time of failure and does not contain any additional information. Therefore, this type of repair is sometimes referred to as *statistical* or *black box* minimal repair [10, 25]. However, to comply with tradition, we will use the term *minimal repair* (without adding "statistical") for the operation described by Definition 2.6.

Comparison of (2.26) with (2.17) results in the important conclusion that the process of minimal repairs is a nonhomogeneous Poisson process with rate $\lambda_r(t) = \lambda(t)$. Therefore, in accordance with Eq. (2.15), the intensity process $\lambda_t, t \geq 0$ that describes the process of minimal repairs that is 'performed on an item' with the failure rate $\lambda(t)$ is also deterministic, i.e., $\lambda_t = \lambda(t)$.

There are two popular interpretations of minimal repair. The first one was introduced to mimic the behavior of a large system of many components when one of the components is perfectly repaired (replacement). It is clear that in this case the performed repair operation can be approximately qualified as a minimal repair. We must assume additionally that the input of the failure rate of this component in the failure rate of the system is sufficiently small.

The second interpretation describes the situation where a failed system is replaced by a statistically identical one, which was operating in the same environment but did not fail. The following example interprets in terms of minimal repairs the meaningful notion of a deprivation of life that is used in demographic literature.

Example 2.3 Let us think of any death in $[t, t + \mathrm{d}t)$, whether from accident, heart disease, or cancer, as an 'accident' that deprives the person involved of the remainder of his expectation of life [33], which in our terms is the MRL function $m(t)$, defined by Eq. (2.6). Suppose that everyone is saved from death once but thereafter is unprotected and is subject to the usual mortality in the population. Then the average deprivation can be calculated as

$$D = \int\limits_{0}^{\infty} f(u)m(u)\mathrm{d}u,$$

where $f(t)$ is the density which corresponds to the Cdf $F(t)$. In our terms, D is the mean duration of the second cycle in the process of minimal repair with rate $\lambda(t)$. Note that the mean duration of the first cycle is $m(0) = m$. The case of several additional life chances or, equivalently, subsequent minimal repairs is considered in Vaupel and Yashin [47]. These authors show that the mortality (failure) rate with a possibility of n minimal repairs is

$$\lambda_n(t) = \lambda(t)\frac{\Lambda^n(t)}{n!\sum_{r=0}^{n}\frac{\Lambda^r(t)}{r!}},$$

where $\lambda(t)$ is the mortality rate without possibility of minimal repairs.

Example 2.3 deals with the limited number of minimal repairs. Another option is to consider the situations when this number is limited in some probabilistic way, e.g., in terms of relevant expectations. The meaningful example of this is the Brown-Proschan model. As it was already stated, real-life repair is neither perfect nor minimal. It is usually intermediate in some suitable sense. Note that it can even be worse than a minimal repair (e.g., correction of a software bug can result in new bugs).

One of the first imperfect repair models was suggested by Beichelt and Fischer [9] (see also [13]). This model combines minimal and perfect repairs in the following way. An item is put into operation at $t = 0$. Each time it fails, a repair is performed, which is perfect with probability p and is minimal with probability $1 - p$. Thus, there can be $k = 0, 1, 2, \ldots$ imperfect repairs between two successive perfect repairs. The sequence of i.i.d. times between consecutive perfect repairs $X_i, i = 1, 2, \ldots$, as usual, forms a renewal process.

The Brown–Proschan model was extended by Block et al. [11] to an age-dependent probability $p(t)$, where t is the time since the last perfect repair. Therefore, each repair is perfect with probability $p(t)$ and is minimal with probability $1 - p(t)$. Denote by $F_p(t)$ the Cdf of the time between two consecutive perfect repairs. Assume that

$$\int_0^\infty p(u)\lambda(u)\mathrm{d}u = \infty, \tag{2.27}$$

where $\lambda(t)$ is the failure rate of our item. Then

$$F_p(t) = 1 - \exp\left\{-\int_0^t p(u)\lambda(u)\mathrm{d}u\right\}. \tag{2.28}$$

Note that Condition (2.27) ensures that $F_p(t)$ is a proper distribution ($F_p(\infty) = 1$). Thus, the failure rate $\lambda_p(t)$ that corresponds to $F_p(t)$ is given by the following meaningful and simple relationship:

$$\lambda_p(t) = p(t)\lambda(t). \tag{2.29}$$

The formal proof of (2.28–2.29) can be found in Beichelt and Fischer [9] and Block et al. [11]. On the other hand, the following simple general reasoning leads to the same result. Let an item start operating at $t = 0$ and let T_p denote the time to the first perfect repair. We will now 'construct' the failure rate $\lambda_p(t)$ in a direct way. Owing to the properties of the process of minimal repairs, we can *reformulate* the described model in a more convenient way that will be frequently used in the next chapter. Assume that events are arriving in accordance with the NHPP with rate $\lambda(t)$. Each event independently from the history 'stays in the process' with probability $q(t) = 1 - p(t)$ and *terminates* the process with probability $p(t)$. Therefore, the random variable T_p can now be interpreted as the time to termination of our point process. The intensity process that corresponds to the NHPP is equal to its rate and does not depend on the history H_{t-} of the point process of minimal repairs. Moreover, owing to our assumption, the probability of termination also does not depend on this history. Therefore,

$$\lambda_p(t)\mathrm{d}t = P(T_p \in [t, t + \mathrm{d}t)|\mathrm{H}_{t-}, T_p \geq t) = p(t)\lambda(t)\mathrm{d}t. \tag{2.30}$$

On the other hand, as we will frequently use the similar reasoning (in more advanced settings) in the next chapter, it is reasonable to present the formal, detailed proof of Eqs. (2.28–2.29). We will derive the distribution of time to termination of the process. As it was stated, the process of minimal repairs (before termination) is the nonhomogeneous Poisson process, $\{N(t), t \geq 0\}$ with rate $\lambda(t)$. Thus, denoting the arrival times by $T_i, i = 1, 2, \ldots$, the cumulative rate by $\Lambda(t) = E[N(t)] = \int_0^t \lambda(u)\mathrm{d}u$ and conditioning on this process (in each realization) gives

$$P(T_p \geq t|N(s), 0 \leq s < t) = \prod_{i=0}^{N(t)} q(T_i),$$

where $q(T_0) \equiv 1$ corresponds to the case when $N(t) = 0$. Then the corresponding expectation is

$$P(T_p \geq t) = E[\prod_{i=1}^{N(t)} q(T_i)].$$

Define $N^*(t) \equiv N(\Lambda^{-1}(t)),\ t \geq 0$, and $T_j^* \equiv \Lambda(T_j), j \geq 1$. As follows from Theorem 2.5, $\{N^*(t),\ t \geq 0\}$. is a stationary Poisson process with rate 1 and $T_j^*, j \geq 1$, are the times of occurrence of events in the new time scale. Let $s = \Lambda(t)$.

Then

$$E[\prod_{i=1}^{N(t)} q(T_i)] = E[\prod_{i=1}^{N^*(s)} q(\Lambda^{-1}(T_i^*))] = E[E[\prod_{i=1}^{N^*(s)} q(\Lambda^{-1}(T_i^*))|N^*(s)]].$$

The joint distribution of $(T_1^*, T_2^*, \cdots, T_n^*)$ given $N^*(s) = n$ is the same as the joint distribution of $(V_{(1)}, V_{(2)}, \cdots, V_{(n)})$, where $V_{(1)} \le V_{(2)} \le \cdots \le V_{(n)}$ are the order statistics of i.i.d. random variables $V_1, V_2, \cdots, V_n$ which are uniformly distributed in the interval $[0, s] = [0, \Lambda(t)]$. Thus

$$\begin{aligned}
&E\left[\prod_{i=1}^{N^*(s)} q(\Lambda^{-1}(T_i^*))|N^*(s) = n\right] \\
&= E\left[\prod_{i=1}^{n} q(\Lambda^{-1}(T_i^*))|N^*(s) = n\right] \\
&= E\left[\prod_{i=1}^{n} q(\Lambda^{-1}(V_{(i)}))\right] \\
&= E\left[\prod_{i=1}^{n} q(\Lambda^{-1}(V_i))\right] \\
&= \left(E[q(\Lambda^{-1}(V_1))]\right)^n = \left(E[q(\Lambda^{-1}(sU))]\right)^n,
\end{aligned}$$

where $U \equiv V_1/s = V_1/\Lambda(t)$ is a random variable uniformly distributed in the unit interval [0,1]. Therefore,

$$E[q(\Lambda^{-1}(sU))] = \int_0^1 q(\Lambda^{-1}(su))\,\mathrm{d}u = \int_0^1 q(\Lambda^{-1}(\Lambda(t)u))\mathrm{d}u = \frac{1}{\Lambda(t)}\int_0^t q(x)\lambda(x)\mathrm{d}x.$$

Hence,

$$E\left[\prod_{i=1}^{N^*(s)} q(\Lambda^{-1}(T_i^*)|N^*(s) = n\right] = \left(\frac{1}{\Lambda(t)}\int_0^t q(x)\lambda(x)\mathrm{d}x\right)^n$$

And, finally,

$$H^*(s) = \int_0^\infty \exp\{-st\}H(t)\mathrm{d}t \quad \text{and} \quad f^*(s) = \int_0^\infty \exp\{-st\}f(t)\mathrm{d}t$$

the Laplace transforms of $H(t)$ and $f(t)$, respectively.

Applying the Laplace transform to both sides of (2.18) and using the fact that the Laplace transform of a convolution of two functions is the product of the Laplace transforms of these functions, we arrive at the following equation:

$$H^*(s) = \frac{1}{s}\sum_{k=1}^{\infty}(f^*(s))^k = \frac{f^*(s)}{s(1-f^*(s))}. \tag{2.19}$$

As the Laplace transform uniquely defines the corresponding distribution, (2.19) implies that the renewal function is uniquely defined by the underlying distribution $F(t)$ via the Laplace transform of its density.

The functions $H(t)$ and $h(t)$ satisfy the following integral equations:

$$H(t) = F(t) + \int_0^t H(t-x)f(x)\mathrm{d}x, \tag{2.20}$$

$$h(t) = f(t) + \int_0^t h(t-x)f(x)\mathrm{d}x. \tag{2.21}$$

Let us prove Eq. (4.10) by conditioning on the time of the first renewal, i.e.,

$$\begin{aligned} H(t) &= \int_0^t E[N(t)|X_1 = x]f(x)\mathrm{d}x = \int_0^t [1 + H(t-x)]f(x)\mathrm{d}x \\ &= F(t) + \int_0^t H(t-x)f(x)\mathrm{d}x. \end{aligned} \tag{2.22}$$

If the first renewal occurs at time $x \le t$, then the process simply restarts and the expected number of renewals after the first one in the interval $(x, t]$ is $H(t-x)$. Note that Eq. (2.19) can also be obtained by applying the Laplace transform to both parts of Eq. (2.20). In a similar way, the equation

$$h(t) = \int_0^t \frac{\mathrm{d}}{\mathrm{d}t}(E[N(t)|X_1 = x])f(x)\mathrm{d}x \tag{2.23}$$

eventually results in (2.21).

Denote, as usual, the failure rate of the underlying distribution $F(t)$ by $\lambda(t)$. The intensity process, which corresponds to the renewal process, is

$$
\begin{aligned}
P(T_p \geq t) &= E[\prod_{i=1}^{N(t)} q(T_i)] \\
&= \sum_{n=0}^{\infty} \left(\frac{1}{\Lambda(t)} \int_0^t q(x)\lambda(x)\mathrm{d}x \right)^n \cdot \frac{(\Lambda(t))^n}{n!} e^{-\Lambda(t)} \\
&= \exp\left\{ - \int_0^t p(x)\lambda(x)\mathrm{d}x \right\} \cdot \sum_{n=0}^{\infty} \frac{\left(\int_0^t q(x)\lambda(x)\mathrm{d}x \right)^n}{n!} \cdot \\
& \exp\left\{ - \int_0^t q(x)\lambda(x)\mathrm{d}x \right\} = \exp\left\{ - \int_0^t p(x)\lambda(x)\mathrm{d}x \right\}.
\end{aligned}
\tag{2.31}
$$

Thus, the time till the first perfect repair is distributed in accordance with Eq. (2.31). Moreover, this setting can be considered more generally (not necessarily with termination), when each event from the original NHPP with rate $\lambda(t)$ is classified with probability $p(t)$ as an event of the Type 1 and with probability $q(t) = 1 - p(t)$, as an event of the Type 2. Then we arrive at the sum of two NHPP processes with rates

$$
p(t)\lambda(t) \quad \text{and} \quad q(t)\lambda(t),
$$

respectively. More discussion on this classification can be found in Chap. 4, where more general point processes will be also considered.

2.6 General (Imperfect) Repair

The conventional models for burn-in of repairable items usually deal with minimally repaired items. However, this assumption is often violated in practice. Therefore, a more general type of repair should be considered. As was discussed in the previous section, minimal repair is the specific case of *imperfect* or *general* repair (we will use these terms interchangeably). After imperfect repair, the system is usually in the intermediate state (between the state that corresponds to perfect repair and the state that corresponds to minimal repair). However, the situation when this state is 'worse' than that after the minimal repair sometimes can also occur in practice. In order to deal effectively with models of imperfect repair we must refer to the concept of virtual age [25].

2.6.1 Virtual Age

Consider a degrading item that operates in a baseline environment (regime) and denote the corresponding Cdf of time to failure by $F_b(t)$. Let another statistically identical item be operating in a more severe environment with the Cdf of time to failure denoted by $F_s(t)$. Denote by $\lambda_b(t)$ and $\lambda_s(t)$ the failure rates in two environments, respectively. We want to establish an age correspondence between the systems in two regimes by considering the baseline as a reference. It is reasonable to assume that degradation in the second regime is more intensive and, therefore, the time for accumulating the same amount of degradation or wear is smaller than in the baseline regime. Therefore, assume that the lifetimes in two environments are ordered as (see Sect. 2.7 for the description of the main stochastic orders)

$$\overline{F}_s(t) < \overline{F}_b(t), \quad t \in (0, \infty). \tag{2.32}$$

Inequality (2.32) implies the following equation:

$$F_s(t) = F_b(W(t)), \qquad W(0) = 0, \; t \in (0, \infty). \tag{2.33}$$

Equation (2.33) can be interpreted as a general Accelerated Life Model (ALM) ([17, 24, 39], to name a few) with a time-dependent scale-transformation function $W(t)$.

Definition 2.7 Let t be the calendar age of a degrading item operating in a baseline environment. Assume that ALM (2.33) describes the lifetime of another statistically identical item, which operates in a more severe environment for the same duration t.

Then the function $W(t) > t$ defines the *statistical* virtual age of the second item, or, equivalently, the inverse function $W^{-1}(t) < t$ defines the *statistical* virtual age of the first item when a more severe environment is set as the baseline environment.

The ALM defined by (2.33) can be viewed as an equation for obtaining $W(t)$, i.e.,

$$\begin{gathered} \exp\left\{-\int_0^t \lambda_s(u)\mathrm{d}u\right\} = \exp\left\{-\int_0^{W(t)} \lambda_b(u)\mathrm{d}u\right\} \\ \Rightarrow \int_0^t \lambda_s(u)\mathrm{d}u = \int_0^{W(t)} \lambda_b(u)\mathrm{d}u. \end{gathered} \tag{2.34}$$

Hence, the statistical virtual age $W(t)$ is uniquely defined by Eq. (2.34). Assume that $W(t)$ is differentiable. Then $W(t) = \int_0^t w(u)\mathrm{d}u$ and $w(t)$ can be interpreted as the *rate of degradation*.

Example 2.4 Let the failure rates in both regimes be increasing, positive power functions (the Weibull distributions), which are often used for lifetime modeling of degrading objects, i.e.,

$$\lambda_b(t) = \alpha t^{\beta}, \; \lambda_s(t) = \mu t^{\eta}, \; \alpha, \beta, \mu, \eta > 0.$$

The statistical virtual age $W(t)$ is defined by Eq. (2.34) as

$$W(t) = \left(\frac{\mu(\beta+1)}{\alpha(\eta+1)}\right)^{\frac{1}{\beta+1}} t^{\frac{\eta+1}{\beta+1}}.$$

In order for the inequality $W(t) > t$ to hold, the following restrictions on the parameters are sufficient: $\eta \geq \beta,\ \ \mu(\beta+1) > \alpha(\eta+1)$.

As follows from Eq. (2.33), the failure rate that corresponds to the Cdf $F_s(t)$ is

$$\lambda_s(t) = \frac{F_b'(W(t))}{\overline{F}_b(W(t))} = w(t)\lambda_b(W(t)). \tag{2.35}$$

Let an item start now operating in a baseline regime at $t = 0$, which is switched at $t = x$ to a more severe regime. In accordance with Definition 2.7, the statistical virtual age immediately after the switching is $V_x = W^{-1}(x)$, where the new notation V_x is used for convenience. Assume now that the governing Cdf after the switching is $F_s(t)$ and that the Cdf of the remaining lifetime is $F_s(t|V_x)$, i.e.,

$$F_s(t|V_x) = 1 - \frac{\overline{F}_s(t+V_x)}{\overline{F}_s(V_x)}. \tag{2.36}$$

Thus, an item starts operating in the second regime with a starting age V_x defined with respect to the Cdf $F_s(t)$. Note that the form of the lifetime Cdf after the switching given by Eq. (2.36) is our assumption and that it does not follow directly from ALM (2.33). Alternatively, we can proceed starting with ALM (2.33) and obtain the Cdf of an item's lifetime for the whole interval $[0,\infty)$, and this will be performed in what follows.

According to our interpretation the rate of degradation is 1 in $t \in [0,x)$. Assume that the switching at $t = x$ results in the rate $w(t) > 1$ in $[x,\infty)$, where $w(t) = W'(t)$. Under the stated assumptions, the item's lifetime Cdf in $[0,\infty)$, to be denoted by $F_{\mathrm{bs}}(t)$, can be written as [25]

$$F_{\mathrm{bs}}(t) = \begin{cases} F_b(t), & 0 \leq t < x, \\ F_b\left(x + \int\limits_x^t w(u)\mathrm{d}u\right), & x \leq t < \infty. \end{cases} \tag{2.37}$$

Transformation of the second row on the right-hand side of this equation results in

$$F_b\left(x + \int\limits_x^t w(u)\mathrm{d}u\right) = F_b(W(t) - W(\tau(x))), \tag{2.38}$$

where $\tau(x) < x$ is uniquely defined from the equation

$$x = \int_{\tau(x)}^{x} w(u)du = W(x) - W(\tau(x)). \tag{2.39}$$

Thus, the cumulative degradation in $[\tau(x), x)$ in the second regime is equal to the cumulative degradation in the baseline regime in $[0, x)$, which is x. Therefore, the age of an item just after switching to a more severe regime can be defined as $\tilde{V}_x = x - \tau(x)$. Let us call it the *recalculated virtual age.*

Definition 2.8 Let a degrading item start operating at $t = 0$ in the baseline regime and be switched to a more severe regime at $t = x$. Assume that the corresponding Cdf in $[0, \infty)$ is given by Eq. (2.37), which follows from the ALM (2.33). Then the recalculated virtual age $\tilde{V}_x$ after switching at $t = x$ is defined as $x - \tau(x)$, where $\tau(x)$ is the unique solution to Eq. (2.39).

Equation (2.39) has the solution:

$$\tau(x) = W^{-1}(W(x) - x).$$

As $V_x = W^{-1}(x)$, the equation $V_x = \tilde{V}_x$ can be written in the form of the following functional equation:

$$x - W^{-1}(x) = W^{-1}(W(x) - x).$$

Applying operation $W(\cdot)$ to both parts of this equation gives

$$W(x - W^{-1}(x)) = W(x) - x.$$

It is easy to show that the linear function $W(t) = wt$ is a solution to this equation. It is also clear that it is the unique solution, as the functional equation $f(x + y) = f(x) + f(y)$ has only a linear solution. Therefore, the recalculated virtual age in this case is equal to the statistical virtual age. When $W(t)$ is a nonlinear function, the statistical virtual age $V_x = W^{-1}(x)$ is not equal to the recalculated virtual age $\tilde{V}_x = x - \tau(x)$ and this should be taken into account.

2.6.2 Models of General Repair

The virtual age concept can also be applied to repairable systems. Keeping the notation but not the literal meaning, assume that initially the lifetime of a repairable item is characterized by the Cdf $F_b(t)$ and the imperfect repair changes it to $F_s(t|V_x)$ defined by Eq. (2.36), where V_x is the virtual age just after repair at $t = x$. This will be our definition for the virtual age for repairable systems, whereas the terms "statistical" and "recalculated" virtual age refer to nonrepairable objects. The important special case $F_s(t) = F_b(t)$ will be also considered. Thus, we have two factors that define a distribution after repair. First, the imperfect repair changes the Cdf from $F_b(t)$ to $F_s(t)$. As an option, parameters of the Cdf $F_b(t)$ can be changed by the repair action. Second, the model includes the virtual age V_x as the starting (initial) age for an item described by the Cdf $F_s(t)$, which was called in Finkelstein [22] "the hidden age of the Cdf after the change of parameters".

Example 2.5 Suppose that a component with an absolutely continuous Cdf $F(t)$ is supplied with an infinite number of 'warm standby' components with Cdfs $F(qt)$, where $0 < q \leq 1$ is a constant. This system starts operating at $t = 0$. The first component operates in a baseline regime, whereas the standby components operate in a less severe regime. Upon each failure in the baseline regime, the component is instantaneously replaced by a standby one, which is switched into operation in the baseline regime. Thus, the virtual age (which was called the recalculated virtual age previously) V_x of a standby component that had replaced the operating one at $t = x$ is qx. The corresponding remaining lifetime Cdf, in accordance with Eq. (2.3), is

$$F(t|V_x) = F(t|qx) = \frac{F(t+qx) - F(qx)}{\bar{F}(qx)}. \tag{2.40}$$

Note that Eq. (2.40) is obtained using the age recalculation approach of Sect. 2.6.1, which is based on the specific linear case of Eq. (2.33). When $q = 1$, (2.40) defines minimal repair; when $q = 0$, the components are in cold standby (perfect repair).

The age recalculation in this model is performed upon each failure. The corresponding sequence of interarrival times $\{X_i\}_{i\geq 1}$ forms a generalized renewal (g-renewal) process. Recall that the cycles of the ordinary renewal process are i.i.d. random variables. In the g-renewal process, the duration of the $(n+1)$th cycle, which starts at $t = s_n \equiv x_1 + x_2 + \ldots + x_n$, $n = 0, 1, 2\ldots$, $s_0 = 0$, is defined by the following conditional distribution:

$$P(X_{n+1} \leq t) = F(t|qs_n),$$

where s_n is a realization of the arrival time S_n.

We will now generalize this example to the case of nonlinear ALM (2.33). Let a failure, not necessarily the first one, occur at $t = x$. It is instantaneously imperfectly repaired and the virtual age after the repair is $V_x = W^{-1}(x) \equiv q(x)$, where $q(x)$ is a continuous increasing function, $0 \leq q(x) \leq x$. Thus the Cdf of the time to the next failure is $F(t|V_x)$. The most important feature of the model is that $F(t|V_x)$ depends only on the time x and not on the other elements of the history of the corresponding point process. This property makes it possible to generalize renewal equations (2.20) and (2.21) to the case under consideration. The point process of imperfect repairs $N(t)$, $t \geq 0$, as in the case of an ordinary renewal process, is characterized by the corresponding renewal function $H(t) = E[N(t)]$ and the renewal density function $h(t) = H'(t)$:

$$H(t) = F(t) + \int_0^t h(x)F(t - x|q(x))\mathrm{d}x, \tag{2.41}$$

$$h(t) = f(t) + \int_0^t h(x)f(t - x|q(x))\mathrm{d}x, \tag{2.42}$$

where $f(t - x|q(x))$ is the density that corresponds to the Cdf $F(t - x|q(x))$.

The strict proof of these equations and the sufficient conditions for the corresponding unique solutions can be found in Kijima and Sumita [35].

Example 2.6 Let $q(x) = 0$. Then $f(t - x|q(x)) = f(t - x)$ and we arrive at ordinary renewal equations (2.20) and (2.21).

Example 2.7 Let $q(x) = x$ (minimal repair). Equations (2.41) and (2.42) can be explicitly solved in this case. However, we will only show that the rate of the nonhomogeneous Poisson process $\lambda_r(t)$, which is equal to the failure rate $\lambda(t)$ of the governing Cdf is a solution to Eq. (2.42). As

$$f(t - x|x)) = f(t)/\bar{F}(x),$$
$$(1/\bar{F}(x))' = \lambda(x)/\bar{F}(x),$$

the right-hand side of Eq. (2.42) is equal to $\lambda(t)$, i.e.,

$$f(t) + \int_0^t h(x)f(t - x|q(x))\mathrm{d}x = f(t) + f(t)\int_0^t \frac{\lambda(x)}{\bar{F}(x)}\mathrm{d}x = \lambda(t),$$

as the process of minimal repairs is the NHPP.

Each cycle of this renewal-type process is defined by the same governing Cdf $F(t)$ with the failure rate $\lambda(t)$ and only the starting age for this distribution is given by the virtual age $V_x = q(x)$. Therefore, the cycle duration after the repair at $t = x$ is described by the Cdf $F(t|V_x)$. The formal definition of the g-renewal process can now be given via the corresponding intensity process [compare with (2.24)].

Definition 2.9 The g-renewal process is defined by the following intensity process:

$$\lambda_t = \lambda(t - S_{N(t)} + q(S_{N(t)})), \tag{2.43}$$

where, as usual, $S_{N(t)}$ denotes the random time of the last renewal.

The function $q(x)$ is usually continuous and increasing and $0 \le q(x) \le x$. Thus, as in the case of an ordinary renewal process, the intensity process is defined by the same failure rate $\lambda(t)$, only the cycles now start with the initial failure rate $\lambda(q(S_{n(t)}))$, $n(t) = 1, 2, \ldots$.

One of the important restrictions of this model is the assumption of the 'fixed' shape of the failure rate. However, this assumption is well motivated, e.g., for the spare-parts setting. Therefore, we will keep the 'sliding along the $\lambda(t)$ curve' reasoning and will generalize it to a more complex case than the g-renewal case dependence on a history of the point process of repairs.

Assume that each imperfect repair reduces the virtual age of an item in accordance with some recalculation rule to be defined for specific models. As the shape of the failure rate is fixed, the virtual age at the start of a cycle is uniquely defined by the 'position' of the corresponding point on the failure rate curve after the repair. Therefore, Eq. (2.43) for the intensity process can be generalized to

$$\lambda_t = \lambda(t - S_{N(t)} + V_{S_{N(t)}}), \tag{2.44}$$

where $V_{S_{N(t)}}$ is the virtual age of an item immediately after the last repair before t. From now on, for convenience, the capital letter V will denote a random virtual age, whereas v will denote its realization. Equation (2.44) gives a general definition for the models with a fixed failure rate shape. It should be specified by the corresponding virtual age model. It follows from Eq. (2.44) that the intensity process between consecutive repairs can be 'graphically' described as *horizontally parallel* to the initial failure rate $\lambda(t)$ as all corresponding shifts are in the argument of the function $\lambda(t)$ [21]. We will consider now a specific but very meaningful and important for practical applications general repair model.

Let an item start operating at $t = 0$. Therefore, the first cycle duration is described by the Cdf $F(t)$ with the corresponding failure rate $\lambda(t)$. Let the first failure (and the instantaneous imperfect repair) occur at $X_1 = x_1$. Assume that the imperfect repair decreases the age of an item to $q(x_1)$, where $q(x)$ is an increasing continuous function and $0 \le q(x) \le x$. Thus, the second cycle of the point process starts with the virtual age $v_1 = q(x_1)$ and the cycle duration X_2 is distributed as $F(t|v_1)$ with the failure rate $\lambda(t + v_1),\ t \ge 0$. Therefore, the virtual age of an item just before the second repair is $v_1 + x_2$ and it is $q(v_1 + x_2)$ just after the second repair, where we assume for simplicity that the function $q(x)$ is the same at each cycle. The sequence of virtual ages after the ith repair $\{v_i\}_{i\ge 0}$ at the start of the $(i+1)$th cycle in this model is defined for realizations x_i as

$$v_0 = 0,\quad v_1 = q(x_1),\quad v_2 = q(v_1 + x_2), \ldots, v_i = q(v_{i-1} + x_i) \tag{2.45}$$

or, equivalently,

$$V_n = q(V_{n-1} + X_n),\quad n \ge 1.$$

For the specific linear case, $q(x) = qx,\ 0 < q < 1$, this model was considered on a descriptive level in Brown et al. [14] and Bai and Yun [5]. Following the publication of the paper by Kijima [34] it usually has been referred to as the Kijima II model, whereas the Kijima I model describes a somewhat simpler version of age reduction when only the duration of the last cycle is reduced by the corresponding imperfect repair [6, 46]. The Kijima II model and its probabilistic analysis was also independently suggested in Finkelstein [23] and later considered in numerous subsequent publications. The term 'virtual age' in connection with imperfect repair models was probably used for the first time in Kijima et al. [36], but the corresponding meaning was already used in a number of publications previously.

When $q(x) = qx$, the intensity process λ_t can be defined in the explicit form. After the first repair the virtual age v_1 is qx_1, after the second repair $v_2 = q(qx_1 + x_2) = q^2x_1 + qx_2,\ldots$, and after the nth repair the virtual age is

$$v_n = q^n x_1 + q^{n-1}x_2 + \ldots + qx_n = \sum_{i=0}^{n-1} q^{n-i} x_{i+1}, \tag{2.46}$$

where $x_i, i \geq 1$ are realizations of interarrival times X_i in the point process of imperfect repairs. Therefore, in accordance with the general Eq. (2.44), the intensity process for this specific model with a linear $q(x) = qx$ is

$$\lambda_t = \lambda\left(t - S_{N(t)} + \sum_{i=0}^{N(t)-1} q^{n-i} X_{i+1}\right). \tag{2.47}$$

Example 2.8 Whereas the repair action in the Kijima II model depends on the whole history of the corresponding stochastic process, the dependence in the Kijima I model is simpler and takes into account the reduction of the last cycle increment only. Similar to (2.45),

$$v_0 = 0, \quad v_1 = qx_1, \quad v_2 = v_1 + qx_2, \ldots, v_n = v_{n-1} + qx_n. \tag{2.48}$$

Therefore,

$$v_n = q(x_1 + x_2 + \ldots + x_n), \quad V_n = q(X_1 + X_2 + \ldots + X_n),$$

and we arrive at the important conclusion that this is exactly the same model as the one defined by the g-renewal process of the previous section [36]. These considerations give another motivation for using the Kijima I model for obtaining the required number of aging spare parts. In accordance with Eqs. (2.44) and (2.48), the intensity process for this model is

$$\begin{aligned}\lambda_t &= \lambda(t - S_{N(t)} + V_{S_{N(T)}}) = \lambda(t - S_{N(t)} + qS_{N(t)}) \\ &= \lambda(t - (1 - q)S_{N(t)}).\end{aligned}$$

The obtained form of the intensity process suggests that the calendar age t is decreased in this model by an increment proportional to the calendar time of the last imperfect repair. Therefore, Doyen and Gaudoin [21] call it the "arithmetic age reduction model".

The two types of the considered models represent two marginal cases of history for the corresponding stochastic repair processes, i.e., the history that 'remembers' all previous repair times and the history that 'remembers' only the last repair time, respectively. Intermediate cases are analyzed in Doyen and Gaudoin [21]. Note that, as q is a constant, the repair quality does not depend on calendar time, or on the repair number.

The original models in Kijima [34] were, in fact, defined for a more general setting when the reduction factors q_i, $i \geq 1$ are different for each cycle (the case of independent random variables $Q_i, i \geq 1$ was also considered). The quality of repair that is deteriorating with i can be defined as $0 < q_1 < q_2 < q_3, \ldots$, which is a natural ordering in this case. Equation (2.47) then becomes

$$v_n = x_1 \prod_{i=1}^{n} q_i + x_2 \prod_{i=2}^{n} q_i + \ldots + q_n x_n = \sum_{i=1}^{n} x_i \prod_{k=i}^{n} q_k, \tag{2.49}$$

and the corresponding intensity process is

$$\lambda_t = \lambda\left(t - S_{N(t)} + \sum_{i=1}^{N(t)} X_i \prod_{k=i}^{N(t)} q_k\right). \tag{2.50}$$

The virtual age in the Kijima I model is

$$v_n = v_{n-1} + q_n x_n = \sum_{1}^{n} q_i x_i,$$

and the corresponding intensity process is defined by

$$\lambda_t = \lambda\left(t - S_{N(t)} + \sum_{i=1}^{N(t)} q_i X_i\right). \tag{2.51}$$

The practical interpretation of (2.49) is quite natural, as the degree of repair at each cycle can be different and usually deteriorates with time. The practical application of Model (2.51) is not so evident. Substitution of a random Q_i instead of a deterministic q_i in (2.50) and (2.51) results in general relationships for the intensity processes in this case.

Note that, when $Q_i \equiv Q$, $i = 1, 2, \ldots$ are i.i.d. Bernoulli random variables, the Kijima II model can be interpreted via the Brown–Proschan model (2.27–2.28). In this model, the repair is perfect with probability p and is minimal with probability $1 - p$. [25].

2.7 Multivariate Accelerated Life and Proportional Hazards Models

The Accelerated Life Model (ALM) and the proportional hazards (PH) model are very popular in reliability theory and applications as convenient tools for modeling, e.g., an impact of a more severe environment on reliability characteristics of items defined for some baseline environment. These models were extensively studied in the literature for single items or systems (see, e.g., Bagdonavicius and Nikulin [4] and references therein).

The univariate ALM is defined by Eq. (2.33), whereas the time-dependent PH model can be defined as

$$\lambda_s(t) = k(t)\lambda_b(t),\ t \in [0, \infty), \tag{2.52}$$

where $\lambda_b(t),\ \lambda_s(t)$ are the failure rates of an item in the baseline and a more severe environment, respectively and $k(t) > 1$.

It should be noted that generalizations of the ALM and the PH models to the case of possibly dependent items, which can be meaningful for reliability practice, are not trivial and, therefore, challenging. We will be mostly interested in the corresponding competing risks problem for possibly dependent items and start, for the presentation sake, with the independent items case.

Survival functions of a series system of n statistically independent items under the baseline and a more severe environment, in accordance with (2.33), are [25]:

$$\overline{F}_b(t) = \prod_1^n \overline{F}_{bi}(t);\ \overline{F}_s(t) = \prod_1^n \overline{F}_{bi}(W_i(t)), \tag{2.53}$$

respectively, where $W_i(t)$ is the scale transformation function for the ith item. Thus $W(t)$ for the system can be obtained from the following equation

$$\overline{F}_b(W(t)) = \prod_1^n \overline{F}_{bi}(W_i(t)) \tag{2.54}$$

or, equivalently, using relationships similar to (2.34):

$$\int_0^{W(t)} \sum_1^n \lambda_{bi}(u)\mathrm{d}u = \sum_1^n \int_0^{W_i(t)} \lambda_{bi}(u)\mathrm{d}u. \tag{2.55}$$

Example 2.9 Let $n = 2$ and $W_1(t) = t$, $W_2(t) = 2t$, which can be interpreted by assuming that the first component is somehow protected from the more severe environment. Then Eq. (2.55) can be transformed to

$$\int_0^{W(t)} (\lambda_{b1}(u) + \lambda_{b2}(u))\mathrm{d}u = \int_0^t \lambda_{b1}(u)\mathrm{d}u + \int_0^{2t} \lambda_{b2}(u)\mathrm{d}u.$$

Assume further that the failure rates are linear, $\lambda_{b1}(t) = \lambda_1 t$, $\lambda_{b2}(t) = \lambda_2 t$, $\lambda_1, \lambda_2 > 0$. Then

$$W(t) = \left(\sqrt{\frac{\lambda_1 + 4\lambda_2}{\lambda_1 + \lambda_2}}\right)t$$

If the components are statistically identical in the baseline environment ($\lambda_1 = \lambda_2$), then $W(t) = \sqrt{5/2}\,t \approx 1.6t$.

It obviously follows from (2.52) that, due to independence (PH model), for each item

$$\lambda_{si}(t) = k_i(t)\lambda_{bi}(t),\ t \in [0, \infty),$$

whereas for the series system, assuming the time-independent impact of a more severe environment on the baseline failure rates of items, we have:

$$\lambda_s(t) = \sum_1^n k_i \lambda_{bi}(t). \tag{2.56}$$

What happens when our items are statistically dependent? We will consider for simplicity of notation the case of two components, $n = 2$. Before generalizing the ALM to this case, we first describe the dependence of components via the concept of *copulas*. A formal definition and numerous properties of copulas can be found, e.g., in Nelsen [42]. Copulas create a convenient way of representing multivariate distributions. In a way, they 'separate' marginal distributions from the dependence structure. It is more convenient for us to consider the survival copulas based on marginal survival functions. In order to deal with the series system (competing risks), we must first consider a general bivariate ($n = 2$) case. For $n > 2$, the discussion is similar.

Let $T_{b1} \geq 0$, $T_{b2} \geq 0$ be the possibly dependent lifetimes of items in the baseline environment and let

$$F_b(t_1, t_2) = P(T_{b1} \leq t_1,\ T_{b2} \leq t_2),$$
$$F_{bi}(t_i) = P(T_{bi} \leq t_i),\ i = 1, 2$$

be the absolutely continuous bivariate and univariate (marginal) Cdfs, respectively (in the baseline environment). The similar notation with the sub index "s" is for the more severe environment. Denote the bivariate (joint) survival function by

$$S_b(t_1, t_2) \equiv P(T_{b1} > t_1,\ T_{b2} > t_2) = 1 - F_{b1}(t_1) - F_{b2}(t_2) + F_b(t_1, t_2) \tag{2.57}$$

and the univariate (marginal) survival functions with the corresponding failure rates $\lambda_{bi}(t_i),\ i = 1, 2$ by

$$S_{b1}(t_1) \equiv P(T_{b1} > t_1,\ T_{b2} > 0) = P(T_{b1} > t_1) = S_b(t_1, 0),$$
$$S_{b2}(t_2) \equiv P(T_{b1} > 0,\ T_{b2} > t_2) = P(T_{b2} > t_2) = S_b(0, t_2).$$

It is well-known [42] that the bivariate survival function can be represented as a function of $S_{bi}(t_i),\ i = 1, 2$ in the following way:

$$S_b(t_1, t_2) = C(S_{b1}(t_1), S_{b2}(t_2)), \tag{2.58}$$

where the survival copula $C(u, v)$ is a bivariate function in $[0, 1] \times [0, 1]$. Note that, such function always exists when the inverse functions for $S_i(t_i),\ i = 1, 2$ exist:

$$S_b(t_1, t_2) = S_b(S_{b1}^{-1} S_{b1}(t_1), S_{b1}^{-1} S_{b2}(t_2)) = C(S_{b1}(t_1), S_{b2}(t_2)).$$

When the lifetimes are independent, the following obvious relationship holds:

$$S_b(t_1, t_2) = S_{b1}(t_1) S_{b2}(t_2) \Leftrightarrow C(u, v) = uv. \tag{2.59}$$

Thus, when the copula and marginal distributions are known the solution of our competing risks problem ($t_1 = t_2 = t$) for the baseline regime is the following survival function:

$$S_b(t) = C(S_{b1}(t), S_{b2}(t)). \tag{2.60}$$

Let the statistically identical system of two items operate now in a more severe environment. All foregoing relationships obviously hold with the substitution of the sub index "b" by the sub index "s" where appropriate. However, (2.58) and (2.60) should be discussed in more detail. For that we need to make the following crucial assumption [26]:

Assume that the copula that defines the dependence structure of a system do not depend on the environment.

Taking into account (2.33), it means that Eqs. (2.58) and (2.60) can be now written as

$$S_s(t_1, t_2) = C(S_{s1}(t_1), S_{s2}(t_2)) = C(S_{b1}(W_1(t_1), S_{b2}(W_2(t_2)), \tag{2.61}$$

$$S_s(t) = C(S_{s1}(t), S_{s2}(t)) = C(S_{b1}(W_1(t), S_{b2}(W_2(t)), \tag{2.62}$$

respectively. Thus, under the stated assumption:

Definition 2.10 The ALM for the series system of two possibly dependent items is defined by Eq. (2.62), where $C(u, v)$ is the survival copula that describes the corresponding dependence structure of the system.

Our assumption seems to be rather natural at many instances, but need to be justified by some 'physical properties' of a system or by the corresponding data, as obviously, it is not a 'universal law', as, e.g., illustrated by the Example 2.10.

Basically, it means that the environment can impact the processes of deterioration in items but cannot influence the dependence properties. The simplest illustrative case is when the items are independent in the baseline environment and the corresponding copula is a product given by Eq. (2.59). It is natural to assume that the independence is preserved under a more severe regime and, therefore, the same product holds. Thus, in this case, a more severe regime does not ruin the property of independence. However, this can happen theoretically when, e.g., the stress defining the severe environment is sufficiently large. Another meaningful example is as follows:

Example 2.10 Consider a system of two components in series. Each component is subject to its own (independent) homogeneous Poisson shock process with rate λ_b. Assume that the shocks constitute the only cause of failure: each shock results in failure of a component with probability $p_{bi}, i = 1, 2$ and is survived (without any consequences) with the complementary probability $q_{bi} = 1 - p_{bi}$, where, as previously, the sub index "b" stands for "baseline" (environment). Then, obviously, the survival probability for the series system is the following product:

$$\begin{aligned} S_b(t) =& (\exp\{-\lambda_b t\} \sum_0^\infty \frac{(\lambda_b t)^i}{i!} (q_{b1})^i)(\exp\{-\lambda_b t\} \sum_0^\infty \frac{(\lambda_b t)^i}{i!} (q_{b2})^i) \\ =& \exp\{-p_{b1}\lambda_b t\} \exp\{-p_{b2}\lambda_b t\}. \end{aligned}$$

Let the HPP of shocks with rate λ_b be the only one now and let it affect both components with given above probabilities. However, the components are not independent now (on the contrary, they are dependent via the mutual shock process). Therefore, the probability of survival of a system under a single shock is $q_{b1}q_{b2}$, whereas the probability of failure is $1 - q_{b1}q_{b2} = p_{b1} + p_{b2} - p_{b1}p_{b2}$ and the corresponding survival probability is:

$$\tilde{S}_b(t) = \exp\{-(1 - q_{b1}q_{b2})\lambda_b t\} = \exp\{-p_{b1}\lambda_b t\} \exp\{-p_{b2}\lambda_b t\} \exp\{p_{b1}p_{b2}\lambda_b t\}.$$

Comparison of $S_b(t)$ and $\tilde{S}_b(t)$ suggests that the term $\exp\{p_{b1}p_{b2}\lambda_b t\}$ is responsible for the described dependence. Thus, the corresponding copula can be written as

$$C(u, v) = uvv^{-p_{b1}} = uvu^{-p_{b2}} = uv^{q_{b1}} = u^{q_{b2}}v.$$

Let a more severe environment be modeled by the shock process with a larger rate, i.e., $\lambda_s > \lambda_b$, whereas the probabilities of failure $p_{bi}, i = 1, 2$ do not change. As we can see, this does not have any effect on the form of the copula as a function of the corresponding marginals. Therefore, the copula in the described setting *is invariant with respect to environment!* The same conclusion can be made when one of the components experiences the increased probability of failure under a more severe shock, whereas the other one retains the same probability (a kind of "protection"). On the other hand, it can be easily seen that if both components experience the increased probability of failure under a more severe shock, then the corresponding copula is not invariant.

Example 2.11 The widely used (especially in survival analysis) Clayton bivariate distribution [19, 20] is given by the following survival copula

$$C(u,v) = (u^{-\theta} + v^{-\theta} - 1)^{-1/\theta},$$

where $\theta > 0$. Therefore,

$$\begin{aligned} S_b(t) &= ((S_{b1}(t))^{-\theta} + (S_{b2}(t))^{-\theta})^{-1/\theta}, \\ S_s(t) &= ((S_{b1}(W_1(t))^{-\theta} + (S_{b2}(W_2(t)))^{-\theta})^{-1/\theta}. \end{aligned}$$

We see that if parameter θ is the same for both environments, then this case complies with our definition of the ALM. The best way to check it is to conduct the corresponding hypothesis testing (given the data).

Example 2.12 The similar reasoning obviously holds for the Farlie–Gumbel–Morgenstern distribution. This bivariate distribution is defined as [32]

$$S(t_1,t_2) = S_1(t_1)S_2(t_2)(1 + \alpha(1 - S_1(t_1))(1 - S_2(t_2))),$$

where $-1 \leq \alpha \leq 1$.

As in the univariate case defined by Eq. (2.52), the PH model for the bivariate case can constitute the alternative to the ALM while modeling the impact of a more severe environment [26]. The environment in this case 'acts directly' on the failure rate. The problem is, however, that now, in contrast to the univariate setting where $S(t) = \exp\{-\int_0^t \lambda(u)du\}$, the single failure rate that defines the corresponding distribution function does not exist. Moreover, it was proved in Finkelstein [25] that the following exponential representation holds (for the baseline environment) in this case:

$$\begin{aligned} S_b(t_1,t_2) = \exp&\left\{-\int_0^{t_1} \lambda_{b1}(u)du\right\} \exp\left\{-\int_0^{t_2} \lambda_{b2}(u)du\right\} \\ &\times \exp\left\{\int_0^{t_1}\int_0^{t_2} (\lambda_b(u,v) - \bar{\lambda}_{b1}(u,v)\bar{\lambda}_{b2}(u,v))dudv\right\}, \end{aligned} \tag{2.63}$$

where $\lambda_{bi}(u)$, $i - 1,2$ arc the failure rates of marginal distributions and the failure rates $\lambda_b(u,v)$, $\bar{\lambda}_{bi}(u,v)$ are defined by the following equations, respectively:

$$\begin{aligned}\lambda_b(t_1,t_2) &= \lim_{\Delta t_1,\Delta t_2\to 0}\frac{\Pr[t_1\le T_{b1}<t_1+\Delta t_1, t_2\le T_{b2}<t_2+\Delta t_2|T_{b1}<t_1, T_{b2}>t_2]}{\Delta t_1\Delta t_2}\\ &=\frac{f_b(t_1,t_2)}{S_b(t_1,t_2)},\end{aligned} \tag{2.64}$$

$$\begin{aligned}\bar{\lambda}_{bi}(t_1,t_2) &= \lim_{\Delta t\to 0}\frac{1}{\Delta t}\Pr[t_i\le T_{bi}<t_i+\Delta t|T_{b1}>t_1, T_{b2}>t_2]\\ &= -\frac{\partial}{\partial t_i}\ln S(t_1,t_2);\quad i=1,2,\end{aligned} \tag{2.65}$$

Thus, $\lambda_b(t_1,t_2)\mathrm{d}t_1\mathrm{d}t_2 + o(\mathrm{d}t_1\mathrm{d}t_2)$ can be interpreted as the probability of failure of both items in intervals of time $[t_1, t_1+\mathrm{d}t_1)$, $[t_2, t_2+\mathrm{d}t_2)$, respectively, on condition that they did not fail before. Similar, e.g., $\bar{\lambda}_{b1}(t_1,t_2)\mathrm{d}t$ can be interpreted as the probability of failure of the first item in $(t_1, t_1+\mathrm{d}t]$ on condition that it did not fail in $[0,t_1]$ and that the second item also did not fail in $[0,t_2]$.

For the series system, (2.63) is obviously modified to:

$$\begin{aligned}S_b(t) &= \exp\left\{-\int_0^t \lambda_{b1}(u)\mathrm{d}u\right\}\exp\left\{-\int_0^t \lambda_{b2}(u)\mathrm{d}u\right\}\\ &\quad\times\exp\left\{\int_0^t\int_0^t(\lambda_b(u,v)-\bar{\lambda}_{b1}(u,v)\bar{\lambda}_{b2}(u,v))\mathrm{d}u\mathrm{d}v\right\},\end{aligned} \tag{2.66}$$

A natural generalization of the univariate PH model, $\lambda_s(t)=k\lambda_b(t)$, $k>0$ to the case of a series system of two possibly dependent components would be to consider multiplying each failure rate in (2.66) by its own multiplier, i.e.,

$$\begin{aligned}S_s(t) &= \exp\left\{-\alpha_1\int_0^t \lambda_{b1}(u)\mathrm{d}u\right\}\exp\left\{-\alpha_2\int_0^t \lambda_{b2}(u)\mathrm{d}u\right\}\\ &\quad\times\exp\left\{\int_0^t\int_0^t(\beta_1\lambda_b(u,v)-\beta_2\bar{\lambda}_{b1}(u,v)\bar{\lambda}_{b2}(u,v))\mathrm{d}u\mathrm{d}v\right\},\end{aligned} \tag{2.67}$$

where $\alpha_i>0$, $\beta_i\ge 0$; $i=1,2$. Thus a more severe environment acts directly on each type of the failure rate.

It can be proved [25] that the sufficient conditions for $S_s(t)$ to be a survival function are:

- $\beta_2\ge\beta_1$;
- $\alpha_i-\beta_2\ge 0,\ i=1,2$;
- $\frac{\lambda(u,v)}{\lambda_1(u,v)\lambda_2(u,v)}\ge\frac{\beta_2}{\beta_1};\ u,v\ge 0.$

Thus under these assumptions, (2.67) defines the bivariate competing risks PH model. The generalization to $n > 2$ can be performed, but it is much more cumbersome. The following example will help to understand the meaning of the quantities involved.

Example 2.13 As a specific case, we will consider the Clayton survival function of Example 2.11, but now we can define parameter $\theta > 0$ explicitly via the failure rates as it should be done in the PH-type reasoning. Let

$$\frac{\lambda_b(u,v)}{\bar{\lambda}_{b1}(u,v)\bar{\lambda}_{b2}(u,v)} = 1+\theta.$$

Thus,

$$\lambda_b(u,v) - \bar{\lambda}_{b1}(u,v)\bar{\lambda}_{b2}(u,v) = \frac{\theta}{1+\theta}\lambda_b(u,v).$$

Constructing the PH model for this case results in:

$$\begin{aligned}&\beta_1\lambda_b(u,v) - \beta_2\bar{\lambda}_{b1}(u,v)\bar{\lambda}_{b2}(u,v)\\ &\quad = \beta(\lambda_b(u,v) - \bar{\lambda}_{b1}(u,v)\bar{\lambda}_{b2}(u,v)),\end{aligned}$$

where β denotes $\frac{1+\theta}{\theta}\left(\beta_1 - \frac{\beta_2}{1+\theta}\right)$. Thus we have reduced the number of parameters of proportionality to 3. Eventually, the corresponding survival function for a more severe regime can be written as

$$S_s(t) = S_{b1}^{\alpha_1-\beta}(t)S_{b2}^{\alpha_2-\beta}(t)\left(S_{b1}^{-\theta}(t) + S_{b2}^{-\theta}(t) - 1\right)^{-\beta\theta^{-1}},$$

which generalizes the Clayton survival function.

2.8 Simplest Stochastic Orders

Throughout this book, we will extensively use several *simplest* stochastic orders for random variables of interest that will be briefly defined in this section. For the comprehensive theory of stochastic ordering, the reader should refer to Shaked and Shanthikumar [45].

Let X and Y be the lifetimes (non-negative random variables) with distribution functions $F(t)$ and $G(t)$, respectively. Assume that the corresponding means are finite. The simplest and one of the weakest stochastic orders is the order with *respect to the means*. Thus, we say that X is larger than Y in this sense, if

$$E[X] \geq E[Y]. \tag{2.68}$$

The first moment is a useful characteristic, but usually more information is needed for better characterization of random variables. Therefore, we say that the random variable X is *stochastically larger* than the random variable Y and write [44]

$$X \geq_{st} Y$$

if $F(t) \leq G(t), \ \forall t \geq 0$, or equivalently,

$$\overline{F}(t) \geq \bar{G}(t) \quad \forall t \geq 0. \tag{2.69}$$

Sometimes in the literature, the terms "*usual stochastic ordering*" or "stochastic dominance" are also used. It is obvious that (2.52) follows from (2.53) as, in accordance with (2.1),

$$E[X] = \int_0^\infty \overline{F}(u)\mathrm{d}u \geq \int_0^\infty \bar{G}(u)\mathrm{d}u = E[Y].$$

The next type of ordering is defined via the corresponding failure rates. The failure rate is a crucial characteristic for reliability and survival analysis and, therefore, this type of ordering is used very often. Assume that the failure rates $\lambda_X(t)$ and $\lambda_Y(t)$ exist. We say that X is larger than Y in the sense of the *hazard (failure) rate ordering*, if

$$\lambda_X(t) \leq \lambda_Y(t), \quad \forall t \geq 0. \tag{2.70}$$

It is clear that Inequality (2.69) follows from Inequality (2.70) as

$$\overline{F}(t) = \exp\left\{-\int_0^t \lambda_X(u)\mathrm{d}u\right\} \geq \exp\left\{-\int_0^t \lambda_Y(u)\mathrm{d}u\right\} = \bar{G}(t).$$

Thus, the hazard rate ordering is obviously stronger than the usual stochastic ordering.

Denote by $f(t)$ and $g(t)$ the probability density functions that correspond to $F(t)$ and $G(t)$, respectively. We say that X is larger than Y in the sense of the *likelihood ratio ordering* and write

$$X \geq_{LR} Y$$

if

$$\frac{f(x)}{g(x)} \leq \frac{f(y)}{g(y)} \quad \text{for all } x \leq y, \tag{2.71}$$

which means that the ratio of the densities $f(x)/g(x)$ is increasing in x. We will use this ordering extensively in Chap. 5. It turns out that (2.71) is a natural ordering for lifetimes in heterogeneous populations. It can be easily proved [44] that ordering in the sense of the likelihood ratio is stronger than the hazard rate ordering.

Sometimes we need to compare the 'variability' of random variables. Assume that $E[X] = E[Y]$ and that

$$E[h(X)] \geq E[h(Y)] \text{ for all convex } h(x). \tag{2.72}$$

Then intuitively, it is clear that X will be more variable than Y. For instance, when $h(x) = x^2$, it is easy to see that $Var(X) \geq Var(Y)$.

It can be proved that (2.72) is equivalent to the following inequality that can be already effectively analyzed:

$$\int_t^{\infty} \overline{F}(u)\mathrm{d}u \geq \int_t^{\infty} \bar{G}(u)\mathrm{d}u \quad \forall t \geq 0. \tag{2.73}$$

When $t = 0$, (2.73) obviously reduces to (2.68).

References

1. Aalen OO, Borgan O, Gjessing HK (2008) Survival and event history analysis. Springer, New York
2. Anderson PK, Borgan O, Gill RD, Keiding N (1993) Statistical models based on counting processes. Springer, New York
3. Aven T, Jensen U (1999) Stochastic models in reliability. Springer, New York
4. Bagdonavicius V, Nikulin M (2002) Accelerated life models. Modelling and statistical analysis. Chapman & Hall, Boca Raton
5. Bai DS, Yun WJ (1986) An age replacement policy with minimal repair cost limit. IEEE Trans Reliab 31:452–459
6. Baxter LA, Kijima M, Tortorella M (1996) A point process model for the reliability of the maintained system subject to general repair. Stoch Models 12:37–65
7. Banevich D (2009) Remaining useful life in theory and practice. Metrica 69:337–349
8. Barlow RE, Hunter LC (1960) Optimal preventive maintenance policies. Oper Res 8:90–100
9. Beichelt FE, Fischer K (1980) General failure model applied to preventive maintenance policies. IEEE Trans Reliab 29:39–41
10. Bergman B (1985) Reliability theory and its applications. Scand J Stat 12:1–41
11. Block HW, Borges WS, Savits TH (1985) Age-dependent minimal repair. J Appl Probab 22:370–386
12. Block HW, Li Y, Savits TH (2003) Initial and final behavior of failure rate functions for mixtures and systems. J Appl Probab 40:721–740
13. Brown M, Proschan F (1983) Imperfect repair. J Appl Probab 20:851–859
14. Brown J, Mahoney J, Sivazlian B (1983) Hysteresis repair in discounted replacement problems. IIE Trans 15:156–165

15. Cox DR, Isham V (1980) Point processes. University Press, Cambridge
16. Cox DR, Lewis PAW (1966) The statistical analysis of series of events. Methuen, London
17. Cox DR, Oakes D (1984) Analysis of survival data. Chapman and Hall, London
18. Cinlar E (1975) Introduction to stochastic processes. Prentice Hall, Englewood Cliffs
19. Clayton DG (1978) A model of association in bivariate life tables and its application in epidemiological studies of familial tendency in chronic disease incidence. Biometrika 65:141–151
20. Clayton DG, Cusick J (1985) Multivariate generalizations of the proportional hazards model. J Roy Stat Soc 148:82–117
21. Doyen L, Gaudoin O (2004) Classes of imperfect repair models based on reduction of failure intensity or virtual age. Reliab Eng Syst Safe 84:45–56
22. Finkelstein M (1997) A concealed age of distribution functions and the problem of general repair. J Stat Plan Inference 65:315–321
23. Finkelstein (1989) Perfect, minimal and imperfect repair (In Russian). Reliab Qual Control 3:17–21
24. Finkelstein M (1999) Wearing-out components in variable environment. Reliab Eng Syst Safe 66:235–242
25. Finkelstein M (2008) Failure rate modelling for reliability and risk. Springer, London
26. Finkelstein M (2013) On dependent items in series in different environments. Reliab Eng Syst Safe 109:119–122
27. Guess F, Proschan F (1988) Mean residual life: theory and applications. In: Krishnaiah PR, Rao CR (eds) Handbook of statistics, vol 9. Elsevier, Amsterdam, pp 215–224
28. Gupta RC (2001) Nonmonotonic failure rates and mean residual life functions. In: Hayakawa Y, Irony T, Xie M (eds) System and Bayesian reliability: essays in honour of Professor R.E. Barlow. Series on quality, reliability and engineering statistics, World Scientific Press, Singapore, pp 147–163
29. Gupta RC, Akman HO (1995) Mean residual life functions for certain types of nonmonotonic aging. Commun Stat-Stoch Models 11:219–225
30. Gupta RC, Warren R (2001) Determination of change points of nonmonotonic failure rates. Commun Stat-Theor Methods 30:1903–1920
31. Glaser RE (1980) Bathtub and related failure rate characterizations. J Am Stat Assoc 75:667–672
32. Johnson NL, Kotz S (1975) A vector multivariate hazard rate. J Multivar Anal 5:53–66
33. Keyfitz N (1985) Applied mathematical demography. Springer, New York
34. Kijima M (1989) Some results for repairable systems with general repair. J Appl Probab 26:89–102
35. Kijima M, Sumita U (1986) A useful generalization of renewal theory: counting processes governed by non-negative Markovian increments. J Appl Probab 23:72–78
36. Kijima M, Morimura H, Suzuki Y (1988) Periodical replacement problem without assuming minimal repair. Eur J Oper Res 37:194–203
37. Lai CD, Xie M (2006) Stochastic ageing and dependence for reliability. Springer, New York
38. Marshall AW, Olkin I (2007) Life distributions. Springer, New York
39. Meeker WQ, Escobar LA (1998) Statistical methods for reliability data. Wiley, New York
40. Meilijson I (1972) Limiting properties of the mean residual lifetime function. Ann Math Stat 43:354–357
41. Mi J (1995) Bathtub failure rate and upside-down bathtub mean residual life. IEEE Trans Reliab 44:388–391
42. Nelsen RB (2006) Introduction to copulas. Springer, New York
43. Navarro J, Hernandez PJ (2004) How to obtain bathtub-shaped failure rate models from normal mixtures. Probab Eng Inform Sci 18:511–531

44. Ross SM (1996) Stochastic processes, 2nd edn. Wiley, New York
45. Shaked M, Shanthikumar J (2007) Stochastic orders. Springer, New York
46. Stadje W, Zuckerman D (1991) Optimal maintenance strategies for repairable systems with general degree of repair. J Appl Probab 28:384–396
47. Vaupel JW, Yashin AI (1987) Repeated resuscitation: how life saving alters life tables. Demography 4:123–135
48. Zahedi H (1991) Proportional mean remaining life model. J Stat Plan Inference 29:221–228

Chapter 3
Shocks and Degradation

This chapter is mostly devoted to basic shock models and their simplest applications. Along with discussing some general approaches and results, we want to present the necessary material for describing our recent findings on shocks modeling of the next chapter. As in the other chapters of this book, we do not intend to perform a comprehensive literature review of this topic, but rather concentrate on notions and results that are vital for further presentation.

We understand the term "shock" in a very broad sense as some instantaneous, potentially harmful event (e.g., electrical impulses of large magnitude, demands for energy in biological objects, insurance claims in finance, etc.). Shock models are widely used in practical and theoretical reliability and in the other disciplines as well. They can also constitute a useful framework for studying aging properties of distributions [2, 3]. It is important to analyze the consequences of shocks to a system (object) that can be *basically* two fold. First, under certain assumptions, we can consider shocks that can either 'kill' a system, or be successfully survived without any impact on its future performance. The corresponding models are usually called the *extreme shock models*, whereas the setting when each shock results in an additive damage (wear) to a system is often described in terms of the *cumulative shock models* ([18–20] to name a few). In the latter case, the failure occurs when the cumulative effect of shocks reaches some deterministic or random level, and therefore, this setting is useful for modeling of degradation (wear) processes. The combination of these two basic models has been also considered in the literature [5, 6, 19].

In Sect. 3.1, we first briefly discuss several simplest stochastic models of wear that are helpful in describing basic cumulative shock models. In the rest of this chapter, we mostly consider the basic results with respect to the extreme and cumulative shock models, and also describe several meaningful modifications, and applications of the extreme shock model. For instance, in Sect. 3.8, a meaningful safety at sea application is considered and in Sect. 3.9, the famous in demography Strehler–Mildvan model of human mortality is discussed from our view point.

M. Finkelstein and J. H. Cha, *Stochastic Modeling for Reliability*,
Springer Series in Reliability Engineering, DOI: 10.1007/978-1-4471-5028-2_3,

3.1 Degradation as Stochastic Process

Stochastic degradation in engineering, ecological, and biological systems is naturally modeled by increasing (decreasing) stochastic processes. The additive nature of the cumulative shock models implies that the corresponding degradation should be strictly monotone. However, it is well-known (e.g., [3] that, for example, the Wiener process with drift (see Definition 3.1) with the nonmonotone realizations under certain assumptions can be also considered as a useful tool for modeling the monotone degradation. In the previous chapter, several point processes were discussed that can be used for modeling degradation induced by shocks in the corresponding cumulative shock models. We will consider now the simplest *continuous-time* stochastic processes, and will be interested in modeling stochastic degradation as such and in obtaining the corresponding distributions for the first passage times when this degradation reaches the predetermined or random level D for the first time. When D defines some critical safety boundary, the latter interpretation can be useful for risk and safety assessment. For instance, when degradation in some structures results in the decreasing resistance to loads, it can result not just in an 'ordinary' failure, but in a severe catastrophic event.

We will briefly define now several approaches, which are most often used in engineering practice for degradation modeling. The simplest and the widely used one is the path model. Its stochastic nature is described either by the additive or by the multiplicative random variable in the following way:

$$W_t = \eta(t) + Z, \tag{3.1}$$

$$W_t = \eta(t)Z, \tag{3.2}$$

where $\{W_t,\ t \geq 0\}$ denotes our stochastic process, $\eta(t)$ is an increasing, continuous function $(\eta(0) = 0,\ \lim_{t\to\infty} \eta(t) = \infty)$ and Z is a nonnegative random variable with the Cdf $G(z)$. Therefore, the sample paths (realizations) for these models are monotonically increasing. The 'nature' of this stochastic process is simple and meaningful: let the failure (catastrophe) be defined as reaching by $\{W_t, t \geq 0\}$ the degradation threshold $D > 0$ and T_D be the corresponding time to failure random variable with the Cdf $F_D(t)$. It follows, e.g., for the model (3.2) that:

$$F_D(t) = P(W_t \geq D) = \Pr\left(Z \geq \frac{D}{\eta(t)}\right) = 1 - G\left(\frac{D}{\eta(t)}\right). \tag{3.3}$$

Example 3.1 Let $\eta(t) = t$ and assume that Z is described by the Weibull distribution, i.e., $G(z) = 1 - \exp\left\{-(\lambda z)^k\right\}$, $\lambda, k > 0$. Then, in accordance with (3.3),

$$F_D(t) = \exp\left\{-\left(\frac{\lambda D}{t}\right)^k\right\},$$

which is often called the Inverse-Weibull distribution [1]. Specifically, when $\lambda = 1, k = 1$:

$$F_D(t) = \exp\left\{-\frac{D}{t}\right\}.$$

It is clear that the value at $t = 0$ for this distribution should be understood as

$$F_D(0) = \lim_{t \to 0} F_D(t) = 0.$$

The Inverse-Weibull distribution is a convenient simple tool for describing threshold models with a linear function $\eta(t)$.

Assume now that the threshold D is a random variable with the Cdf $F_0(d) = \Pr(D \leq d)$ and let, at first, degradation be modeled by the deterministic, increasing function $W(t)$ ($W(0) = 0$, $\lim_{t\to\infty} W(t) = \infty$). Equivalently, the problem can be reformulated in terms of the fixed threshold and random initial value of degradation. Denote by T the random time to failure. As events $T \leq t$ and $W(t)$ are equivalent, similar to (3.3) [12],

$$F(t) \equiv P(T \leq t) = P(D \leq W(t)) = F_0(W(t)), \tag{3.4}$$

where the last equality is due to the fact that the Cdf of D is $F_0(d)$. Substituting d by $W(t)$, finally results in (3.4).

Let now the deterministic degradation $W(t)$ in (3.4) be replaced by a stochastic process $W_t, t \geq 0$. In order to derive the corresponding distribution of the time to failure in this case we must obtain the expectation of $F_0(W_t)$ with respect to the process $W_t, t \geq 0$:

$$F(t) = E\left[F_0(W)_t\right]. \tag{3.5}$$

This equation is too general, as the stochastic process is not specified. The following example considers the multiplicative path model for $W_t, t \geq 0$.

Example 3.2 Let, e.g., $F_0(d) = 1 - \exp\{-\lambda d\}$ and $W_t = \eta(t)Z$, where Z is also exponentially distributed with parameter μ. Direct integration in (3.5) gives:

$$\begin{aligned} F(t) &= E[1 - \exp\{-\lambda\eta(t)Z\}] \\ &= \int_0^\infty (1 - \exp\{-\lambda\eta(t)z\})\mu\exp\{\ \mu z\} \\ &= 1 - \frac{\mu}{\mu + \lambda\eta(t)}. \end{aligned}$$

The path model can be very useful for illustration. However, obviously, the real life stochastic processes are much more complex. Probably, the most popular in applications and well investigated from the formal point of view stochastic process is the Wiener process. The Wiener process with drift is often used for modeling wear although its sample paths are not monotone (but the mean of the process is a monotonically increasing function).

Definition 3.1 Stochastic process $\{W_t, t \geq 0\}$ is called the Wiener process with drift

$$W_t = \mu t + X(t),$$

where $\mu > 0$ is a drift parameter and $X(t)$ is a standard Wiener process: for the fixed $t \geq 0$, the random variable $X(t)$ is normally distributed with zero mean and variance $\sigma^2 t$.

It is well-known (see, e.g., Cox and Miller [8] that the first passage time T_D, i.e.,

$$T_D = \inf_t \{t, W_t > D\}$$

for this process is described by the inverse Gaussian distribution:

$$\overline{F}_D(t) = \Pr(T_D > t) = \Phi\left(\frac{D - \mu t}{\sqrt{t}\,\sigma}\right) - \exp\{-2D\mu\}\Phi\left(\frac{D + \mu t}{\sqrt{t}\,\sigma}\right) \tag{3.6}$$

and

$$E[T_D] = \frac{D}{\mu}, \quad Var(T_D) = \frac{D\sigma^2}{\mu^3},$$

where, as usual, $\Phi(t)$, denotes the Cdf of the standard normal random variable.

Another popular process for modeling degradation is the gamma process (see, e.g., the perfect survey by Van Nortwijk [30]). Although, parameter estimation for the degradation models driven by the gamma process is usually more complicated than for the Wiener process, it better captures the desired monotonicity.

Definition 3.2 The gamma process is a stochastic process $(W_t, t \geq 0)$, $W_0 = 0$ with independent nonnegative increments having a gamma Cdf with identical scale parameters. The increment $W_t - W_\tau$ has a gamma distribution with a shape parameter $v(t) - v(\tau)$ and a scale parameter u, where $v(t)$ is an increasing function $(v(0) = 0)$.

Thus W_t for each fixed t is gamma-distributed with shape parameter $v(t)$ and scale parameter u, whereas

$$E[W_t] = \frac{v(t)}{u}, \quad Var(W_t) = \frac{v(t)}{u^2}.$$

The first passage time T_D, is described in this case by the following distribution [30]

$$F_D(t) = \Pr(T_D \leq t) = \Pr(W_t \geq D) = \frac{\Gamma(v(t), Du)}{\Gamma(v(t))},$$

where $\Gamma(a,x) = \int_x^\infty t^{a-1}e^{-t}dt$ is an incomplete gamma function for $x > 0$. Thus, deterioration with independent increments can be often modeled by the gamma process.

3.2 Shocks and Shot Noise Process

A natural way of modeling additive degradation is via the sum of random variables, which represent the degradation increments:

$$W_t = \sum_1^n X_i,$$

where $X_i, i = 1, 2, \ldots, n$ are positive i.i.d. random variables with a generic variable denoted by X, and n is an integer.

The next step to a more real stochastic modeling is to view n as a random variable N (the compound random variable) or a point process $\{N_t, t \geq 0\}$. The latter is counting the point events of interest in $[0, t), t \geq 0$ (the compound point process):

$$W_t = \sum_1^{N_t} X_i. \tag{3.7}$$

Denote by $Y_i, i = 1, 2, \ldots$ a sequence of inter-arrival times for $\{N_t, t \geq 0\}$. If $Y_i, i = 1, 2, \ldots$ are i.i.d (and this case will be considered in what follows) with a generic variable Y, then the Wald's equation [26] immediately yields

$$E[W_t] = E[N_t]E[X],$$

where, specifically for the compound Poisson process with rate m: $E[N_t] = mt$. Note that [9] under certain assumptions the stationary gamma process ($v(t) = vt$) can be viewed as a limit of a specially constructed compound Poisson process.

Relationship (3.7) has a meaningful interpretation via shocks, as $X_i, i = 1, 2, \ldots$ can be interpreted as an amount of damage caused by the ith shock. An important modification of this additive model is given by the shot noise process [25, 26]. In a shot noise point process, an additive input of a shock of magnitude X_i is decreased in accordance with some decreasing (nonincreasing) response function $h(t - s)$. Therefore, Eq. (3.7) turns to

$$W_t = \sum_1^{N_t} X_i h(t - \tau_i), \tag{3.8}$$

where $\tau_1 < \tau_2 < \tau_3, \ldots$ is the sequence of the corresponding arrival (waiting) times in the point process. This setting has a lot of applications in electrical engineering, materials science, health sciences, risk, and safety analysis. For instance, cracks due to fatigue in some materials tend to close up after the material has borne a load, which has caused the cracks to grow. Another example is the human heart muscle's tendency to heal after a heart attack [27]. Thus, the inputs of each shock in the accumulated damage decrease with time.

Equivalently, (3.8) can be written as:

$$W_t = \int_0^t Xh(t-u)dN_u,$$

where $dN_u = N(u, u + \mathrm{d}u)$ denotes the number of shocks in $[u, u + \mathrm{d}u)$.

First, we are interested in the mean of the defined process. Assume that $E[X] < \infty$. As $X_i, i = 1, 2, \ldots$ are independent from the point process $\{N_t, t \geq 0\}$,

$$E[W_t] = E[X] \int_0^t h(t-u)dN_u = E[X] \int_0^t h(t-u)m(u)\mathrm{d}u, \tag{3.9}$$

where $m(u) = dE[N_u]/\mathrm{d}u$ is the rate (intensity) of the point process. For the Poisson process, $m(u) = m$ and:

$$E[W_t] = mE[X] \int_0^t h(u)\mathrm{d}u. \tag{3.10}$$

Therefore, asymptotically the mean accumulative damage is finite, when the response function has a finite integral, i.e.,

$$\lim_{t\to\infty} E[W_t] < \infty \ , \ if \ \int_0^\infty h(u)\mathrm{d}u < \infty.$$

This property has an important meaning in different engineering and biological applications. It can be shown directly that, if $E[X^2] < \infty$:

$$Cov(W_{t_1}, W_{t_2}) = mE[X^2] \int_0^{t_1} h(t_1-u)h(t_2-u)\mathrm{d}u; \quad t_1 \geq t_2.$$

The central limit theorem for the sufficiently large m also takes place in the following form [23, 24]:

$$\frac{W_t - E[W_t]}{(Var(W_t))^{1/2}} \to^D N(0,1), \, t \to \infty \ , \tag{3.11}$$

where the sign "D" means convergence in distribution and $N(0,1)$ denotes the standard normal distribution. The renewal case with the interarrival time denoted by X gives similar results

$$\lim_{t\to\infty} E[W_t] = \frac{1}{E[X]}\int_0^\infty h(u)\mathrm{d}u.$$

Example 3.3 Consider a specific exponential case of the response function $h(u)$ and the Poisson process of shocks with rate m:

$$W_t = \sum_1^{N_t} X_i \exp\{\alpha(t-\tau_i)\}.$$

By straightforward calculations [26], using the technique of the moment generating functions, it can be shown that the stationary value of W_t for t sufficiently large is described by the gamma distribution with mean $m/\lambda\alpha$ and variance $m/\lambda^2\alpha$. Moreover, the distribution of the first passage time is given by

$$F_D(t) = \Pr(T_D \le t) = \Pr(W_t \ge D) = \frac{\Gamma(m/\alpha, D\lambda)}{\Gamma(m/\alpha)}.$$

It is well-known from the properties of the gamma distribution that as m/λ increases, it converges to the normal distribution and, therefore, there is no contradiction between this result and asymptotic relation (3.11).

In the next chapter, we will consider another shot noise model where the shot-noise process models the failure rate of an object. Some meaningful generalizations will be also considered.

3.3 Asymptotic Properties

In many applications, the number of shocks in the time interval of interest is large, which makes it possible to apply the corresponding asymptotic methods.

Consider a family of nonnegative, i.i.d, two-dimensional random vectors $\{(X_i, Y_i), i \ge 0\}, X_0 = 0, Y_0 = 0$, where $\sum_1^n X_i$ is the accumulated damage after n shocks and $Y_i, i = 1, 2, \ldots$ is the sequence of the i.i.d inter-arrival times of the corresponding renewal process. Recall that the renewal process is defined by the sequence of the i.i.d inter-arrival times. Specifically, when these times are exponentially distributed, the renewal process 'reduces' to the Poisson process. We will assume for simplicity that X and Y are independent, although the case of dependent variables can be also considered [19]. Let $0 < E[X]$, $E[Y] < \infty$, $0 < Var(X)$, $Var(Y) < \infty$. It follows immediately from (3.7) and the elementary renewal theorem [26] that

$$\lim_{t\to\infty}\frac{E[W_t]}{t}=\lim_{t\to\infty}\frac{E[N_t]E[X]}{t}=\frac{E[X]}{E[Y]}. \tag{3.12}$$

The corresponding central limit theorem can be proved using the theory of stopped random walks [19]

$$\frac{W_t-(E[X]/E[Y])t}{(E[Y])^{-3/2}\sigma\, t^{1/2}}\to N(0,1),\ t\to\infty. \tag{3.13}$$

where $\sigma=\sqrt{\mathrm{var}(E[Y]X-E[X]Y)}$.

Relationship (3.13) means that for large t, the random variable W_t is approximately normally distributed with expected value $(E[X]/E[Y])t$ and variance $(E[Y])^{-3}\,\sigma^2(E[X])^2\,t$. Therefore, we need only $E[X]$, $E[Y]$ and σ for the corresponding asymptotic analysis, which is very convenient in practice.

Similar to (3.12),

$$\lim_{t\to\infty}\frac{E[T_D]}{D}=\lim_{D\to\infty}\frac{E[N_D]E[Y]}{D}=\frac{E[Y]}{E[X]}, \tag{3.14}$$

where N_D denotes a random number of shocks to reach the cumulative value D. Equation (3.13) can be now rewritten for the distribution of the first passage time T_D as [19]

$$\frac{T_D-(E[Y]/E[X])D}{(E[X])^{-3/2}\,\sigma\, D^{1/2}}\to N(0,1),\ D\to\infty.$$

This equation means that for large threshold D the random variable T_D can be approximately described by a normal distribution with expected value $(E[Y]/E[X])D$, and variance $(E[X])^{-3}\,\sigma^2\,D$. Therefore, the results of this section can be easily and effectively used in safety and reliability analysis.

3.4 Extreme Shock Models

Let the shocks occur in accordance with a renewal process or a nonhomogeneous Poisson process. Each shock *independently of the previous history* leads to a failure of a system with probability p and is survived with the complementary probability $q=1-p$. Assume, that a shock is the only cause of failure. We see that there is no accumulation of damage and the fatal 'damage' can be a consequence of a single shock. Numerous problems in reliability, risk, and safety analysis can be interpreted by means of this model. This setting is often referred to as an *extreme shock model* [12, 18]. Our main interest in the rest of this chapter will be in different settings, and applications that are described within the framework of the extreme shock model. We will use these results and reasoning in the rest of this book.

Consider first, a general point process $\{T_n\}; T_0 = 0, T_{n+1} > T_n, n = 0, 1, 2, \ldots$, where T_n is the time to the nth arrival of an event with the corresponding cumulative distribution function $F^{(n)}(t)$. Therefore, $F^{(n)}(t) - F^{(n+1)}(t)$ is the probability of exactly n events in $[0, t)$; $F^{(0)}(t) \equiv 1$, $F^{(1)}(t) \equiv F(t)$. Let G be a geometric variable with parameter p (independent of $\{T_n\}_{n \geq 0}$) and denote by T a random variable with the following survival function

$$P(t) = \sum_{k=0}^{\infty} q^k \left(F^{(k)}(t) - F^{(k+1)}(t)\right). \tag{3.15}$$

Thus $P(t)$ is the system's survival probability for the described extreme shock model. We can also interpret the setting in terms of the terminating point process when $1 - P(t)$ is the probability of its termination in $[0, t)$.

Obtaining probability $P(t)$ is an important problem in various reliability and safety assessment applications. It is clear that in this general form, Eq. (3.15) does not allow for explicit results that can be used in practice, and therefore, assumptions on the type of the point process of shocks should be made. Two specific point processes are mostly used in reliability applications, i.e., the Poisson process and the renewal process. For the homogeneous Poisson process with rate λ, the derivation is trivial

$$P(t) = \sum_{0}^{\infty} q^k \exp\{-\lambda t\} \frac{(\lambda t)^k}{k!} = \exp\{-p\lambda t\}. \tag{3.16}$$

It follows from (3.16) that the corresponding constant failure rate, which describes the lifetime of our system T, is given by a simple and meaningful relationship

$$\lambda_S = p\,\lambda. \tag{3.17}$$

Thus, the rate of the underlying Poisson process λ is decreased by the factor $p \leq 1$.

This result can be generalized to the case of the NHPP with rate $\lambda(t)$ and time-dependent probability $p(t)$. It is clear that the Brown–Proschan model of Chap. 2 described by Eqs. (2.17–2.19) can be interpreted in terms of our extreme shock model, and therefore,

$$P(t) = 1 - \exp\left\{-\int_0^t p(u)\lambda(u)\mathrm{d}u\right\} \tag{3.18}$$

with the corresponding failure rate

$$\lambda_S(t) = p(t)\lambda(t).$$

Numerous generalizations of these results under the assumption of the underlying NHPP of shocks will be considered further in this chapter and in the next chapter as well. In spite of its relative simplicity, the renewal process of shocks does not allow for the similar explicit relationships. However, it is well-known (see, e.g., [21]) that, as $p \to 0$, the following convergence in distribution takes place:

$$P(t) \to \exp\left\{-\frac{pt}{\mu}\right\}, \quad \forall t \in (0, \infty) , \tag{3.19}$$

where μ is the mean that corresponds to the governing distribution. Thus, (3.19) constitutes a very simple asymptotic exponential approximation. In practice, however, parameter p is not usually sufficiently small for using effectively this approximation, and therefore, the corresponding bounds for $P(t)$ can be very helpful.

The simplest and useful in practice but a rather crude bound for the survival function can be obtained via the following identity:

$$E\left[q^{N_t}\right] = \sum_{k=0}^{\infty} q^k \left(F^{(k)}(t) - F^{(k+1)}(t)\right).$$

Finally, using Jensen's inequality [12]:

$$P(t) = E\left[q^{N_t}\right] \geq q^{E[N_t]}.$$

In the next three sections, the extreme shock model with the *homogeneous* Poisson process of shocks will be generalized to different settings that can occur in practice [13]. For instance, the probability of a failure of an operable system under a shock, which is in conventional models either a constant or depends only on chronological time t, can depend also on a state of a system. This is a natural assumption, as resistance to shocks, e.g., in multistate systems (discrete or continuous) often depends on the current state of a system. Another extension of conventional models to be considered is when the failure occurs if two successive shocks 'are too close' to each other. A system in this case cannot recover from the consequences of the previous shock. This setting is similar to that of the σ -shock model considered in the literature [22, 28], however, our method allows for more general and flexible results. The main analytical tool allowing for the *explicit solutions* for all mentioned settings is the method of integral equations developed in Finkelstein [12]. These equations can be effectively solved in terms of the Laplace transform and explicitly inverted for the sufficiently simple cases.

3.5 State-Dependent Probability of Termination

Consider first, the Poisson process of shocks with rate λ and probability of failure (termination) on each shock, p. In this case, the survival probability is given by Eq. (3.16). In order, to illustrate the *method of integral equations* to be used further [13] we will describe how it works for this simplest case. It is easy to see that the following integral equation with respect to $P(t)$ holds

$$P(t) = e^{-\lambda t} + \int_0^t \lambda e^{-\lambda x} q P(t-x) \mathrm{d}x . \tag{3.20}$$

The first term, on the right hand side is the probability that there are no shocks in $[0, t)$ and the integrand defines the probability that the first shock that have occurred in $[x, x + \mathrm{d}x)$ was survived and then the system have survived in $[x, t)$. Due to the properties of the homogeneous Poisson process, the probability of the latter event is $P(t - x)$.

We have now a simple integral equation with respect to the unknown function $P(t)$. Applying the Laplace transform to both sides of Eq. (3.20) results in

$$\widetilde{P}(s) = \frac{1}{s+\lambda} + \frac{\lambda q}{s+\lambda}\widetilde{P}(s) \Rightarrow \widetilde{P}(s) = \frac{1}{s+\lambda p},$$

where $\widetilde{P}(s)$ denotes the Laplace transform of $P(t)$. The corresponding inversion results in $\exp\{-p\lambda t\}$.

Consider now a repairable system with instantaneous, perfect repair that starts functioning at $t = 0$. Let its lifetime be described by the Cdf $F(t)$, which is a governing distribution for the corresponding renewal process with the renewal density function to be denoted by $h(t)$. Assume, that the quality of performance of our system is characterized by some deterministic for simplicity function of performance $Q(t)$ to be called the quality function. The considered approach can be generalized to the case of a random $Q(t)$. It is often a decreasing function of time, and this assumption is quite natural for degrading systems. In applications, the function $Q(t)$ can describe some key parameter of a system, e.g., the decreasing in time accuracy of the information measuring system or effectiveness (productivity) of some production process. As repair is perfect, the quality function is also restored to its initial value $Q(0)$. It is clear that the quality function of our system at time t is now random and equal to $Q(Y)$, where Y is a random time since the last (before t) repair.

The system is subject to the Poisson process of shocks with rate λ. As previously, each shock can terminate the performance of the repairable system and we are interested in obtaining the survival probability $P(t)$. Note, that the repaired failure of the system does not terminate the process and only a shock can result in termination. Assume, that the probability of termination depends on the system's quality at the time of a shock. This is a reasonable assumption meaning that the larger value of quality implies the smaller probability of termination. Let the first shock arrive before the first failure of the system. Denote by $p^*(Q(t))$ the corresponding probability of termination in this case. Now we are able to obtain $p(t)$—the probability of termination of the operating system *by the first shock* at time instant t. Using the standard 'renewal-type reasoning' [13], the following relationship for $p(t)$ can be derived

$$p(t) = p^*(Q(t))\overline{F}(t) + \int_0^t h(x)\overline{F}(t-x)p^*(Q(t-x))\mathrm{d}x , \tag{3.21}$$

where $\overline{F}(t) \equiv 1 - F(t)$.

The first term on the right-hand side of Eq. (3.21) gives the probability of termination during the first cycle of the renewal process, whereas $h(x)\overline{F}(t-x)\mathrm{d}x$ defines the probability that the last failure (renewal) of the system before t had occurred in $[x, x+\mathrm{d}x)$ (as $h(x)\mathrm{d}x$ is the probability that a failure (renewal) had occurred in $[x, x+\mathrm{d}x)$ and $\overline{F}(t-x)$ is the probability that no failure had occurred in $[x+\mathrm{d}x, t]$. Therefore, the corresponding probability of termination at t is equal to $p^*(Q(t-x))$.

Thus, the probability of termination under the first shock $p(t)$, which is now time-dependent, has been derived. Assume, now that the survived shock can be interpreted as an instantaneous, perfect repair of the system (the 'repaired shock' is survived, the 'non-repaired' results in termination). Therefore, the instants of survived shocks can be also considered as the renewal points for the system. Having this in mind, we can now proceed with obtaining the survival probability $P(t)$. Using the similar reasoning as when deriving Eq. (3.20)

$$P(t) = e^{-\lambda t} + \int_0^t \lambda e^{-\lambda x} q(x)\, P(t-x)\mathrm{d}x\,, \tag{3.22}$$

where $q(x) \equiv 1 - p(x)$.

Applying the Laplace transform to Eq. (3.22):

$$\begin{aligned} \widetilde{P}(s) &= \frac{1}{s+\lambda} + \lambda\widetilde{q}(s+\lambda)\widetilde{P}(s) \\ \Rightarrow \widetilde{P}(s) &= \frac{1}{(s+\lambda)(1-\lambda\widetilde{q}(s+\lambda))}. \end{aligned} \tag{3.23}$$

Given the functions $F(t)$ and $p^*(Q(t))$, Eqs. (3.21) and (3.23) can be solved numerically, but we can still proceed with the Laplace transforms under an additional assumption that the underlying distribution is exponential, i.e., $F(t) = 1 - \exp\{-ht\}$. In this case, $h(x) = h$ and the Laplace transform of Eq. (3.21) results in [13]

$$\widetilde{p}(s) = \widetilde{p}^*(s+h)\left(1 + \frac{h}{s}\right), \tag{3.24}$$

where $\widetilde{p}^*(s) = \int_0^\infty e^{-sx} p^*(Q(x))\mathrm{d}x$ denotes the Laplace transform of the function $p^*(Q(t))$. Substituting (3.24) into (3.23) and taking into account that $\widetilde{q}(s) = (1/s) - \widetilde{p}(s)$

$$\widetilde{P}(s) = \frac{1}{s + \lambda\widetilde{p}^*(s+h+\lambda)(s+h+\lambda)}. \tag{3.25}$$

To proceed further with inversion, we must make some assumptions on the form of the function $p^*(Q(t))$. Let $p^*(Q(t)) = 1 - \exp\{-\alpha t\}$, $\alpha \geq 0$. This is a reasonable assumption (as the probability of termination increases as $Q(t)$ decreases with t) that allows for a simple Laplace transform. Then

$$\widetilde{P}(s) = \frac{s + h + \lambda + \alpha}{s^2 + s(\lambda + h + \alpha) + \alpha\lambda}$$

and the inversion gives

$$P(t) = \frac{s_1 + \lambda + \alpha}{s_1 - s_2}\exp\{s_1 t\} - \frac{s_2 + \lambda + \alpha}{s_1 - s_2}\exp\{s_2 t\},$$

where

$$s_{1,2} = \frac{-(h + \lambda + \alpha) \pm \sqrt{(h + \lambda + \alpha)^2 - 4\lambda\alpha}}{2}.$$

An important specific case is when the system is absolutely reliable ($h = 0$) but is characterized by the quality function $Q(t)$. Then $s_1 = -\lambda,\ s_2 = -\alpha;\ \alpha \neq \lambda$ and

$$P(t) = \frac{\lambda}{\lambda - \alpha}\exp\{-\alpha t\} - \frac{\alpha}{\lambda - \alpha}\exp\{-\lambda t\}. \tag{3.26}$$

If, for instance, $p^*(Q(t)) = 1$, which means that $\alpha \to \infty$, then $P(t) = \exp\{-\lambda t\}$ as expected, the probability that there are no shocks in $[0, t)$. On the contrary, if $\alpha = 0$, which means that $p^*(Q(t)) = 0$, the survival probability is equal to 1. Another marginal case is defined by the value of the rate λ. If $\lambda = 0$, then again, as expected, $P(t) = 1$. On the other hand, it follows from (3.26) that as $\lambda \to \infty$,

$$P(t) \to \exp\{-\alpha t\}, \tag{3.27}$$

which can be confusing at first sight, as one would expect that when the rate of a shock process tends to infinity, the probability of survival in $[0, t)$ should tend to 0, but this is not the case because the function $p^*(Q(t)) = 1 - \exp\{-\alpha t\}$ is close to 0 for small t and each survived shock is the renewal point for our system. Therefore, as the number of shocks increases, due to the properties of exponential function, relationship (3.27) holds.

3.6 Termination with Recovery Time

In the previous sections, the only source of termination was an immediate effect of a shock. Consider now another setting that can be often encountered in practical reliability and safety analysis. Let, as previously, each shock from the Poisson process with rate λ terminate the process with probability p and be survived with probability $q = 1 - p$. Assume, now that termination additionally can also occur when the consecutive shocks are 'too close', which means that the system cannot recover from the consequences of a previous shock. Therefore, the time for recovering should be taken into account. It is natural to assume that it is a random variable τ with the Cdf $R(t)$ (different values of damage need different time of recovering and this fact is described by $R(t)$). Thus, if the shock occurs while the system still has not recovered from the previous non-terminating shock, it terminates the process. It is the simplest criterion of termination of this kind. Other criterions can be also considered. As previously, we want to derive $P(t)$—the probability of survival of our system in $[0,t)$.

First, assume that a shock had occurred at $t=0$ and has been survived. Denote the probability of survival under this condition by $P^*(t)$. Then the corresponding supplementary integral equation is

$$P^*(t) = e^{-\lambda t} + \int_0^t \lambda e^{-\lambda x} q R(x) P^*(t-x)\mathrm{d}x \,, \tag{3.28}$$

where the multiplier $R(x)$ in the integrand is the probability that the recovery time after the first shock at $t=0$ (and before the next one at $t=x$) is sufficient (smaller than x).

Applying, the Laplace transform to both sides of (3.28) results in the following relationship for the Laplace transform of $P^*(t)$:

$$\widetilde{P}^*(s) = \frac{1}{(s+\lambda)(1-\lambda q\widetilde{R}(s+\lambda))}, \tag{3.29}$$

where $\widetilde{R}(s)$ is the Laplace transform of the Cdf $R(t)$.

Using probability $P^*(t)$, we can derive now the following equation:

$$P(t) = e^{-\lambda t} + \int_0^t \lambda e^{-\lambda x} q P^*(t-x)\mathrm{d}x. \tag{3.30}$$

As previously, the first term on the right-hand side of this equation is the probability of shocks absence in $[0,t)$, $\lambda e^{-\lambda x} q\mathrm{d}x$ is the probability that the first shock has occurred and was survived in $[x, x+\mathrm{d}x)$. Finally, $P^*(t-x)$ is the probability that the system survives in $[x,t)$.

We can obtain $P(t)$, applying the Laplace transform to both sides of (3.30), i.e.,

$$\widetilde{P}(s) = \frac{1}{s+\lambda} + \frac{\lambda q}{s+\lambda}\widetilde{P}^*(s) \,,$$

where $\widetilde{P}^*(s)$ is defined by (3.29). This gives the general solution of the problem under the stated assumptions in terms of the Laplace transforms. In order to be able to invert $\widetilde{P}(s)$, assume additionally that the Cdf $R(t)$ is exponential, i.e., $R(t) = 1-\exp\{-\gamma t\}$, $\gamma > 0$. Performing simple algebraic transformations

$$\widetilde{P}(s) = \frac{s+2\lambda+\gamma-p\lambda}{s^2+s(\gamma+2\lambda)+\lambda^2+\gamma\lambda p}. \tag{3.31}$$

Inversion of (3.31) gives

$$P(t) = \frac{s_1 + \gamma + 2\lambda - p\lambda}{s_1 - s_2} \exp\{s_1 t\} - \frac{s_2 + \gamma + 2\lambda - p\lambda}{s_1 - s_2} \exp\{s_2 t\} , \qquad (3.32)$$

where

$$s_{1,2} = \frac{-(\gamma + 2\lambda) \pm \sqrt{(\gamma + 2\lambda)^2 - 4(\lambda^2 + \gamma\lambda p)}}{2} .$$

Equation (3.32) presents the exact solution for $P(t)$. In applications, it is convenient to use simple approximate formulas. Consider the following meaningful assumption [13]:

$$\frac{1}{\lambda} \gg \bar{\tau} \equiv \int_0^\infty (1 - R(x))\mathrm{d}x , \qquad (3.33)$$

where $\bar{\tau}$ denotes the mean time of recovery.

Relationship (3.33) means that the mean inter-arrival time in the shock process is much larger than the mean time of recovery, and this is often the case in practice. In the study of repairable systems, the similar case is usually called the *fast repair* condition. Using this assumption, the equivalent rate of termination for our process for $\lambda\bar{\tau} \to 0$, $\lambda t \gg 1$ can be written as

$$\lambda(t) = B\,\lambda(1 + o(1)), \qquad (3.34)$$

where B is the probability of termination for the *occurred shock* due to two causes, i.e., the termination immediately after the shock and the termination when the next shock occurs before the recovery is completed. Therefore, for sufficiently large t $(t \gg \bar{\tau})$ the integration in the following integral can be performed to ∞ and the approximate value of B is

$$B = \theta + (1 - \theta) \int_0^\infty \lambda e^{-\lambda x}(1 - R(x))\mathrm{d}x.$$

Assuming, as previously, that $R(t) = 1 - \exp\{-\gamma t\}$, $\gamma > 0$ gives

$$B = \frac{\lambda + \theta\gamma}{\lambda + \gamma} .$$

Finally, the fast repair approximation for the survival probability is

$$P(t) \approx \exp\left\{ -\frac{\lambda + p\gamma}{\lambda + \gamma} \lambda t \right\}. \qquad (3.35)$$

It can be easily seen that when $\gamma \to \infty$ (instant recovery), Relationship (3.35) reduces to Eq. (3.16). The accuracy of the fast repair approximation (3.35) with respect to the time of recovery can be analyzed similar to Finkelstein and Zarudnij [14].

3.7 Two Types of Shocks

Assume now that there are two types of shocks [13]. As in the previous section, potentially harmful shocks (to be called *redshocks*) result in termination of the process when they are 'too close', i.e., when the time between two consecutive red shocks is smaller than a recovery time with the Cdf $R(t)$. Therefore, in this case, the system does not have enough time to recover from the consequences of the previous red shock. Assume for simplicity that the probability of immediate termination on red shock's occurrence is equal to 0 ($p = 0$). The model can be easily generalized to the case when $p \neq 0$. On the other hand, our system is subject to the process of 'good' (*blue*) shocks. If the blue shock follows the red shock, termination cannot happen no matter how soon the next red shock will occur. Therefore, the blue shock can be considered as a kind of an additional recovery action.

Denote by λ and β the rates of the independent Poisson processes of red and blue shocks, respectively. First, assume that the first red shock has already occurred at $t = 0$. An integral equation for the probability of survival in $[0, t)$, $P^*(t)$ for this case is as follows:

$$\begin{aligned} P^*(t) = e^{-\lambda t} + \int_0^t \beta e^{-\beta x} e^{-\lambda x} \int_0^{t-x} \lambda e^{-\lambda y} P^*(t-x-y)\mathrm{d}y\mathrm{d}x \\ + \int_0^t e^{-\beta x} \lambda e^{-\lambda x} R(x) P^*(t-x)\mathrm{d}x, \end{aligned} \tag{3.36}$$

where

- The first term on the right-hand side is the probability that there are no other red shocks in $[0, t)$;
- $\beta e^{-\beta x} e^{-\lambda x}\mathrm{d}x$ is the probability that a blue shock occurs in $[x, x + \mathrm{d}x)$ and no red shocks occur in $(0, x)$;
- $\lambda e^{-\lambda y}\mathrm{d}y$ is the probability that the second red shock occurs in $[x + y, x + y + \mathrm{d}y)$;
- $P^*(t - x - y)$ is the probability that the system survives in $[x + y, t)$ given the red shock has occurred at time $x + y$;
- $e^{-\beta x} \lambda e^{-\lambda x}\mathrm{d}x$ is the probability that there is one red shock (the second) in $(0, t)$ and no blue shocks in this interval of time;

- $R(x)$ is the probability that the recovery time x is sufficient and, therefore, the second red shock does not terminate the process;
- $P^*(t-x)$ is the probability that the system survives in $[x, t)$ given the red shock has occurred at time x.

Using $P^*(t)$ that can be obtained from Eq. (3.36), as previously, we can now construct an equation with respect to $P(t)$—the probability of survival without assuming occurrence of the red shock at $t = 0$. Thus

$$P(t) = e^{-\lambda t} + \int_0^t \lambda e^{-\lambda x} P^*(t-x)\mathrm{d}x \,. \tag{3.37}$$

Applying the Laplace transform to Eq. (3.36) results in

$$\widetilde{P}^*(s) = \frac{s+\beta+\lambda}{(s+\beta+\lambda)(s+\lambda) - \beta\lambda - \lambda(s+\beta+\lambda)(s+\lambda)\widetilde{R}(s+\beta+\lambda)}. \tag{3.38}$$

Applying the Laplace transform to Eq. (3.38) gives

$$\widetilde{P}(s) = \frac{1}{s+\lambda} + \frac{\lambda}{s+\lambda}\widetilde{P}^*(s).$$

This equation gives a general solution of the problem under the stated assumptions in terms of the Laplace transforms. In order to be able to invert $\widetilde{P}(s)$, as in the previous section, assume that the Cdf $R(t)$ is exponential $R(t) = 1 - \exp\{-\gamma t\}$, $\gamma > 0$. Performing simple algebraic transformations

$$\widetilde{P}(s) = \frac{s+\gamma+\beta+2\lambda}{s^2 + s(\gamma+\beta+2\lambda) + \lambda^2}. \tag{3.39}$$

Inversion of (3.39) results in

$$P(t) = \frac{s_1+\gamma+\beta+2\lambda}{s_1 - s_2}\exp\{s_1 t\} - \frac{s_2+\gamma+\beta+2\lambda}{s_1 - s_2}\exp\{s_2 t\}\,, \tag{3.40}$$

where

$$s_{1,2} = \frac{-(\gamma+2\lambda+\beta) \pm \sqrt{(\gamma+\beta)^2 + 4\lambda(\gamma+\beta)}}{2}.$$

When $\gamma = 0$, there is no recovery time and the process is terminated when two consecutive red shocks occur.

Equation (3.40) gives an exact solution for $P(t)$. Similar to the previous section, it can be simplified under certain assumptions. Assume that the fast repair condition (3.33) holds. The first red shock cannot terminate the process. The probability that the subsequent shock can result in termination is

$$B = \int_0^t \lambda e^{-\lambda x} \int_0^{t-x} \lambda e^{-\lambda y} e^{-\beta y}(1 - R(y))\mathrm{d}y\mathrm{d}x.$$

For the exponentially distributed time of recovery

$$B = \frac{\lambda}{\lambda+\beta+\gamma} - \frac{\lambda}{\beta+\gamma} e^{-\lambda t} + \frac{\lambda^2}{(\lambda+\beta+\gamma)(\beta+\gamma)} e^{-(\lambda+\beta+\gamma)t}.$$

For the sufficiently large t, $B \approx \lambda/\lambda + \beta + \gamma$ and this approximate value can be used for subsequent shocks as well. Therefore, the relationship

$$P(t) \approx \exp\left\{-\frac{\lambda^2}{\lambda+\beta+\gamma} t\right\}.$$

is the fast repair approximation in this case.

The considered in Sects. 3.5–3.7 method of integral equations, which is applied to deriving the survival probability for different shock models is an effective tool for obtaining probabilities of interest in situations where the object under consideration has renewal points. As the considered process of shocks is the homogeneous Poisson process, each shock (under some additional assumptions) constitutes these renewal points. When a shock process is the NHPP, there are no renewal points, but the integral equations usually can also be derived. For the illustration, consider the corresponding generalization of Eq. (3.20). Denote by $P(t-x, x)$ the survival probability in $[x, t)$, $x < t$ for the 'remaining shock process' that has started at $t = 0$ and was not terminated by the first shock at time x. Note that this probability depends now not only on $x - t$ as in the homogeneous case, but on x as well. Equation (3.20) is modified now to

$$P(t) = \exp\left\{-\int_0^t \lambda(u)\mathrm{d}u\right\} + \int_0^t \lambda(x)\exp\left\{-\int_0^x \lambda(u)\mathrm{d}u\right\} qP(t-x, x)\mathrm{d}x.$$

It can be seen by substitution that

$$P(t-x, x) = \exp\left\{-p\int_x^t \lambda(u)\mathrm{d}u\right\}, \quad 0 \le x, t$$

is the solution to this equation.

One can formally derive integral equations for other models (with the NHPP process of shocks) considered in this section, however, the corresponding solutions can be obtained only numerically, as the explicit inversions of the Laplace transforms are not possible in these cases.

The method of integral equations can be also obviously applied to the renewal process of shocks, as in this case we also have 'pure renewal points'. For instance, the simplest Eq. (3.20) turns into

$$P(t) = (1 - F(t)) + \int_0^t f(x)\, qP(t - x)\mathrm{d}x\,,$$

where $F(t)$ and $f(t)$ are the Cdf and the pdf of the inter-arrival times, respectively. Applying the Laplace transform gives

$$\widetilde{P}(s) = \frac{1 - \widetilde{f}(s)}{s(1 - q\widetilde{f}(s))},$$

which is a formal solution to our problem in terms of the Laplace transforms. Note that it can be usually inverted only numerically.

3.8 Spatial Extreme Shock Model

In this section, we consider a two-dimensional model of spatial survival [10, 12]. It is a meaningful generalization of the univariate extreme shock model to the case of the spatial Poisson process of shocks. The random obstacles along the route of a moving object will play the role of these shocks. Although the initial setting is bivariate, the constructed failure rate is an univariate function and, therefore, our previous one-dimensional results can be used.

The setting of the problem is as follows: a sufficiently small normally or tangentially oriented interval is moving along a fixed route in the plane, crossing points of the spatial Poisson random process. Each crossing leads to a termination of the process (failure, accident) with a predetermined probability. As previously, the probability of passing the route without termination is of interest. An immediate application of the method to be considered is the *safety at sea* assessment. Our approach takes into account the fixed obstacles (e.g., shallows), which can lead to foundering and the moving obstacles (e.g., other ships), which can lead to collisions. The latter setting is not considered in this section and can be found in Finkelstein [12].

The field of fixed obstacles is considered to be random. In this application, there are two types of fixed obstacles: obstacles with known coordinates, marked in the corresponding navigational sea charts (and, therefore, not random), and obstacles with unknown coordinates, which following the subjective approach can be considered random. It turns out that, owing to the accuracy of navigation and motion control systems of a ship, weather influences, currents, etc., the obstacles with the known coordinates can also be modeled as random points in the plane. The 'geometric densities' of these obstacles, which can be obtained from the navigational charts, define the rates of the corresponding planar point processes to be used in the model [12].

The values of probabilities of accidents in "safety at sea" analysis are usually in the range 10^{-4} to 10^{-6}. Such estimates are often meaningless since there are not enough data to justify them. Therefore, simple relations for comparison of these probabilities can be very helpful in practice.

The developed approach can also be used for obtaining solutions that are optimal, for example, for finding a route with maximal probabilities of safe performance with or without specific restrictions (time on the route, fuel consumption, etc.). In what follows we consider the two-dimensional setting, but the generalization to $n = 3$ is straightforward and can be applied to assessing air traffic safety.

Denote by $\{N(B)\}$ an orderly point process in the plane, where $N(B)$ is a number of points in some domain $B \subset \Re^2$. We shall consider points of the process as prospective point influences (shocks) on our system (shallows for a ship, for instance). Similar to (2.12), the rate of this process $\lambda_f(\xi)$ can be formally defined as

$$\lambda_f(\xi) = \lim_{S(\delta(\xi)) \to 0} \frac{E[N(\delta(\xi))]}{S(\delta(\xi))}, \tag{3.41}$$

where $B = \delta(\xi)$ is the neighborhood of ξ with the area $S(\delta(\xi))$ and the diameter tending to zero. The subscript f stands for "fixed" obstacles.

Definition 3.3 The spatial nonhomogeneous Poisson process is defined similar to the one-dimensional case by the following relations [7]:

$$P\big(N(\delta(\xi)) = 1 | H_{\delta(\xi)}\big) = \lambda_f(\xi) S(\delta(\xi)) + o(S(\delta(\xi))),$$

$$P\big(N(\delta(\xi)) > 1 | H_{\delta(\xi)}\big) = o(S(\delta(\xi))),$$

where $H_{\delta(\xi)}$ denotes the configuration of all points outside $\delta(\xi)$.

It can be shown for an arbitrary B that $N(B)$ has a Poisson distribution with mean

$$\int_B \lambda_f(\xi) d\xi$$

and that the numbers of points in nonoverlapping domains are mutually independent random variables [7].

Our goal is to obtain a generalization of Eq. (3.18) to the bivariate case. The main feature of this generalization is a suitable parameterization allowing us to reduce the problem to the one-dimensional case [12]. Assume for simplicity that $\lambda_f(\xi)$ is a continuous function of ξ in an arbitrary closed circle in $\Re^2$. Let R_{ξ_1,ξ_2} be a fixed continuous curve connecting two distinct points in the plane, ξ_1 and ξ_2. We will call R_{ξ_1,ξ_2} a route. A point (a ship in our application) is moving in one direction along the route. Every time it 'crosses the point' of the process $\{N(B)\}$ (see later the corresponding regularization), an accident (failure) can happen with a given probability. We are interested in assessing the probability of moving along R_{ξ_1,ξ_2} without accidents. Let r be the distance from ξ_1 to the current point of the route (coordinate) and $\lambda_f(r)$ denote the corresponding rate. Thus, the one-dimensional parameterization is considered. For defining the corresponding

Poisson measure, the dimensions of objects under consideration should be taken into account.

Let $\left(\gamma_n^+(r), \gamma_n^-(r)\right)$ be a small interval of length $\gamma_n(r) = \gamma_n^+(r) + \gamma_n^-(r)$ in a normal direction to R_{ξ_1,ξ_2} at the point with the coordinate r, where the upper index denotes the corresponding direction ($\gamma_n^+(r)$ is on one side of R_{ξ_1,ξ_2}, whereas $\gamma_n^-(r)$ is on the other). Let $\overline{R} \equiv |R_{\xi_1\xi_2}|$ be the length of R_{ξ_1,ξ_2} and assume that the interval is small compared with the length of the route, i.e.,

$$\overline{R} > > \gamma_n(r), \forall r \in [0, \overline{R}].$$

The interval $\left(\gamma_n^+(r), \gamma_n^-(r)\right)$ is moving along R_{ξ_1,ξ_2}, crossing points of a random field. For "safety at sea" applications, it is reasonable to assume the symmetrical $\left(\gamma_n^+(r) = \gamma_n^-(r)\right)$ structure of the interval with length $\gamma_n(r) = 2\delta_s + 2\delta_o(r)$, where $2\delta_s$, $2\delta_o(r)$ are the diameters of the ship and of an obstacle, respectively. For simplicity, we assume that all obstacles have the same diameter. Thus, the ship's dimensions are already 'included' in the length of our equivalent interval. There can be other models as well, e.g., the diameter of an obstacle can be considered a random variable.

Taking Eq. (3.41) into account, the *equivalent rate* of occurrence of points, $\lambda_{e,f}(r)$ is defined as

$$\lambda_{ef}(r) = \lim_{\Delta r \to 0} \frac{E[N(B(r, \Delta r, \gamma_n(r)))]}{\Delta r}, \tag{3.42}$$

where $N(B(r, \Delta r, \gamma_n(r))$ is the random number of points crossed by the interval $\gamma_n(r)$ when moving from r to $r + \Delta r$. Thus, the specific domain in this case is defined as an area covered by the interval moving from r to $r + \Delta r$.

When $\Delta r \to 0$, $\gamma_n(r) \to 0$, and taking into account that $\lambda_f(\xi)$ is a continuous function [12],

$$\begin{aligned} E[N(B(r, \Delta r, \gamma_n(r)))] &= \int\limits_{B(r,\Delta r,\gamma_n(r))} \lambda_f(\xi) dS(\delta(\xi)) \\ &= \gamma_n(r)\lambda_f(r)\mathrm{d}r[1 + o(1)], \end{aligned}$$

which leads to the relationship for the equivalent rate of the corresponding one-dimensional nonhomogeneous Poisson process, i.e.,

$$\lambda_{ef}(r) = \gamma_n(r)\lambda_f(r)[1 + o(1)], \quad \Delta r \to 0, \ \gamma_n(r) \to 0.$$

As the radius of curvature of the route $R_c(r)$ is sufficiently large compared with $\gamma_n(r)$, i.e.,

$$\gamma_n(r) \ll R_c(r),$$

the domain covered by the interval $\left(\gamma_n^+(r), \gamma_n^-(r)\right)$ when it moves from r to $r + \Delta r$ along the route, is asymptotically ($\Delta r \to 0$) rectangular with area $\gamma_n(r)\Delta r$. Hence, the performed r -parameterization along the fixed route reduces the problem to the one-dimensional setting.

Assume now, as in the previous sections of this chapter, that the crossing of a point with a coordinate r leads to an accident (termination) with probability $p_f(r)$ and to the survival with the complementary probability $q_f(r) = 1 - p_f(r)$. Denote by R the random distance from the initial point of the route ξ_1 to a point of the route where an accident has occurred. Similar to (3.18), the probability of passing the route R_{ξ_1,ξ_2} without accidents can be derived in the following way:

$$P(R > \overline{R}) = \exp\left\{ -\int_0^{\overline{R}} \lambda_{af}(r)\mathrm{d}r \right\}, \tag{3.43}$$

where

$$\lambda_{af}(r) \equiv \theta_f(r)\lambda_{ef}(r) \tag{3.44}$$

is the corresponding failure (accident) rate. As previously, Eq. (3.43) and (3.44) constitute a simple and convenient tool for obtaining probabilities of safe (reliable) performance of our object. Thus, the univariate extreme shock model can be effectively applied to this initially two-dimensional setting.

3.9 Shock-Based Theory of Biological Aging

As a remarkable application to health sciences, we will show how the extreme shock model 'works' for obtaining the law of mortality of human populations. For this reason, we discuss and generalize the famous result by Strehler and Mildvan [29]. Our reasoning will mostly follow Finkelstein [15]. In this section, in accordance with the demographic and actuarial terminology, we will use the term "the force of mortality" (mortality rate) instead of the failure rate.

The Strehler–Mildvan [29] model suggests the justification of an exponential increase in the force of mortality $\mu(t)$, and describes some formal properties of the Gompertz mortality curve [17]:

$$\mu(t) = ae^{bt}. \tag{3.45}$$

The conventional generalization is the Gompertz–Makeham model, which adds a constant term c to the right-hand side of (3.45) in order to account for the 'background' mortality. In the current section, as in the original publication, we will assume that this term is negligible. Equation (3.45) usually provides a satisfactory fit to human mortality data for ages since maturity to the upper limit of around 90–100 years.

The goal of this section is to *discuss the underlying assumptions* of the Strehler–Mildvan (SM) *shock* model and the SM-correlation, which defines a negative correlation between parameters a and b. For several decades, the SM-correlation was believed to be a universal demographic law valid both for period and cohort mortality data [32].

The SM-model relies on the notion of vitality, i.e., an organism is characterized by its vitality function $V(t),\ V(0) \equiv V_0$, which decreases with age t. In the rest of this book, we will come back several times to the notion of vitality or its equivalents and will suggest a more mathematically advanced modeling of the vitality-related problems. Specifically, several strength–stress models will be considered when the failure (death) occurs if the magnitude of the stress (shock) exceeds the value of the strength (vitality).

According to Strehler and Mildvan [29], an organism is subject to stresses of internal or external nature that cause demands for energy. Those are shocks in our terminology. Let $(T_i,\ Y_i), i = 1, 2, \ldots$ be the sequence of pairs of i.i.d. random variables (therefore, the notation will be $(T,\ Y)$), characterizing the times at which stress events (demands for energy) occur, and the value of the demand for energy that is needed to recover from these stresses, respectively. Let $K(t)$ be the rate of the corresponding counting process describing arrival times of stress events. The following assumptions were made in the original paper:

Assumption 1 Y_i are exponentially distributed:

$$P(Y > y) = e^{-\frac{y}{D}}, \tag{3.46}$$

where, D is the mean value of this demand.

Assumption 2 An organism is characterized by its vitality function $V(t),\ V(0) \equiv V_0$ which decreases with age t. Yashin et al. [33], as in the original paper, called this function the maximum capacity of energy supply for an organism at age t. It can be also obviously interpreted as the stress resistance of an organism. Death occurs at age t when, for the first time, $Y > V(t)$. We discuss this assumption in conjunction with the last one.

Assumption 3 The rate $K(t) = K$ is a constant and the force of mortality is defined as [compare with Eq. (3.18)]

$$\mu(t) = KP(Y > V(t)) = Ke^{-\frac{V(t)}{D}}. \tag{3.47}$$

Equation (3.47) is called "a postulate" in Strehler and Mildvan [29]. However, it follows from the theory of point processes that (3.47) (see Chap. 2 and Sect. 3.4) is true *only* when the underlying point process $\{T_i\}_{i \geq 1}$ is the homogeneous Poisson process and, therefore, that the inter-arrival times of events (stresses) are exponentially distributed. This is a rather stringent condition, which was not pointed out in the original and subsequent papers discussing the SM-model. It should also be noted that, while (3.47), similar to (3.18), can be generalized to the case of the *nonhomogeneous Poisson process* with the age-dependent rate $K(t)$, the Poisson property of the underlying process is crucial for the product in the right-hand side of (3.47).

The following remark should be also made: as the force of mortality is a population characteristic, the vitality $V(t)$ should also be understood in this sense. However, it is obviously introduced by Assumption 1 as an individual (stochastic) characteristic. Therefore, we cannot simply substitute it with the corresponding expectation, as the exponential function is not linear:

$$E[e^{-\frac{V(t)}{D}}] \neq e^{-\frac{E[V(t)]}{D}}.$$

Thus, while there are a few important deficiencies in the original formulation of the model, it *formally* leads to the justified in practice properties of mortality rates.

Now we are ready to equate (3.45) and (3.47). As in the original paper, we will show using elementary derivations that $V(t)$ is *linearly declining* with age. It should be noted that this 'shape' is in consensus with the current understanding of the decline in the essential biological markers and the corresponding data, at least, for the human middle-age span [16]. Thus

$$\mu(t) = ae^{bt} = Ke^{-\frac{V(t)}{D}} \tag{3.48}$$

and taking logarithms of both sides ($V(0) \equiv V_0$):

$$V(t) = V_0(1 - (b/V_0)t) = V_0(1 - Bt), \tag{3.49}$$

where formally, $B = b/\ln(K/a) = Db/V_0$, and this quantity is usually called the individual rate of aging (in contrast with the population rate of aging b). Substituting (3.49) into (3.48):

$$\mu(t) = ae^{bt} = Ke^{-\frac{V_0(1-Bt)}{D}} = Ke^{-\frac{V_0}{D}}e^{\frac{V_0Bt}{D}} \tag{3.50}$$

and thus

$$a = Ke^{-\frac{V_0}{D}}; \quad b = V_0B/D. \tag{3.51}$$

Comparing two equations for the force of mortality, we see the dependence between a and b (negative correlation): the larger a results in the smaller b. From (3.51), this dependence can be written as

$$\ln a = \ln K - \frac{1}{B}b, \tag{3.52}$$

which is known in the literature as *SM-correlation*. This correlation has been observed empirically in various human populations. It follows from (3.52) that

$$\ln \mu(t) = \ln a + bt = \ln K + b(t - 1/B)), \tag{3.53}$$

meaning that the logarithms of mortality rates for different populations (e.g., with different a) intersect in one point with coordinates $(\ln K,\ 1/B)$. This has been experimentally observed and reported in the literature, although some criticism and violations of this rule were also discussed (see e.g., [32, 33]).

At first sight, it seems intriguing that the SM-correlation, which is derived using some general, partially unjustified assumptions, complies with the real mortality data. However, recently a certain departure from this pattern has been observed. A possible explanation is in consideration of the *vitality-independent* approach. It is based on the concept of lifesaving: i.e., that the environment not only supplies additional energy under stress, but due to the crucial advances in healthcare in recent decades, saves lives that previously would have been lost. The stochastic 'lifesaving model' (with a discussion of necessary assumptions) was developed in Finkelstein [11, 12]. It should be noted that Vaupel and Yashin [31] assumed that there can be a finite number of lifesavings, whereas we are dealing with a random number of these events.

Consider a lifetime that is characterized by the force of mortality $\mu(t)$ and the corresponding Cdf $F(t)$. Assume that a stress event affecting an organism, which occurs in accordance with this Cdf at age t_1 is fatal with probability $p(t_1)$ and is 'cured' with probability $1-p(t_1)$. The next stress occurs at age $t_2 > t_1$ in accordance with the Cdf $(F(t+t_1)-F(t))\overline{F}(t_1)$ and it is fatal with probability $p(t_2)$ and 'is cured' with probability $1-p(t_2)$, etc. It should be noted that the decreasing in age vitality of an organism can be still part of this model, if we assume that $1-p(t)$ is a decreasing function of age. In this case, $1-p(t)$ has a meaning of probability that the magnitude of a stress is smaller then the value of vitality at age t (probability of survival under a single shock). Therefore, in accordance with the lifesaving model [11], the initial nonhomogeneous Poisson process of stress events with rate $\mu(t)$ is terminated (*i.e.*, each event terminates the process with probability $p(t)$ and is 'harmless' with probability $1-p(t)$) and the Cdf of time to termination is characterized by the force of mortality $p(t)\mu(t)$. Thus, we again arrive at our extreme shock model (3.18)!

In order to explain the departures from the Srtehler–Mildvan correlation that were observed in recent decades, assume now that probability $p(t)$ in the described lifesaving model is not age-dependent any more, *i.e.*, $p(t) \equiv p$. Obviously, the state of an organism (vitality) can 'affect' this probability, However, today it is mostly defined by the new 'technical' abilities of treating, e.g., medical conditions that could not be treated before or performing medical operations that were not possible before. Therefore, we can consider this probability as approximately constant. Our assumption also means that the proportion of conditions that can be now cured does not depend on age. Thus, the resulting force of mortality $p\mu(t)$ follows the proportional hazards (PH) model. In order to illustrate our further reasoning, consider the following example. Let Eq. (3.45) define the baseline force of mortality for a developed country at, e.g., chronological time $x_b = 1950$. Then it can be modified for time $x > x_b$ to

$$\mu_\tau(t) = p_\tau a e^{bt}, \tag{3.54}$$

where $\tau = x - x_b$ and p_τ is constant in age for the fixed τ. Thus, the environment, due to lifesaving and in accordance with the extreme shock model, 'decreases' only parameter a without affecting the slope of the logarithmic mortality rate b. This perfectly complies with the Gompertz shift model of Bongaarts and Feeney [4] and with other experimental studies. It also can explain the change in the rectangularization pattern (that is usually attributed to the Strehler–Mildvan correlation) to shifts in the corresponding survival curves (which can be explained by the PH model). The mortality data for developed countries in recent decades support these claims. It should be noted that the assumption of the underlying Gompertz law is essential for the described change in the pattern, which can be easily seen from Eq (3.54), as $p_\tau = e^{\ln p_\tau}$ ($\ln p_\tau < 0$) creates shifts in age for the baseline mortality rate. It is also worth mentioning that, although the method of constructing the resulting force of mortality in the SM model, which is captured by Eq (3.47), formally resembles our lifesaving approach, the difference lies in the fact that the corresponding probabilities are 'applied' to each stress event (with a constant rate) in the former case and to events occurring in accordance with the nonhomogeneous Poisson process with rate $\mu(t)$, in the latter case.

References

1. Bae S, Kuo W, Kvam P (2007) Degradation models and implied lifetime distributions. Reliab Eng Syst Saf 92:601–608
2. Barlow RE, Proschan F (1975). Statistical theory of reliability and life testing. Holt, Renerhart & Winston, New York
3. Beichelt FE, Fatti LP (2002) Stochastic processes and their applications. Taylor and Francis, London
4. Bongaarts J, Feeney G (2002) How long do we live? Popul Dev Rev 28:13–29
5. Cha JH, Finkelstein M (2009) On a terminating shock process with independent wear increments. J Appl Probab 46:353–362
6. Cha JH, Finkelstein M (2010) Burn-in by environmental shocks for two ordered subpopulations. Eur J Oper Res 206:111–117
7. Cox DR, Isham V (1980) Point processes. University Press, Cambridge
8. Cox DR, Miller HD (1965) Theory of stochastic processes. Methuen & Co, London
9. Dufresne F, Gerber H, Shiu E (1991) Risk theory with the gamma process. ASTIN Bull 21(2):177–192
10. Finkelstein M (1998) A point process stochastic model with application to safety at sea. Reliab Eng Syst Saf 60:227–234
11. Finkelstein M (2005) Lifesaving explains mortality decline with time. Math Biosci 196:187–197
12. Finkelstein M (2008) Failure rate modelling for reliability and risk. Springer, London
13. Finkelstein M, Marais F (2010) On terminating Poisson processes in some shock models. Reliab Eng Syst Saf 95:874–879
14. Finkelstein M, Zarudnij VI (2002) Laplace transform methods and fast repair approximations for multiple availability and its generalizations. IEEE Trans Reliab 51:168–177
15. Finkelstein M (2012) Discussing the Strehler-Mildvan model of mortality. Demographic Res. 26:191–206
16. Golubev A (2009) How could the Gompertz-Makeham law evolve. J Theor Biol 238:1–17

17. Gompertz B (1825) On the nature of the function expressive of the law of human mortality and on a new mode of determining the value of life contingencies. Philos Trans R Soc 115:513–585
18. Gut A (1990) Cumulated shock models. Adv Appl Probab 22:504–507
19. Gut A, Huslcr J (2005) Realistic variation of shock models. Stat Probab Lett 74:187–204
20. Kahle W, Wendt H (2004) On accumulative damage process and resulting first passage times. Appl Stoch Models Bus Ind 20:17–27
21. Kalashnikov V (1997) Geometric sums: bounds for rare events with applications. Kluwer Academic Publishers, Dordrecht
22. Lam Y, Zhang YL (2004) A shock model for the maintenance problem of a repairable system. Comput Oper Res 31:1807–1820
23. Lund R, McCormic W, Xiao U (2004) Limiting properties of Poisson shot noise processes. J Appl Probab 41:911–918
24. Papoulis A (1971) High density shot noise and Gaussianity. J Appl Probab 8:118–127
25. Rice J (1977) On generalized shot noise. Adv Appl Probab 9:553–565
26. Ross SM (1996) Stochastic processes, 2nd edn. Wiley, New York
27. Singpurwalla N (1995) Survival in dynamic environment. Stat Sci 10:86–103
28. Tang YY, Lam Y (2006) A delta-shock maintenance model for a deteriorating system. Eur J Oper Res 168:541–556
29. Strehler L, Mildvan AS (1960) General theory of mortality and aging. Science 132:14–21
30. Van Nortwijk JM (2009) A survey of the application of gamma processes in maintenance. Reliab Eng and Syst Saf 94:2–21
31. Vaupel JW, Yashin AI (1987) Repeated resuscitation: how life saving alters life tables. Demography 4:123–135
32. Yashin AI, Begun AS, Boiko SI, Ukraintseva SV, Oeppen J (2001) The new trends in survival improvement require a revision of traditional gerontological concepts. Exp Gerontol 37:157–167
33. Yashin AI, Ukraintseva SV, Boiko SI, Arbeev KG (2002) Individual aging and mortality rate: how are they related. Soc Biol 49:206–217

Chapter 4
Advanced Theory for Poisson Shock Models

In this chapter, we extend and generalize approaches and results of the previous chapter to various reliability-related settings of a more complex nature. We relax some assumptions of the traditional models except the one that defines the underlying shock process as the nonhomogeneous Poisson process (NHPP). Only in the last section, we suggest an alternative to the Poisson process to be called *the geometric point process*. It is remarkable that although the members of the class of geometric processes do not possess the property of independent increments, some shock models can be effectively described without specifying the corresponding dependence structure. Most of the contents of this chapter is based on our recent work [5–11] and covers various settings that, we believe, are meaningful both from the theoretical and the practical points of view. The chapter is rather technical in nature, however, general descriptions of results are reasonably simple and illustrated by meaningful examples. As the assumption of the NHPP of shocks is adopted, many of the proofs follow the same pattern by using the time-transformation of the NHPP to the HPP (see the derivation of Eq. (2.31)). This technique will be used often in this chapter. Sometimes the corresponding derivations will be reasonably abridged, whereas other proofs will be presented at full length.

Recall that in extreme shock models, only an impact of the current, possibly fatal shock is usually taken into account, whereas in cumulative shock models, the impacts of the preceding shocks are accumulated as well. In this chapter, we combine extreme shock models with specific cumulative shock models and derive probabilities of interest, e.g., the probability that the process will not be terminated during a 'mission time'. We also consider some meaningful interpretations and examples. We depart from the assumption that the probability of termination does not depend on the history of the process and this makes the modeling more complex on the one hand, but more adequate on the other hand.

M. Finkelstein and J. H. Cha, *Stochastic Modeling for Reliability*,
Springer Series in Reliability Engineering, DOI: 10.1007/978-1-4471-5028-2_4,

4.1 The Terminating Shock Process with Independent Wear Increments

4.1.1 General Setting

Consider a system subject to a NHPP of shocks with rate $\nu(t)$. Let it be 'absolutely reliable' in the absence of shocks. As in Chap. 3, assume that each shock (regardless of its number) results in the system's failure (and, therefore, in the termination of the corresponding Poisson shock process) with probability $p(t)$ and is harmless to the system with probability $q(t) = 1 - p(t)$. Denote the corresponding time to failure of a system by T_S. Then Eq. (3.18) can be written now as

$$P(T_S > t) \equiv \bar{F}_S(t) = \exp\left(-\int_0^t p(u)\nu(u)\,du\right), \tag{4.1}$$

whereas the corresponding failure rate is

$$\lambda_S(t) = p(t)\nu(t).$$

The formal proof of (4.1) can be found in Beichelt and Fisher [3] and Block et al. [4]. A 'non-technical proof', based on the notion of the conditional intensity function (CIF) (see [15]) is given e.g., in Nachlas [25] and Finkelstein [17]. Thus, (4.1) describes an *extreme shock model*, as only the impact of the current, possibly fatal shock is taken into account. For convenience, we shall often call the described model the $p(t) \Leftrightarrow q(t)$ model.

It is clear that the extreme shock model can be easily modified to the case when a system can also fail from causes other than shocks. Denote the corresponding Cdf in the absence of shocks by $F(t)$ and assume that the process of failure from other causes and the shock process are independent. It follows from the competing risks considerations that

$$P(T_S > t) = \bar{F}(t)\exp\left(-\int_0^t p(u)\nu(u)\,du\right). \tag{4.2}$$

A crucial assumption for obtaining Eqs. (4.1) and (4.2) is the assumption that with probability $q(t) = 1 - p(t)$, a shock does not result in any changes in a system. However, in practice, shocks can also increase deterioration, wear, etc. The effect of different shocks is also usually accumulated in some way. Therefore, we start with the following setting [5]:

Let the lifetime of a system in a baseline environment (without shocks) be denoted by R. Thus, $P(R \leq t) = F(t)$. We interpret here R as some initial, random resource, which is 'consumed' by a system (with rate 1) in the process of its operation. Therefore, the age of our system in this case is equal to a calendar time t, and a failure occurs when this age reaches R. It is clear that when the remaining resource decreases with time, our system can be considered as aging (deteriorating).

Let $\{N(t),\ t \geq 0\}$ denote an orderly point process of shocks with arrival times $T_i,\ i = 1, 2, \ldots$ Denote also by $F_S(t)$ the Cdf that describes the lifetime of our system, T_S in the presence of shocks. Assume that the ith shock causes immediate system's failure with probability $p(t)$, but in contrast to the extreme shock model, with probability $q(t)$, it now increases the age of a system by a random increment $W_i \geq 0$. In terms of repair actions, this repair is 'worse than minimal'. In accordance with this setting, a random age of a system at time t (which is similar to the 'virtual age' of Finkelstein [16, 17]) is

$$T_v = t + \sum_{i=0}^{N(t)} W_i,$$

where, formally, $W_0 = 0$ corresponds to the case $N(t) = 0$ when there are no shocks in $[0, t]$. Failure occurs when this random variable reaches the boundary R. Therefore,

$$\begin{aligned} P(T_S > t|N(s),\ 0 \leq s \leq t;\ W_1,\ W_2, \ldots, W_{N(t)};\ R) \\ = \prod_{i=0}^{N(t)} q(T_i) I(T_v \leq R) \\ = \prod_{i=0}^{N(t)} q(T_i) I\left(\sum_{i=0}^{N(t)} W_i \leq R - t\right), \end{aligned} \tag{4.3}$$

where $q(T_0) = 1$ describes the case when $N(t) = 0$ and $I(x)$ is the corresponding indicator. This probability should be understood conditionally on realizations of $N(t),\ W_i, i = 1, 2, \ldots, N(t)$ and R.

Relationship (4.3) is very general and it is impossible to 'integrate out' explicitly $N(t),\ W_i,\ i = 1, 2, \ldots, N(t)$ and R without substantial simplifying assumptions. Therefore, after the forthcoming comment we will consider two important specific cases [5].

The described model can be equivalently formulated in the following way. Let $F(t)$ be the distribution of a lifetime of the wearing item in a baseline environment. Failure occurs when this wear, which in the standardized form is equal to t, reaches the resource (boundary) R. Denote the random wear in a more severe environment by $W_t,\ t \geq 0$. Specifically, for our shock model, $W_t = t + \sum_{i=0}^{N(t)} W_i$, where $W_i,\ i = 1, 2, \ldots, N(t)$, are the random increments of wear due to shocks and $W_0 \equiv 0$ [18]. For convenience, in what follows we will use this wear-based interpretation.

4.1.2 Exponentially Distributed Boundary

In addition to the previous assumptions, we need the following:

Assumption 1. $N(t)$, $t \geq 0$, is the NHPP with rate $v(t)$.

Assumption 2. $W_i, i = 1, 2, \ldots$, are i.i.d. random variables characterized by the moment generating function $M_W(t)$ and the Cdf $G(t)$.

Assumption 3. $N(t)$, $t \geq 0$; W_i, $i = 1, 2, \ldots$ and R are independent of each other.

Assumption 4. R is exponentially distributed with the failure rate λ, i.e., $\overline{F}(t) = \exp\{-\lambda t\}$.

The following result gives the survival function and the failure rate function for T_S [5].

Theorem 4.1 *Let $m(t) \equiv E(N(t)) = \int_0^t v(x)\,\mathrm{d}x$. Suppose that Assumptions 1–4 hold and that the inverse function $m^{-1}(t)$ exists for $t > 0$. Then the survival function for T_S and the corresponding failure rate $\lambda_S(t)$ are given by*

$$P(T_S > t) = \exp\left\{ -\lambda t - \int_0^t v(x)\,\mathrm{d}x + M_W(-\lambda) \cdot \int_0^t q(x)\, v(x)\,\mathrm{d}x \right\}, \; t \geq 0,$$

and

$$\lambda_S(t) = \lambda + (1 - M_W(-\lambda) \cdot q(t))\, v(t), \tag{4.4}$$

respectively.

Proof Given the assumptions, we can directly 'integrate out' the variable R and define the corresponding probability as

$$\begin{aligned} &P(T_S > t \,|\, N(s),\, 0 \leq s \leq t,\, W_1,\, W_2, \cdots,\, W_{N(t)}) \\ &= \left(\prod_{i=0}^{N(t)} q(T_i) \right) \cdot \exp\left\{ - \int_0^{t + \sum_{i=0}^{N(t)} W_i} \lambda\, du \right\} \\ &= \exp\left\{ -\lambda t - \lambda \sum_{i=1}^{N(t)} W_i + \sum_{i=1}^{N(t)} \ln q(T_i) \right\}. \end{aligned}$$

Thus

$$\begin{aligned} &P(T_S > t \,|\, N(s),\, 0 \leq s \leq t) \\ &= \exp\{-\lambda t\} \cdot \exp\left\{ \sum_{i=1}^{N(t)} \ln q(T_i) \right\} \cdot E\left[\exp\left\{ -\sum_{i=1}^{N(t)} \lambda W_i \right\} \right] \\ &= \exp\{-\lambda t\} \cdot \exp\left\{ \sum_{i=1}^{N(t)} \left[\ln q(T_i) + \ln\left(M_W(-\lambda)\right) \right] \right\}. \end{aligned} \tag{4.5}$$

We use now the same reasoning as when deriving Eq. (2.31). Therefore, some evident intermediate transformations are omitted. More details can be found in the original publication [5]. A similar approach is applied to our presentation in the rest of this chapter.

Define $N^*(t) \equiv N(m^{-1}(t))$, $t \geq 0$, and $T_j^* \equiv m(T_j)$, $j \geq 1$. It is well-known that $\{N^*(t), t \geq t\}$ is a stationary Poisson process with intensity one (see, e.g., [14]) and T_j^*, $j \geq 1$, are the times of occurrence of shocks in the new time scale. Let $s = m(t)$. Then

$$\begin{aligned} & E\left[\exp\left\{\sum_{i=1}^{N(t)} [\ln q(T_i) + \ln(M_W(-\lambda))]\right\}\right] \\ & = E\left[E\left[\exp\left\{\sum_{i=1}^{N^*(s)} \left[\ln q\left(m^{-1}(T_i^*)\right) + \ln(M_W(-\lambda))\right]\right\} \mid N^*(s)\right]\right]. \end{aligned} \tag{4.6}$$

The joint distribution of $\left(T_1^*, T_2^*, \ldots, T_n^*\right)$ given $N^*(s) = n$ is the same as the joint distribution of $\left(V_{(1)}, V_{(2)}, \ldots, V_{(n)}\right)$, where $V_{(1)} \leq V_{(2)} \leq \ldots \leq V_{(n)}$ are the order statistics of i.i.d. random variables $V_1, V_2, \ldots, V_n$ which are uniformly distributed in the interval $[0, s] = [0, m(t)]$. Then

$$\begin{aligned} & E\left[\exp\left\{\sum_{i=1}^{N^*(s)} \left(\ln q\left(m^{-1}\left(T_i^*\right)\right) + \ln(M_W(-\lambda))\right)\right\} \mid N^*(s) = n\right] \\ & = E\left[\exp\left\{\sum_{i=1}^{n} \left(\ln q\left(m^{-1}\left(V_{(i)}\right)\right) + \ln(M_W(-\lambda))\right)\right\}\right] \\ & = E\left[\exp\left\{\sum_{i=1}^{n} \left(\ln q\left(m^{-1}(V_i)\right) + \ln(M_W(-\lambda))\right)\right\}\right] \\ & = \left(E\left[\exp\left\{\ln q\left(m^{-1}(sU)\right) + \ln(M_W(-\lambda))\right\}\right]\right)^n, \end{aligned} \tag{4.7}$$

where $U \equiv V_1/s = V_1/m(t)$ is a random variable uniformly distributed in the unit interval [0,1]. Therefore,

$$\begin{aligned} & E\left[\exp\left\{\ln q\left(m^{-1}(sU)\right) + \ln\left(M_W(-\lambda)\right)\right\}\right] \\ & = \int_0^1 \exp\left\{\ln q\left(m^{-1}(m(t)u)\right) + \ln\left(M_W(-\lambda)\right)\right\} \mathrm{d}u \\ & = \frac{M_W(-\lambda)}{m(t)} \int_0^t q(x)v(x)\,\mathrm{d}x. \end{aligned} \tag{4.8}$$

From Eqs. (4.5)–(4.8),

$$\begin{aligned} P(T_S > t) &= \exp\{-\lambda t\} \cdot \sum_{n=0}^{\infty} \left(\frac{M_W(-\lambda)}{m(t)} \int_0^t q(x)v(x)\mathrm{d}x \right)^n \frac{s^n}{n!} e^{-s}. \\ &= \exp\{-\lambda t\} \cdot e^{-s} \cdot \exp\left\{ M_W(-\lambda) \cdot \frac{s}{m(t)} \int_0^t q(x)v(x)\mathrm{d}x \right\} \\ &= \exp\left\{ -\lambda t - \int_0^t v(x)\,\mathrm{d}x + M_W(-\lambda) \cdot \int_0^t q(x)v(x)\,\mathrm{d}x \right\}. \end{aligned}$$

Therefore, the failure rate of the system, $\lambda_S(t)$, is given by

$$\lambda_S(t) = \lambda + (1 - M_W(-\lambda) \cdot q(t))v(t). \qquad \square$$

The following corollary defines the failure rate that describes T_S when W_i's are distributed exponentially with mean μ.

Corollary 4.1 *If the W_i's are distributed exponentially with mean μ then the failure rate $\lambda_S(t)$ is given by*

$$\lambda_S(t) = \lambda + \left(1 - \frac{q(t)}{\lambda\mu + 1}\right)v(t). \tag{4.9}$$

We present now a qualitative analysis of the obtained result. Eq. (4.4) suggests that the failure rate $\lambda_S(t)$ can be interpreted as a failure rate of a series system with dependent (via R) components. When $\mu \to \infty$, from Eq. (4.9), we obtain $\lambda_S(t) \to \lambda + v(t)$, which means that a failure occurs either in accordance with the baseline $F(t)$ or as a result of the first shock (competing risks). Note that, in accordance with the properties of Poisson processes, the rate $v(t)$ is equal to the failure rate, which corresponds to the time to the first shock. Therefore, the two 'components' of the described series system are asymptotically independent as $\mu \to \infty$.

When $\mu = 0$, which means that $W_i = 0,\ i \geq 1$, Eq. (4.9) becomes $\lambda_S(t) = \lambda + p(t)v(t)$. Therefore, this specific case describes the series system with two independent components. The first component has the failure rate λ and the second component has the failure rate $p(t)\,v(t)$.

Let $q(t) = 1$ (there are no 'killing' shocks) and let W_i be deterministic and equal to μ. Then $M_W(-\lambda) = \exp\{-\mu\lambda\}$ and Eq. (4.4) becomes

$$\lambda_S(t) = \lambda + (1 - \exp\{-\mu\lambda\})v(t).$$

Assume for simplicity of notation that there is no baseline wear and all wear increments come from shocks. Then from Theorem1

$$P(T_S > t) = \exp\left\{ -\int_0^t \nu(x)\,\mathrm{d}x + M_W(-\lambda) \cdot \int_0^t q(x)\nu(x)\mathrm{d}x \right\}.$$

The form of this equation suggests the following probabilistic interpretation [6]. A system can fail from (i) the critical shock or (ii) the accumulated wear caused by the shocks. Suppose that the system has survived until time t. Then, as the distribution of the random boundary R is exponential, the accumulated wear until time t, $\sum_{i=0}^{N(t)} W_i$, does not affect the failure process of the component after time t. That is, on the next shock, the probability of the system's failure due to the accumulated wear given that a critical shock has not occurred, is just $P(R \le W_{N(t)+1})$. This probability does not depend on the wear accumulation history, that is,

$$\begin{aligned} &P(R \ge W_1 + W_2 + \ldots + W_n \,|\, R > W_1 + W_2 + \ldots + W_{n-1}) \\ &\quad = P(R > W_n), \quad \forall n = 1, 2, \ldots, W_1, W_2, \ldots, \end{aligned}$$

where $W_1 + W_2 + \ldots + W_{n-1} \equiv 0$ when $n = 1$. Finally, each shock results in the immediate failure with probability $p(t) + q(t)\,P(R \le W_1)$; otherwise, the system survives with probability $q(t)\,P(R > W_1)$. Although we have two (independent) causes of failure in this case, the second cause also does not depend on the history of the process and, therefore, our initial $p(t) \Leftrightarrow q(t)$ model can be applied after an obvious modification. In accordance with (4.1), the corresponding failure rate can then be immediately obtained as

$$\begin{aligned} \lambda_S(t) &= (p(t) + q(t)\,P(R \le W_1))\,\nu(t) \\ &= (1 - q(t)\,P(R > W_1))\,\nu(t) \\ &= (1 - q(t)\,M_W(-\lambda))\nu(t). \end{aligned}$$

The validity of the above reasoning and interpretation can be verified by comparing this failure rate function with that directly derived in (4.4) ($\lambda = 0$).

It is clear that this reasoning can be applied due to the specific, exponential distribution of the boundary R, which implies the Markov property for the wear 'accumulation'. In the next section, the case of a deterministic boundary will be considered and, obviously, the foregoing interpretation 'does not work' for this case.

4.1.3 Deterministic Boundary

Let $R = b$ be the deterministic boundary. Let other assumptions of Sect. 4.3.1 hold. We consider the case when $t < b$, which means that a failure cannot occur without shocks. The following result gives the survival function for T_S.

Theorem 4.2 *Suppose that Assumptions 1–3 of Sect. 4.3.1 hold and that the inverse function $m^{-1}(t)$ exists for $t > 0$. Furthermore, let the W_i's be i.i.d. exponential with mean $1/\eta$. Then the survival function for T_S is given by*

$$\begin{aligned} P(T_S > t) = \sum_{n=0}^{\infty} \left(\sum_{j=n}^{\infty} \frac{(\eta(b-t))^j}{j!} \exp\{-\eta(b-t)\} \right) \\ \times \left(\frac{1}{m(t)} \int_0^t q(x)v(x)\,\mathrm{d}x \right)^n \cdot \frac{m(t)^n}{n!} \exp\{-m(t)\}, \; 0 \le t < b. \end{aligned} \tag{4.10}$$

Proof Similar to the previous subsection,

$$\begin{aligned} &P(T_S > t | N(s), \quad 0 \le s \le t, \quad W_1, W_2, \ldots, W_{N(t)}) \\ &\quad = \left(\prod_{i=1}^{N(t)} q(T_i) \right) \cdot I\left(t + \sum_{i=1}^{N(t)} W_i \le b \right). \end{aligned}$$

Thus, we have

$$\begin{aligned} &P(T_S > t | \, N(s), \quad 0 \le s \le t) \\ &\quad = \left(\prod_{i=1}^{N(t)} q(T_i) \right) P\left(\sum_{i=1}^{N(t)} W_i \le b - t \right) \\ &\quad = \left(\prod_{i=1}^{N(t)} q(T_i) \right) G^{(N(t))}(b-t), \end{aligned}$$

where $G^{(n)}(t)$ is the n-fold convolution of $G(t)$ with itself.

As a special case, when the W_i's are i.i.d. exponential with mean $1/\eta$,

$$P(T_S > t | \, N(s), 0 \le s \le t) = \left(\prod_{i=1}^{N(t)} q(T_i) \right) \cdot \Psi(N(t)),$$

where

$$\Psi(N(t)) \equiv \sum_{j=N(t)}^{\infty} \frac{(\eta(b-t))^j}{j!} \exp\{-\eta(b-t)\},$$

and

$$\begin{aligned} P(T_S > t) &= E\left[\left(\prod_{i=1}^{N(t)} q(T_i) \right) \cdot \Psi(N(t)) \right] \\ &= E\left[E\left[\left(\prod_{i=1}^{N(t)} q(T_i) \right) \cdot \Psi(N(t)) | N(t) \right] \right], \end{aligned}$$

where

$$E\left[\left(\prod_{i=1}^{N(t)} q(T_i)\right) \cdot \Psi(N(t)) | N(t) = n\right]$$
$$= \Psi(n) \cdot E\left[\left(\prod_{i=1}^{N(t)} q(T_i)\right) | N(t) = n\right].$$

Using the same notation and properties as those of the previous subsection, we have

$$E\left[\left(\prod_{i=1}^{N(t)} q(T_i)\right) | N(t) = n\right] = \left[E\left(q\left(m^{-1}(sU)\right)\right)\right]^n$$

and

$$E\left(q\left(m^{-1}(sU)\right)\right) = \frac{1}{m(t)} \int_0^t q(x) v(x)\, \mathrm{d}x.$$

Therefore,

$$E\left[\left(\prod_{i=1}^{N(t)} q(T_i)\right) \cdot \Psi(N(t)) | N(t) = n\right]$$
$$= \Psi(n) \cdot \left(\frac{1}{m(t)} \int_0^t q(x) v(x)\, \mathrm{d}x\right)^n.$$

Finally, we obtain a rather cumbersome Eq. (4.10).

□

It can be easily shown that the survival function in (4.10) can be written in the following compact form [6]:

$$P(T_s > t) = \exp\left\{-\int_0^t p(x)\, v(x)\, \mathrm{d}x\right\} \cdot \sum_{n=0}^{\infty} P(Z_1 \geq n) \cdot P(Z_2 = n), \qquad (4.11)$$

where Z_1 and Z_2 are two Poisson random variables with parameters $\eta(b \quad t)$ and $\int_0^t q(x)\, v(x)\, \mathrm{d}x$, respectively. The following presents a qualitative analysis for two marginal cases of Eq. (4.11) for each fixed $t < b$.

When $\eta = 1/\mu \to \infty$, which means that the mean of increments W_i tends to 0, Eq. (4.11) 'reduces' to (4.1). Indeed, as $\eta \to \infty$,

$$\sum_{n=0}^{\infty} P(Z_1 \geq n)\, P(Z_2 = n) \to \sum_{n=0}^{\infty} P(Z_2 = n) = 1,$$

because $P(Z_1 \geq n) \to 1$ for $\forall n \geq 1$ and $P(Z_1 \geq 0) = 1$. From 'physical considerations', it is also clear that as increments vanish, their impact on the model also vanishes.

When $\eta \to 0$, the mean of the increments tends to infinity and, therefore, the first shock will kill the system with probability tending to one as $\eta \to 0$. The infinite sum in the right-hand side in the following equation vanishes in this case:

$$\sum_{n=0}^{\infty} P(Z_1 \geq n)\, P(Z_2 = n) = P(Z_1 \geq 0)\, P(Z_2 = 0) + \sum_{n=1}^{\infty} P(Z_1 \geq n)\, P(Z_2 = n) \to P(Z_2 = 0),$$

as $P(Z_1 \geq 0) = 1$ and $P(Z_1 \geq n) \to 0$ for $\forall n \geq 1$ when $\eta \to 0$. Therefore, finally

$$P(T_S > t) \to \exp\left\{ -\int_0^t p(x)\nu(x)\,\mathrm{d}x \right\} \exp\left\{ -\int_0^t q(x)\nu(x)\,\mathrm{d}x \right\} = \exp\left\{ -\int_0^t \nu(x)\,\mathrm{d}x \right\},$$

which is the probability that no shocks have occurred in $[0, t]$. This is what we also expect from general considerations for $\eta \to 0$, as the system can survive for $t < b$ only without shocks.

4.2 History-Dependent Termination Probability

Consider first, the orderly point process with the conditional (complete) intensity function (CIF) $\nu(t|H(t))$ [2, 15], where $H(t)$ is the history of the process up to t. This notion is similar to the intensity process defined in (2.12). Whereas the intensity process is considered as a stochastic process defined by filtration H_{t-}, the CIF is usually a realization of this process defined by the realization of filtration $H(t)$. We will use these terms in our book interchangeably. Accordingly, let the probability of termination under a single shock be adjusted in a similar way and, therefore, also depend on this history, i.e., $p(t|H(t))$. Denote, as previously, by T_S the corresponding lifetime. It is clear that in accordance with our assumptions, the conditional probability of termination in the infinitesimal interval of time can be written in the following *simplified* form [17]:

$$P[T_S \in [t, t + \mathrm{d}t) | T_S \geq t, H(t)] = p(t|H(t))\, \nu(t|\mathrm{H}(t))\, \mathrm{d}t.$$

The only way for $p(t|H(t))\, \nu(t|\mathrm{H}(t))$ to become a 'full-fledged' failure rate that corresponds to the lifetime T_S is when there is no dependence on $H(t)$ for both multipliers in the right-hand side. It is obvious that elimination of this dependence for the second multiplier uniquely leads to the NHPP. In what follows, we will consider this case. However, specific types of dependence on history in the first multiplier *will be retained* and this will give rise to the new classes of extreme shock models.

Model A. We will consider the NHPP of shocks with rate $v(t)$ and with the history-dependent termination probability

$$p(t|H(t)) = p(t|N(s),\, 0 \le s < t).$$

Let this be the simplest history case, i.e., the number of shocks, $N(t)$ that our system has experienced in $[0, t)$. This seems to be a reasonable assumption, as each shock can contribute to 'weakening' of the system by increasing the probability $p(t|H(t)) \equiv p(t, N(t))$ and, therefore, the function $p(t, N(t))$ is usually increasing in $n(t)$ (for each realization, $N(t) = n(t)$). To obtain the following result, we must assume the specific form of this function. It is more convenient to consider the corresponding probability of survival. Let

$$q(t, n(t)) \equiv 1 - p(t, n(t)) = q(t)\,\rho(n(t)), \tag{4.12}$$

where $\rho(n(t))$ is a decreasing function of its argument (for each fixed t). Thus the survival probability at each shock decreases as the number of survived shocks in $[0, t)$ increases. The multiplicative form of (4.12) will be important for us as it will be 'responsible' for the vital independence to be discussed later.

The survival function of the system's lifetime T_S is given by the following theorem.

Theorem 4.3 *Let* $m(t) \equiv E(N(t)) = \int_0^t v(x)\,\mathrm{d}x$ *and* $\Psi(n) \equiv \prod_{i=0}^{n} \rho(i)$ $(\rho(0) \equiv 1)$. *Suppose that the inverse function* $m^{-1}(t)$ *exists for* $t > 0$. *Then*

$$P(T_S \ge t) = E\left[\Psi\left(N_{qv}(t)\right)\right] \cdot \exp\left\{-\int_0^t p(x)\,v(x)\,\mathrm{d}x\right\}, \tag{4.13}$$

where $\{N_{qv}(t),\, t \ge 0\}$ *follows the NHPP with rate* $q(t)v(t)$.

Proof Obviously, conditioning on the process (in each realization) gives

$$P(T_S \ge t|N(s),\, 0 \le s < t) = \prod_{i=0}^{N(t)} q(T_i)\rho(i),$$

where formally $q(T_0) \equiv 1$ and $\rho(0) \equiv 1$ corresponds to the case when $N(t) = 0$. Also, by convention, $\prod_{i=1}^{n} (\cdot)_i \equiv 1$ for $n = 0$. Then the corresponding expectation is

$$P(T_S \ge t) = E\left[\prod_{i=1}^{N(t)} q(T_i)\,\rho(i)\right].$$

As previously, define the stationary Poisson process with rate 1: $N^*(t) \equiv N(m^{-1}(t))$, $t \geq 0$, and $T_j^* \equiv m(T_j)$, $j \geq 1$ are the times of occurrence of shocks in the new time scale. Let $s = m(t)$. Then

$$E\left[\prod_{i=1}^{N(t)} q(T_i)\,\rho(i)\right] = E\left[E\left[\prod_{i=1}^{N^*(s)} q\left(m^{-1}(T_i^*)\right)\rho(i)|N^*(s)\right]\right].$$

The joint distribution of $\left(T_1^*, T_2^*, \ldots, T_n^*\right)$ given $N^*(s) = n$ is the same as the joint distribution of $\left(V_{(1)}, V_{(2)}, \ldots, V_{(n)}\right)$, where $V_{(1)} \leq V_{(2)} \leq \cdots \leq V_{(n)}$ are the order statistics of i.i.d. random variables $V_1, V_2, \ldots, V_n$ which are uniformly distributed in the interval $[0, s] = [0, m(t)]$. Thus omitting derivations that are similar, to those in the proofs of Theorems 4.1 and 4.2 (see [6] for more details):

$$E\left[\prod_{i=1}^{N^*(s)} q\left(m^{-1}(T_i^*)\right)\rho(i)|N^*(s) = n\right] = \prod_{i=1}^{n}\rho(i)\left(E\left[q\left(m^{-1}(sU)\right)\right]\right)^n,$$

where $U \equiv V_1/s = V_1/m(t)$ is a random variable uniformly distributed in the unit interval [0,1]. Therefore,

$$\begin{aligned} E\left[q(m^{-1}(sU))\right] &= \int_0^1 q\left(m^{-1}(su)\right)\mathrm{d}u = \int_0^1 q\left(m^{-1}(m(t)u)\right)\mathrm{d}u \\ &= \frac{1}{m(t)}\int_0^t q(x)\nu(x)\,\mathrm{d}x. \end{aligned}$$

Hence,

$$E\left[\prod_{i=1}^{N^*(s)} q(m^{-1}(T_i^*)\,\rho(i)|N^*(s) = n\right] = \prod_{i=1}^{n}\rho(i)\cdot\left(\frac{1}{m(t)}\int_0^t q(x)\nu(x)\,\mathrm{d}x\right)^n.$$

Using $\Psi(n) \equiv \prod_{i=1}^{n}\rho(i)$,

$$\begin{aligned} P(T_S \geq t) &= E\left[\prod_{i=1}^{N(t)} q(T_i)\rho(i)\right] \\ &= \sum_{n=0}^{\infty}\Psi(n)\left(\frac{1}{m(t)}\int_0^t q(x)\nu(x)\,\mathrm{d}x\right)^n\cdot\frac{(m(t))^n}{n!}e^{-m(t)} \\ &= \exp\left\{-\int_0^t p(x)\nu(x)\,\mathrm{d}x\right\}\cdot\sum_{n=0}^{\infty}\Psi(n)\cdot\frac{\left(\int_0^t q(x)\nu(x)\,\mathrm{d}x\right)^n}{n!}\cdot \end{aligned}$$

$$\exp\left\{-\int_0^t q(x)\nu(x)\,\mathrm{d}x\right\} = E\left[\Psi(N_{q\nu}(t))\right]\cdot\exp\left\{-\int_0^t p(x)\nu(x)\,\mathrm{d}x\right\},$$

where $\{N_{qv}(t),\ t \geq 0\}$ follows the NHPP with rate $q(t)v(t)$.

□

Example 4.1 Let $\rho(i) = \rho^{i-1}$, $i = 1, 2, \ldots$. Then $\Psi(n) \equiv \rho^{n(n-1)/2}$ and

$$P(T_S \geq t) = \sum_{n=0}^{\infty} \rho^{n(n-1)/2} \cdot \frac{\left(\int_0^t q(x)v(x)\,dx\right)^n}{n!} \cdot \exp\left\{-\int_0^t q(x)v(x)\,dx\right\} \cdot \exp\left\{-\int_0^t p(x)v(x)\,dx\right\}$$

$$= \sum_{n=0}^{\infty} \rho^{n(n-1)/2} \cdot \frac{\left(\int_0^t q(x)v(x)\,dx\right)^n}{n!} \cdot \exp\left\{-\int_0^t v(x)\,dx\right\}. \tag{4.14}$$

The following discussion will help us in the further presentation of our time-dependent results. Let $\{N(t),\ t \geq 0\}$ be the NHPP with rate $v(t)$. If an event occurs at time t, it is classified as a Type I event with probability $p(t)$ and as a Type II event with the complementary probability $1 - p(t)$, as in our initial $p(t) \Leftrightarrow q(t)$ model. Then $\{N_1(t),\ t \geq 0\}$ and $\{N_2(t),\ t \geq 0\}$ are the independent NHPP with rates $p(t)\,v(t)$ and $q(t)\,v(t)$, respectively, and $N(t) = N_1(t) + N_2(t)$. Accordingly, e.g., given that there have been no Type I events in $[0,\ t)$, the process $\{N(t),\ t \geq 0\}$ reduces to $\{N_2(t),\ t \geq 0\}$, as in our specific case when a Type I event (fatal shock) leads to the termination of the process (failure). Therefore, in order to describe the lifetime to termination, it is obviously sufficient to consider $\{N_2(t),\ t \geq 0\}$, and not the original $\{N(t),\ t \geq 0\}$.

We will use a similar reasoning for a more general $p(t|H(t)) \Leftrightarrow q(t|H(t))$ model considered above, although interpretation of the types of events will be slightly different in this case. In the following, in accordance with our previous notation, $N_2(t) = N_{qv}(t)$ and the arrival times of this process are denoted by $T_{(qv)1}, T_{(qv)2}, \ldots$.

The multiplicative form of the specific result in (4.13) indicates that it might be also obtained and interpreted via the following general reasoning, which can be useful for probabilistic analysis of various extensions of standard extreme shock models. Considering the classical $p(t) \Leftrightarrow q(t)$ extreme shock model, assume that there can be other additional causes of termination dependent either directly on a history of the point process (as in Model A) or on some other variables, as for the marked point process, when each event is 'characterized' by some variable (e.g., damage or wear). Just for the sake of definiteness of presentation, let us call this 'initial' cause of failure, which corresponds to the $p(t) \Leftrightarrow q(t)$model, the *main* or the *critical cause of failure* (termination) and the shock that leads to this event—*the critical shock* (Type I event). However, distinct from the $p(t) \Leftrightarrow q(t)$ model, the Type II events, which follow the Poisson process with rate $q(t)\,v(t)$, can now also result in failure.

Let $E_C(t)$ denote the event that there were no critical shocks until time t in the absence of other causes of failures. Then, obviously,

$$P(T_S \geq t | E_C(t)) = \frac{P(T_S \geq t, E_C(t))}{P(E_C(t))} = \frac{P(T_S \geq t)}{P(E_C(t))},$$

and, thus,

$$P(T_S \geq t) = P(T_S \geq t | E_C(t))\, P(E_C(t)),$$

where

$$P(E_C(t)) = P(N_1(t) = 0) = \exp\left\{ - \int_0^t p(x)v(x)\,\mathrm{d}x \right\}. \tag{4.15}$$

Therefore, in accordance with our previous reasoning and notation, we can describe $P(T_S \geq t | E_C(t))$ in terms of the process $\{N_{qv}(t), t \geq 0\}$ (and not in terms of the original process $\{N(t), t \geq 0\}$) in the following general form to be specified for the forthcoming model:

$$P(T_S \geq t | E_C(t)) = E(I(\Psi(N_{qv}(t), \Theta) \in S) | E_C(t)),$$

where $I(\cdot)$ is the corresponding indicator, Θ is a set of random variables that are 'responsible' for other causes of failure (see later), $\Psi(N_{qv}(t), \Theta)$ is a real-valued function of $(N_{qv}(t), \Theta)$ which represents the state of the system at time t (given $E_C(t)$ i.e., no critical shock has occurred), and S is a set of real values which defines the survival of the system in terms of $\Psi(N_{qv}(t), \Theta)$. That is, if the critical shock has not occurred, the system survives when $\Psi(N_{qv}(t), \Theta) \in S$.

In order to apply effectively Model A, we have to reinterpret it as follows. Suppose first, that the system is composed of two parts in series and that each shock affects only one component. If it hits the first component (with probability $p(t)$), it directly causes its (and the systems) failure (the critical shock). On the other hand, if it hits the second component (with probability $q(t)$), then this component fails with probability $1 - \rho(n(t))$ and survives with probability $\rho(n(t))$. This interpretation nicely conforms with the two independent causes of failure model in (4.12). Note that, in fact, we are speaking about the *conditional independence* of causes of failure (on condition that a shock from the Poisson process with rate $v(t)$ has occurred).

Another (and probably more practical) interpretation is as follows. Assume that there are some parts of a system (component 1) that are critical *only* to, e.g., the shock's level of severity, which is assumed to be random. This results in failure with probability $p(t)$. On the other hand, the other parts (component 2) are critical *only* to accumulation of damage (failure with probability $1 - \rho(n(t))$). Assuming the series structure and the corresponding independence, we arrive at the survival (on shock) probability (4.12).

We can define now the function $\Psi(N_{qv}(t), \Theta)$ for Model A. Suppose that there have been no critical shocks in $[0, t)$ and let $\varphi_i = 1$ if the second component survives the ith shock, and $\varphi_i = 0$, $i = 1, 2, 3, \ldots N(t)$ otherwise. Then

$$\Psi(N_{qv}(t), \Theta) = \prod_{i=1}^{N_{qv}(t)} \varphi_i,$$

and $S = \{1\}$. Therefore, as events $E_C(t)$ and $\Psi(N_{qv}(t), \Theta) \in S$ are 'related' only to the first and the second causes of failure, respectively, and these causes of failure are independent, we have:

$$\begin{aligned}
&P(T_S \geq t | E_C(t)) \\
&= E(I(\Psi(N_{qv}(t), \Theta) \in S) | E_C(t)) \\
&= E(I(\Psi(N_{qv}(t), \Theta) \in S)) \\
&= E\left(I\left(\prod_{i=1}^{N_{qv}(t)} \varphi_i = 1\right)\right) = E\left[P\left(\prod_{i=1}^{N_{qv}(t)} \varphi_i = 1 | N_{qv}(t)\right)\right] \\
&= E\left(\prod_{i=1}^{N_{qv}(t)} \rho(i)\right).
\end{aligned}$$

Combining this equation with (4.15), we arrive at the original result in (4.13).

Model B. Consider now another type of extreme shock model, which is, in fact, a generalization of Model A. In model A, the second cause of failure (termination) was due to the number of noncritical shocks, no matter what the severity of these shocks was. Now, we will count only those shocks (to be called 'dangerous') with severity larger than some level κ. Assume that the second cause of failure 'materializes' only when the number of dangerous shocks exceeds some random level M. That is, given $M = m$, in the absence of critical shocks, the system fails as soon as it experiences the $(m + 1)$th dangerous shock.

Assume that the shock's severity is a random variable with the Cdf $G(t)$, and the survival function for M, $P(M > l)$, $l = 0, 1, 2, ..$, is also given. Suppose that there have been no critical shocks until time t and let φ_i be the indicator random variable ($\varphi_i = 1$ if the ith shock is dangerous and $\varphi_i = 0$ otherwise). Then, as previously,

$$\Psi\left(N_{qv}(t), \Theta\right) = I\left(M \geq \sum_{i=1}^{N_{qv}(t)} \varphi_i\right),$$

and $S - \{1\}$. Thus

$$\begin{aligned}
P(T_S \geq t | E_C(t)) &= E(I(\Psi(N_{qv}(t), \Theta) \in S)) = E\left(I\left(M \geq \sum_{i=1}^{N_{qv}(t)} \varphi_i\right)\right) \\
&= P\left(M \geq \sum_{i=1}^{N_{qv}(t)} \varphi_i\right) = E\left[P\left(M \geq \sum_{i=1}^{N_{qv}(t)} \varphi_i | N_{qv}(t)\right)\right],
\end{aligned}$$

where,

$$
\begin{aligned}
&P\left(M \geq \sum_{i=1}^{N_{qv}(t)} \varphi_i | N_{qv}(t) = n\right) \\
&= P(M > n | N_{qv}(t) = n) + \sum_{m=0}^{n} P\left(M \geq \sum_{i=1}^{n} \varphi_i | N_{qv}(t) = n,\, M = m\right) \cdot P(M = m | N_{qv}(t) = n). \\
&= P(M > n) + \sum_{m=0}^{n} \sum_{l=0}^{m} \binom{n}{l} \bar{G}(\kappa)^l G(\kappa)^{n-l} \cdot P(M = m) \\
&= P(M > n) + \sum_{l=0}^{n} \sum_{m=l}^{n} \binom{n}{l} \bar{G}(\kappa)^l G(\kappa)^{n-l} \cdot P(M = m) \\
&= P(M > n) + \sum_{l=0}^{n} \binom{n}{l} \bar{G}(\kappa)^l G(\kappa)^{n-l} \cdot (P(M \geq l) - P(M \geq n+1)) \\
&= \sum_{l=0}^{n} \binom{n}{l} \bar{G}(\kappa)^l G(\kappa)^{n-l} \cdot P(M \geq l).
\end{aligned}
$$

Thus, similar to the derivations of the previous section

$$
\begin{aligned}
P(T_S \geq t | E_C(t)) = \sum_{n=0}^{\infty} &\left[\sum_{l=0}^{n} P(M \geq l) \cdot \binom{n}{l} \bar{G}(\kappa)^l G(\kappa)^{n-l} \right] \\
&\cdot m_q(t)^n \frac{\exp\{-m_q(t)\}}{n!},
\end{aligned}
$$

where $m_q(t) \equiv \int_0^t q(x)v(x)\,dx$, and finally, we have

$$
\begin{aligned}
P(T_S \geq t) = \exp\left\{ -\int_0^t p(x)v(x)\,dx \right\} \cdot \sum_{n=0}^{\infty} \left[\sum_{l=0}^{n} P(M \geq l) \cdot \binom{n}{l} \bar{G}(\kappa)^l G(\kappa)^{n-l} \right] \\
\cdot m_q(t)^n \frac{\exp\{-m_q(t)\}}{n!}
\end{aligned}
$$

Note that, when the expression for $P(T_S \geq t | E_C(t))$ involves not only the number of shocks $N_{qv}(t)$ but also the filtration generated by $(N_{qv}(s),\, 0 \leq s \leq t)$, the computation becomes intensive and the results might not be useful in practice. The corresponding example with numerical results can be found in [6].

4.3 Shot Noise Process for the Failure Rate

4.3.1 Shot Noise Process Without Critical Shocks

Assume that a system is subject to the NHPP of shocks $\{N(t),\, t \geq 0\}$ with rate $v(t)$, which is the only possible cause of its failure. The consequences of shocks are accumulated in accordance with the 'standard' shot noise process $X(t)$, $X(0) = 0$ (see e.g., [26], [27] and the previous chapter). Similar to (3.8), but in a slightly different and more convenient for us here notation, define the level of the cumulative stress (wear) at time t as the following stochastic process:

$$X(t) = \sum_{j=1}^{N(t)} D_j h(t - T_j), \tag{4.16}$$

where T_n is the n-th arrival time in the shock process, D_j, $j = 1, 2, \ldots$ are the i.i.d. magnitudes of shocks and $h(t)$ is a non-negative, nonincreasing for $t \geq 0$, deterministic function ($h(t) = 0$ for $t < 0$). The usual assumption for considering asymptotic properties of $X(t)$ is that $h(t)$ vanishes as $t \to \infty$ and its integral in $[0, \infty)$is finite, however, we formally do not need this rather restrictive assumption here. The shock process $\{N(t),\, t \geq 0\}$ and the sequence $\{D_1, D_2, \ldots\}$ are supposed to be independent.

The cumulative stress eventually results in failures, which can be probabilistically described in different ways. Denote by T_S, as previously, the failure time of our system. Lemoine and Wenocur [23, 24], for example, modeled the distribution of T_S by assuming that the corresponding intensity process is *proportional* to $X(t)$ (see (2.12) for a general definition). As we are dealing with the intensity process, we will rather use the term "stress" instead of "wear". Proportionality is a reasonable assumption that describes the proportional dependence of the probability of failure in the infinitesimal interval of time on the level of stress

$$\lambda_t \equiv k X(t) = k \sum_{j=1}^{N(t)} D_j h(t - T_j), \tag{4.17}$$

where $k > 0$ is the constant of proportionality. Then

$$\begin{aligned} &P(T_S > t | N(s),\, 0 \leq s \leq t,\, D_1, D_2, \ldots, D_{N(t)}) \\ &\quad = \exp\left\{ -k \int_0^t \sum_{j=1}^{N(x)} D_j h(x - T_j) \, dx \right\}, \end{aligned} \tag{4.18}$$

Therefore, it means that the intensity process (4.17) can be also considered as the failure rate process [22]. Probability (4.18) should be understood conditionally on the corresponding realizations of $\{N(s),\, 0 \leq s \leq t\}$ and $D_1, D_2, \ldots, D_{N(t)}$. Therefore, 'integrating them out',

$$P(T_S > t) = E\left[\exp\left\{ -k \int_0^t X(u) \, du \right\} \right].$$

Lemoine and Wenocur [24] had finally derived the following relationship for the survival probability $P(T_S > t)$:

provided that the system is operating at time t [i.e., the event $D_j > g_U(T_j), j = 1, 2, \ldots$ did not happen in $[0, t)$].

Generalizing (4.17), assume that the conditional failure rate process $\hat{\lambda}_t$ (on condition that the event $D_j > g_U(T_j), j = 1, 2, \ldots$ did not happen in $[0, t)$ and $\{N(t), T_1, T_2, \ldots, T_{N(t)}\}$ and $\{D_1, D_2, \ldots, D_{N(t)}\}$ are given) is proportional to $X(t)$

$$\hat{\lambda}_t \equiv kX(t) = k\sum_{n=1}^{N(t)} \xi(T_j, D_j)\, D_j h(t - T_j), \; k > 0. \tag{4.22}$$

It is clear that conditionally on the corresponding history

(i) If $D_j > g_U(T_j)$, for at least one j, then

$$P(T_S > t \mid N(s), \, 0 \le s \le t, \, D_1, D_2, \ldots, D_{N(t)}) \; = \; 0;$$

(ii) If $D_j \le g_U(t)$, for all j, then

$$P(T_S > t \mid N(s), \, 0 \le s \le t, \, D_1, D_2, \ldots, D_{N(t)}) = \exp\left\{ -k \int_0^t \sum_{j=1}^{N(x)} \xi(T_j, D_j)\, D_j h(x - T_j)\, \mathrm{d}x. \right\}.$$

Therefore,

$$\begin{aligned} &P(T_S > t \mid N(s), \, 0 \le s \le t, \, D_1, D_2, \ldots, D_{N(t)}) \\ &= \prod_{j=1}^{N(t)} \gamma(T_j, D_j) \cdot \exp\left\{ -k \int_0^t \sum_{j=1}^{N(x)} \xi(T_j, D_j)\, D_j h(x - T_j)\, \mathrm{d}x \right\}, \end{aligned} \tag{4.23}$$

where

$$\gamma(T_j, D_j) \; = \; \begin{cases} 0, & D_j > g_U(T_j) \\ 1, & D_j \le g_U(T_j) \end{cases}. \tag{4.24}$$

Thus, we have described a rather general model that extends (4.18) to the defined deterioration pattern. Indeed, if $g_U(t) = \infty$; $g_L(t) = 0$, then $\xi(T_j, D_j) \equiv 1$and (4.23) reduces to (4.18) with the corresponding survival probability (4.19). On the other hand, let $g_U(t) = g_L(t) = g(t)$. Then, defining $p(t) = P(D_j > g(t))$ as the probability of failure under a shock at time t $(q(t) = P(D_j \le g(t))$, we obviously arrive at the $p(t) \Leftrightarrow q(t)$ model described by Eq. (4.1).

On the basis of the above described model, we will derive now the (unconditional) survival function and the corresponding failure rate function. First, wc need the following general lemma (see, [13] for the proof):

Lemma 4. 1 *Let $X_1, X_2, \ldots, X_n$ be i.i.d. random variables and $Z_1, Z_2, \ldots, Z_n$ be i.i.d. continuous random variables with the corresponding common pdf. Furthermore, let* $\mathrm{X} = (X_1, X_2, \ldots, X_n)$ *and* $\mathrm{Z} = (Z_1, Z_2, \ldots, Z_n)$ *be independent. Suppose that the function $\varphi(x, z) \, : \, R^n \, \times \, R^n \, \to \, R$ satisfies $\varphi(\mathrm{X}, t) \, =^d \, \varphi(\mathrm{X}, \pi(t))$, for any vector $t \in R^n$ and for any n-dimensional permutation function $\pi(\cdot)$. Then*

$$\varphi(\mathrm{X}, \mathrm{Z}) =^d \varphi(\mathrm{X}, \mathrm{Z}^*),$$

where $\mathrm{Z}^* = (\mathrm{Z}_{(1)}, \mathrm{Z}_{(2)}, \ldots, \mathrm{Z}_{(n)})$ *is the vector of the order statistics of* Z.

We are ready now to prove the following theorem [11].

Theorem 4.4 *Let* $H(t) = \int_0^t h(v)\,\mathrm{d}v$, $m(t) \equiv E(N(t)) = \int_0^t v(x)\,\mathrm{d}x$ *and* $f_D(u)$, $F_D(u)$ *be the pdf and the Cdf of* $D =^d D_j$, $j = 1, 2, \ldots$. *Assume that the inverse function* $m^{-1}(t)$ *exists for* $t > 0$. *Then the survival function that corresponds to the lifetime* T_S *is*

$$P(T_S > t) = \exp\left\{ -\int_0^t \bar{F}_D(g_L(u))\, v(u)\,\mathrm{d}u \right\}$$
$$\exp\left\{ \int_0^t \int_{g_L(s)}^{g_U(s)} \exp\{-kuH(t - s)\} f_D(u)\,\mathrm{d}u\, v(s)\,\mathrm{d}s \right\}, \qquad (4.25)$$

and the corresponding failure rate is

$$\lambda_S(t) = P(D > g_U(t))\,\lambda(t)$$
$$+ \int_0^t \int_{g_L(s)}^{g_U(s)} kuh(t - s)\, \exp\{-kuH(t - s)\} f_D(u)\,\mathrm{d}u\, v(s)\,\mathrm{d}s \qquad (4.26)$$

Proof Observe that

$$P(T_S > t \mid N(s),\ 0 \le s \le t,\ D_1, D_2, \ldots, D_{N(t)})$$
$$= \prod_{j=1}^{N(t)} \gamma(T_j, D_j)\, \exp\left\{ -k \sum_{j=1}^{N(t)} \xi(T_j, D_j)\, D_j\, H(t - T_j) \right\}$$
$$= \exp\left\{ \sum_{j=1}^{N(t)} (\ln \gamma(T_j, D_j) - k\xi(T_j, D_j)\, D_j\, H(t - T_j)) \right\}.$$

Therefore,

$$P(T_S > t) = E\left[\exp\left\{ \sum_{j=1}^{N(t)} (\ln \gamma(T_j, D_j) - k\xi(T_j, D_j)\, D_j H(t - T_j)) \right\} \right]$$
$$= E\left[E\left(\exp\left\{ \sum_{j=1}^{N(t)} (\ln \gamma(T_j, D_j) - k\xi(T_j, D_j)\, D_j\, H(t - T_j)) \right\} \middle| N(t) \right) \right].$$

As previously, if $m^{-1}(t)$ exists, then the joint distribution of $T_1, T_2, \ldots, T_n$, given $N(t) = n$, is the same as the joint distribution of the order statistics $T'_{(1)} \le T'_{(2)} \le \ldots \le T'_{(n)}$ of i.i.d. random variables $T'_1, T'_2, \ldots, T'_n$, where the pdf of the common distribution of T'_j's is given by $v(x)/m(t)$. Thus,

$$E\left(\exp\left\{\sum_{j=1}^{N(t)} (\ln \gamma(T_j, D_j) - k\xi(T_j, D_j)\, D_j\, H(t - T_j))\right\} | N(t) = n\right)$$
$$= E\left(\exp\left\{\sum_{j=1}^{n} (\ln \gamma(T'_{(j)}, D_j) - k\xi(T'_{(j)}, D_j)\, D_j H(t - T'_{(j)}))\right\}\right).$$

Let X $= (D_1, D_2, \ldots, D_n)$, Z $= (T'_1, T'_2, \ldots, T'_n)$ and

$$\varphi(\mathrm{X}, \mathrm{Z}) \equiv \sum_{j=1}^{n} (\ln \gamma(T'_j, D_j) - k\xi(T'_j, D_j)\, D_j H(t - T'_j)). \tag{4.27}$$

Note that, as was mentioned, if $g_U(t) = \infty$; $g_L(t) = 0$, then $\xi(T_j, D_j) \equiv 1$ and our model reduces to the original model of Lemoine and Wenocur [24], where each term in $\varphi(\mathrm{X}, \mathrm{Z})$ is just a simple product of D_j and $H(t - T'_j)$. Due to this simplicity, the rest was straightforward. Now we have a much more complex form of $\varphi(\mathrm{X}, \mathrm{Z})$, as given in (4.27), where the terms in the sum cannot be factorized.

Observe that the function $\varphi(x, z)$ satisfies

$$\varphi(\mathrm{X}, t) =^d \varphi(\mathrm{X}, \pi(t))$$

for any vector $t \in R^n$ and for any n-dimensional permutation function $\pi(\cdot)$. Thus, applying Lemma 4.1,

$$\sum_{j=1}^{n} (\ln \gamma(T'_j, D_j) - k\xi(T'_j, D_j)\, D_j\, H(t - T'_j))$$
$$=^d \sum_{j=1}^{n} (\ln \gamma(T'_{(j)}, D_j) - k\xi(T'_{(j)}, D_j)\, D_j\, H(t - T'_{(j)}))$$

and, therefore,

$$E\left(\exp\left\{\sum_{j=1}^{n} (\ln \gamma(T'_{(j)}, D_j) - k\xi(T'_{(j)}, D_j)\, D_j\, H(t - T'_{(j)}))\right\}\right)$$
$$= E\left(\exp\left\{\sum_{j=1}^{n} (\ln \gamma(T'_j, D_j) - k\xi(T'_j, D_j)\, D_j\, H(t - T'_j))\right\}\right)$$
$$= \left(E\left(\exp\{\ln \gamma(T'_1, D_1) - k\xi(T'_1, D_1)\, D_1\, H(t - T'_1)\}\right)\right)^n.$$

As

$$
\begin{aligned}
&E\left[\exp\{\ln\gamma(T_1',D_1) - k\xi(T_1',D_1)\,D_1\,H(t-T_1')\}\mid T_1'=s\right]\\
&= E\left[\exp\{\ln\gamma(s,D_1) - k\xi(s,D_1)\,D_1\,H(t-s)\}\right]\\
&= \int\limits_{g_L(s)}^{g_U(s)} \exp\{-kuH(t-s)\}f_D(u)\,\mathrm{d}u + P(D_1\le g_L(s)),
\end{aligned}
\tag{4.28}
$$

where for $D_1 > g_U(s)$, $\exp\{\ln\gamma(s, D_1) - k\xi(s,D_1)\,D_1\,H(t-s)\} = 0$, for all $s > 0$, the unconditional expectation is

$$
\begin{aligned}
&E\left[\exp\{\ln\gamma(T_1',D_1) - k\xi(T_1',D_1)\,D_1\,H(t-T_1')\}\right]\\
&= \int\limits_0^t\int\limits_{g_L(s)}^{g_U(s)} \exp\{-kuH(t-s)\}f_D(u)\,\mathrm{d}u\frac{v(s)}{m(t)}\mathrm{d}s + \int\limits_0^t P(D_1\le g_L(s))\,\frac{v(s)}{m(t)}\mathrm{d}s.
\end{aligned}
$$

Let

$$
\alpha(t)\equiv \int\limits_0^t\int\limits_{g_L(s)}^{g_U(s)} \exp\{-kuH(t-s)\}f_D(u)\,\mathrm{d}u\lambda(s)\,\mathrm{d}s + \int\limits_0^t P(D_1\le g_L(s))\,v(s)\,\mathrm{d}s,
$$

and we finally arrive at

$$
\begin{aligned}
P(T_S>t) &= \sum_{n=0}^{\infty}\left(\frac{v(t)}{m(t)}\right)^n\cdot\frac{m(t)^n}{n!}\exp\left\{-\int\limits_0^t v(u)\,\mathrm{d}u\right\}\\
&= \exp\left\{-\int\limits_0^t v(u)\,\mathrm{d}u + \int\limits_0^t\int\limits_{g_L(s)}^{g_U(s)} \exp\{-kuH(t-s)\}f_D(u)\,\mathrm{d}u\,v(s)\,\mathrm{d}s + \int\limits_0^t P(D_1\le g_L(u))v(u)\,\mathrm{d}u\right\},
\end{aligned}
$$

which is obviously equal to (4.25).

The corresponding failure rate can be obtained as

$$
\begin{aligned}
\lambda_S(t) &= -\frac{\mathrm{d}}{\mathrm{d}t}\ln P(T_S>t)\\
&= v(t) - P(g_L(t)\le D_1\le g_u(t))\,v(t)\\
&\quad + \int\limits_0^t\int\limits_{g_L(s)}^{g_U(s)} kuh(t-s)\exp\{-kuH(t-s)\}f_D(u)\,\mathrm{d}u\,v(s)\,\mathrm{d}s - P(D_1\le g_L(t))\,v(t)\\
&= P(D_1>g_U(t))v(t) + \int\limits_0^t\int\limits_{g_L(s)}^{g_U(s)} kuh(t-s)\exp\{-kuH(t-s)\}f_D(u)\,\mathrm{d}u\,v(s)\,\mathrm{d}s,
\end{aligned}
$$

where the Leibnitz rule was used for differentiation of the double integral. □

Relationship (4.26) suggests that (4.25) can be equivalently written as

$$P(T_S > t) = \exp\left\{-\int_0^t \bar{F}_D(g_U(u))\,v(u)\,du\right\} \exp\left\{-\int_0^t \int_{g_L(s)}^{g_U(s)} kuh(t-s)\exp\{-kuH(t-s)\}f_D(u)\,du\,v(s)\,ds\right\}.$$

Therefore, we can again interpret our system as a series one with two independent components: one that fails only because of fatal (critical) shocks and the other that fails because of nonfatal shocks.

Example 4.2 Consider the special case when $g_U(t) = \infty$ and $g_L(t) = 0$. Then the survival function in (4.25) is

$$\begin{aligned} P(T > t) &= \exp\left\{-\int_0^t \bar{F}_D(g_L(u))\,v(u)\,du\right\} \exp\left\{\int_0^t \int_{g_L(s)}^{g_U(s)} \exp\{-kuH(t-s)\}f_D(u)\,du\,v(s)\,ds\right\} \\ &= \exp\{-m(t)\}\exp\left\{\int_0^t L(kH(t-s))\,v(s)\,ds\right\} = \exp\{-m(t)\}\exp\left\{\int_0^t L(kH(u))\,v(t-u)\,du\right\}, \end{aligned}$$

where $L(\cdot)$ is the operator of the Laplace transform with respect to $f_D(u)$. Therefore, we arrive at Eq. (4.19) obtained in [24].

Example 4.3 Suppose that $v(t) = v,\ t \geq 0,\ D_j \equiv d,\ j = 1, 2, \ldots,$ and there exist $t_2 > t_1 > 0$ such that

$g_U(t) > g_L(t) > d$, for $0 \leq t < t_1$ (shocks are harmless);
$d > g_U(t) > g_L(t)$, for $t_2 < t$ (shocks are fatal), and
$g_U(t) > d > g_L(t)$, for $t_1 < t < t_2$; $g_L(t_1) = g_U(t_2) = d$.

Let for the sake of further integration, $h(t) = 1/(1+t),\ t \geq 0$, and $k = 1/d$ (for simplicity of notation). From Eq. (4.28),

$$\begin{aligned} &E[\exp\{\ln\gamma(T_1', D_1) - k\xi(T_1', D_1)\,D_1 H(t - T_1')\} \mid T_1' = s] \\ &= \exp\{\ln\gamma(s,d) - k\xi(s,d)\,dH(t-s)\} \\ &= \begin{cases} 0, & \text{if} \quad g_U(s) > d(s > t_2) \\ \exp\{-H(t-s)\}, & \text{if} \quad g_L(s) < d \leq g_U(s)\,(t_1 < s \leq t_2) \\ 1, & \text{if} \quad d \leq g_L(s)\,(s \leq t_1) \end{cases} \\ &= \exp\{-H(t-s)\}I(g_L(s) > d \leq g_U(s)) + I(d \leq g_L(s)) \\ &= \exp\{-H(t-s)\}I(t_1 > s \leq t_2) + I(s_1). \end{aligned}$$

Thus, 'integrating $T_1' = s$ out':

$$E\left[\exp\{\ln\gamma(T_1',D_1) - k\xi(T_1',D_1)D_1 H(t - T_1')\}\right]$$
$$= \frac{1}{m(t)}\left[\int_0^t \exp\{-H(t-s)\} I(t_1 < s \le t_2)\, v(s)\,\mathrm{d}s + \int_0^t I(s \le t_1)\, v(s)\,\mathrm{d}s\right].$$

Then,

$$P(T_S > t) = \exp\left\{-\int_0^t v(u)\,du + \int_0^t \exp\{-H(t - s)\} I(t_1 < s \le t_2)\, v(s)\,\mathrm{d}s + \int_0^t I(s \le t_1)\, v(s)\,\mathrm{d}s\right\}$$
$$= \exp\left\{-\int_0^t I(s > t_1)\, v(s)\,\mathrm{d}s + \int_0^t \exp\{-H(t - s)\} I(t_1 < s \le t_2)\, v(s)\,\mathrm{d}s\right\}.$$

Thus [11],

(i) For $0 \le t \le t_1$, $P(T > t) = 1$;
(ii) For $t_1 \le t \le t_2$,

$$P(T_S > t) = \exp\left\{-\int_{t_1}^t \lambda \mathrm{d}u\right\} \exp\left\{\lambda \int_{t_1}^t \exp\{-H(t - s)\}\,\mathrm{d}s\right\}$$
$$= \exp\{-v(t - t_1)\} \exp\{v \ln(1 + t - t_1)\}$$
$$= \exp\{-v(t - t_1)\}(1 + t - t_1)^v;$$

(iii) For $t_2 \le t$,

$$P(T_S > t) = \exp\left\{-\int_{t_1}^t v\mathrm{d}u\right\} \exp\left\{v \int_{t_1}^{t_2} \exp\{-H(t - s)\}\,\mathrm{d}s\right\}$$
$$= \exp\{-v(t - t_1)\}(1 + t_2 - t_1)^v,$$

which shows (compared with case (ii)) that if the system has survived in $0 \le t \le t_1$, then the next shock with probability 1 will 'kill' it.

Model 2. We consider now the following useful modification of Model 1:

Let, on each shock, depending on its magnitude D_j, $j = 1, 2, ..$, the following mutually exclusive events occur:

(i) If $D_j > g_U(T_j)$, the shock results in an immediate system failure (as in Model 1)
(ii) If $D_j \le g_L(T_j)$, the shock is harmless (as in Model 1)
(iii) If $g_L(T_j) < D_j \le g_U(T_j)$, then the shock imposes a (constant) effect on the system *lasting for a random time*, which depends on its arrival time and magnitude.

$$P(T_S > t) = \exp\{-m(t)\} \exp\left\{ \int_0^t L(kH(u))\, v(t - u)\, du \right\}, \tag{4.19}$$

where $m(t) = \int_0^t v(u)\, du$, $H(t) = \int_0^t h(u)\, du$ and $L(\cdot)$ is the operator of the Laplace transform with respect to the distribution of the shock's magnitude. In what follows, we generalize the approach of these authors to the case when a system can also fail due to a fatal shock with the magnitude exceeding the time-dependent bound, which is more realistic in practice.

4.3.2 Shot Noise Process with Critical Shocks and Deterioration

Model 1. In addition to the general assumptions of Lemoine and Wenocur [24] stated in the previous subsection, let on each shock, depending on its magnitude D_j, $j = 1, 2\ldots$, the following mutually exclusive events occur [11]:

(i) If $D_j > g_U(T_j)$, then the shock results in an immediate system's failure
(ii) If $D_j \le g_L(T_j)$, then the shock does not cause any change in the system (harmless)
(iii) If $g_L(T_j) < D_j \le g_U(T_j)$, then the shock increases the stress by $D_j\, h(0)$,

where $g_U(t)$, $g_L(t)$ are the decreasing, deterministic functions.

The functions of operating time, $g_U(t)$, $g_L(t)$ define the corresponding upper and lower bounds. Because they are decreasing, this means that the probability that the shock arriving at time t results in the system's failure is increasing in time, whereas the probability that the shock is harmless is decreasing with time. Therefore, obviously, a deterioration of our system is described in this way. The function $g_U(t)$ can also be interpreted as the strength of our system with respect to shocks, whereas the function $g_L(t)$, can be interpreted as the 'sensitivity' to shocks. At many instances, they can be defined from the general 'physical considerations' on the criterion of failure of a system. For instance, the minimum peak voltage that can ruin a new electronic item is usually given in its specifications.

Define the following 'membership function':

$$\xi(T_j, D_j) = \begin{cases} 1, & g_L(T_j) < D_j \le g_U(T_j) \\ 0, & D_j \le g_L(T_j) \end{cases}. \tag{4.20}$$

Using this notation, the cumulative stress, similar to (4.16), can be written as

$$X(t) \equiv \sum_{j=1}^{N(t)} \xi(T_j, D_j)\, D_j\, h(t - T_j), \tag{4.21}$$

In the latter case, assume that the larger are the shock's arrival time and magnitude, the longer this effect lasts. Formally, let the shock increase the system failure rate by η units (constant) for the random time $w(T_j, D_j)$, where $w(t,d)$ is a strictly increasing function of each argument. Thus, along with decreasing functions $g_U(t)$, $g_L(t)$, the increasing function $w(t,d)$ models *deterioration* of our system.

Similar to (4.22) (where for simplicity of notation, we set $k \equiv 1$), the conditional failure rate process (on condition that the event $D_j > g_U(T_j)$, $j = 1, 2, \ldots$ did not happen in $[0,t)$ and $\{N(t), T_1, T_2, \ldots, T_{N(t)}\}$ and $\{D_1, D_2, \ldots, D_{N(t)}\}$ are given) is

$$\hat{\lambda}_t \equiv X(t) = \sum_{j=1}^{N(t)} \xi(T_j, D_j)\, \eta I(T_j \le t < T_j + w(T_j, D_j)).$$

Then, similar to (4.23),

$$\begin{aligned} &P(T_S > t \mid N(s),\ 0 \le s \le t,\ D_1, D_2, \ldots, D_{N(t)}) \\ &= \prod_{j=1}^{N(t)} \gamma(T_j, D_j) \cdot \exp\left\{ -\int_0^t \sum_{j=1}^{N(x)} \xi(T_j, D_j)\, \eta I(T_j \le x < T_j + w(T_j, D_j))\, \mathrm{d}x \right\}. \end{aligned} \tag{4.29}$$

where the functions $\xi(T_j, D_j)$ and $\gamma(T_j, D_j)$ are defined in (4.20) and (4.24), respectively.

Similar to Theorem 4.4, the following result holds.

Theorem 4.5 *Let η be the increment in the system's failure rate due to a single shock that lasts for the random time $w(T_j, D_j)$. Under assumptions of Theorem 4.4, the survival function $P(T_S > t)$ is given by*

$$\begin{aligned} P(T_S > t) &= \exp\left\{ -\int_0^t \bar{F}_D(g_L(u))\, v(u)\, \mathrm{d}u \right\} \\ &\times \exp\left\{ \int_0^t \int_{g_L(s)}^{g_U(s)} \exp\{-\eta \cdot \min\{w(u,s),\ (t - s)\}\} f_D(u)\, \mathrm{d}u\, v(s)\, \mathrm{d}s \right\}. \end{aligned} \tag{4.30}$$

Proof Observe that from (4.29),

$$\begin{aligned} &P(T_S > t \mid N(s),\ 0 \le s \le t,\ D_1, D_2, \ldots, D_{N(t)}) \\ &= \exp\left\{ \sum_{j=1}^{N(t)} (\ln \gamma(T_j, D_j) - \eta \xi(T_j, D_j) \min\{w(T_j, D_j),\ (t - T_j)\}) \right\}. \end{aligned}$$

Therefore,

$$P(T_S > t) = E\left[\exp\left\{\sum_{j=1}^{N(t)} (\ln \gamma(T_j, D_j) - \eta\xi(T_j, D_j) \min\{w(T_j, D_j), (t - T_j)\})\right\}\right]$$

$$= E\left[E\left(\exp\left\{\sum_{j=1}^{N(t)} (\ln \gamma(T_j, D_j) - \eta\xi(T_j, D_j) \min\{w(T_j, D_j), (t - T_j)\})\right\} |N(t)\right)\right].$$

Following straightforwardly the procedure described in the proof of Theorem 4.4, we eventually arrive at (4.30).

□

In contrast to Theorem 4.4 and owing to dependence in (4.30) on the function of minimum, the corresponding failure rate can only be obtained when specific forms of $g_U(t)$, $g_L(t)$, and $w(t, d)$ are given. As in the case of Model 1, when $g_U(t) = g_L(t) = g(t)$, this model also obviously reduces to the $p(t) \Leftrightarrow q(t)$ model (4.1).

Example 4.4 Let $g_L(t) = 0$, $g_U(t) = \infty$, for all $t \geq 0$, and $w(t, d) = d$(no deterioration in time). This means that the shocks are not fatal with probability 1 and that the durations of the shock's effect do not depend on the arrival times but are just given by the i.i.d. random variables D_j. In this case, from (4.30),

$$P(T_S > t) = \exp\left\{-\int_0^t v(u)\,du\right\}$$
$$\times \exp\left\{\int_0^t\int_0^\infty \exp\{-\eta \cdot \min\{w(u, s), (t - s)\}\} f_D(u)\,du\,v(s)\,ds\right\},$$

where

$$\int_0^t\int_0^\infty \exp\{-\eta \cdot \min\{w(u, s), (t - s)\}\} f_D(u)du\,v(s)\,ds$$
$$= \int_0^t\int_0^{t-s} \exp\{-\eta u\} f_D(u)\,du\,v(s)\,ds + \int_0^t\int_{t-s}^\infty \exp\{-\eta(t - s)\} f_D(u)\,du\,v(s)\,ds$$
$$= \int_0^t\int_0^{t-u} v(s)\,ds\,\exp\{-\eta u\} f_D(u)\,du + \int_0^t \exp\{-\eta(t - s)\}\,\overline{F}_D(t - s)\,v(s)\,ds.$$
$$= \int_0^t m(t - u)\exp\{-\eta u\} f_D(u)\,du + \int_0^t \exp\{-\eta(u)\}\,\overline{F}_D(u)\,v(t - u)\,du$$

$$= [-\overline{F}_D(u)\exp\{-\eta u\}\,m(t-u)]_0^t - \int_0^t \overline{F}_D(u)\exp\{-\eta u\}\,v(t-u)\,du$$

$$-\eta\int_0^t \overline{F}_D(u)\exp\{-\eta u\}\,m(t-u)\,du + \int_0^t \exp\{-\eta(u)\}\,\overline{F}_D(u)\,v(t-u)\,du$$

$$= m(t) - \eta\int_0^t \overline{F}_D(u)\exp\{-\eta u\}\,m(t-u)\,du.$$

Therefore,

$$P(T_S > t) = \exp\left\{-\eta\int_0^t \exp\{-\eta u\}\cdot\overline{F}_D(u)\cdot m(t-u)\,du\right\},$$

and thus

$$\lambda_S(t) = \eta\int_0^t \exp\{-\eta u\}\cdot\overline{F}_D(u)\cdot v(t-u)\,du.$$

4.4 Extreme Shock Model with Delayed Termination

Consider an orderly point process (without multiple occurrences) $\{N(t),\ t \geq 0\}$ of some 'initiating' events (IEs) with arrival times $T_1 < T_2 < T_3 < \ldots$. Let each event from this process triggers the 'effective event' (EE), which occurs after a random time (delay) $D_i, i = 1, 2, \ldots,$ since the occurrence of the corresponding IE at T_i. Obviously, in contrast to the initial ordered sequence $T_1 < T_2 < T_3 < \ldots$, the EEs $\{T_i + D_i\}, i = 1, 2, \ldots$ are now not necessarily ordered. This setting can be encountered in many practical situations, when, e.g., initiating events start the process of developing the non-fatal faults in a system and we are interested in the number of these faults in $[0, t)$. Alternatively, effective events can result in fatal, terminating faults (failures) and then we are interested in the survival probability of our system. Therefore, the latter setting means that the *first EE ruins our system*. When there are no delays, each shock (with the specified probability) results in the failure of the survived system and the described model obviously reduces to the classical extreme shock model ([17]; [19]) considered in the previous section of this chapter and in Chap. 3.

The IEs can often be interpreted as some external shocks affecting a system, and for convenience and in the spirit of the current chapter, we will often use this term (interchangeably with the "IE"). We will consider the case of the NHPP of the IEs.

The approach can, in principle, be applied to the case of renewal processes, but the corresponding formulas are too cumbersome. However, the obtained results for the NHPP case are in simple, closed forms that allow intuitive interpretations and proper analyses. Our presentation in this and the subsequent section will mostly follow Cha and Finkelstein[7].

Thus, a system is subject to the NHPP of IEs, $\{N(t),\ t \geq 0\}$ to be called shocks. Let the rate of this process be $v(t)$ and the corresponding arrival times be denoted as $T_1 < T_2 < T_3 \ldots$. Assume that the ith shock is 'harmless' to the system with probability $q(T_i)$, and with probability $p(T_i)$ it *triggers the failure process* of the system which results in its failure after a random time $D(T_i)$, $i = 1, 2, \ldots$, where $D(t)$ is a non-negative, semicontinuous random variable with the point mass at "0" (at each fixed t). Note that, this 'point mass' at 0 opens the possibility of the 'immediate failure' of the system on a shock's occurrence, which is practically very important. Furthermore, the case of the 'full point mass' of $D(t)$ at 0 reduces to the ordinary 'extreme shock model'. Obviously, without the point mass at 0, we arrive at an absolutely continuous random variable. The distributions of $D(t)$ having point masses at other values of time could be considered similarly.

Let $G(t,x) \equiv P(D(t) \leq x)$, $\bar{G}(t,x) \equiv 1 - G(t,x)$, and $g(t,x)$ be the Cdf, the survival function and the pdf for the 'continuous part' of $D(t)$, respectively. Then, in accordance with our terminology, the failure in this case is the EE.

First of all, we are interested in describing the lifetime of our system T_S. The corresponding conditional survival function is given by

$$P(T_S > t \mid N(s),\ 0 \leq s \leq t;\ D(T_1),\ D(T_2), \ldots, D(T_{N(t)});\ J_1, J_2, \ldots, J_{N(t)})$$
$$= \prod_{i=1}^{N(t)} (J_i + (1 - J_i) I(D(T_i) > t - T_i)), \tag{4.31}$$

where the indicators are defined as

$$I(D(T_i) > t - T_i) = \begin{cases} 1, & \text{if } D(T_i) > t - T_i \\ 0, & \text{otherwise} \end{cases},$$

$$J_i = \begin{cases} 1, & \text{if the ith shock does not trigger the subsequent failure process,} \\ 0, & \text{otherwise.} \end{cases}$$

Assume the following conditions regarding 'conditional independence':

(i) Given the shock process, $D(T_i)$, $i = 1, 2, \ldots$, are mutually independent.
(ii) Given the shock process, J_i, $i = 1, 2, \ldots$, are mutually independent. (It means that whether each shock triggers the failure process of the system or not is 'independently determined').
(iii) Given the shock process, $\{D(T_i),\ i = 1, 2, \ldots\}$ and $\{J_i,\ i = 1, 2, \ldots\}$ are mutually independent.

As in the previous sections, integrating out all conditional random quantities in (4.31) under the basic assumptions described above results in the following theorem.

Theorem 4.6 *Let* $m^{-1}(t),\ t > 0$ *exist (* $m(t) \equiv E(N(T))$*). Then*

$$P(T_S \geq t) = \exp\left\{ -\int_0^t G(x,\, t\ -\ x)\, p(x)\, v(x)\, \mathrm{d}x \right\},\ t \geq 0,$$

and the failure rate function of the system is

$$\lambda_S(t) = \int_0^t g(x, t\ -\ x)\, p(x)\, v(x)\, \mathrm{d}x\ +\ G(t, 0)\, p(t)\, v(t),\ t \geq 0.$$

Proof Given the assumptions, we can directly 'integrate out' J_i's and D_i's and define the corresponding probability in the following way:

$$P(T_S > t \mid N(s),\ 0 \leq s \leq t) = \prod_{i=1}^{N(t)} \left(q(T_i) + p(T_i)\overline{G}(T_i,\ t - T_i)\right).$$

Therefore,

$$\begin{aligned} P(T_S > t) &= E\left[\prod_{i=1}^{N(t)} \left(q(T_i) + p(T_i)\overline{G}(T_i,\ t - T_i)\right)\right] \\ &= E\left[E\left[\prod_{i=1}^{N(t)} \left(q(T_i) + p(T_i)\,\overline{G}(T_i,\ t - T_i)\right) \mid N(t)\right]\right]. \end{aligned} \tag{4.32}$$

As the joint distribution of $T_1, T_2, \ldots, T_n$ given $N(t) = n$ is the same as the joint distribution of order statistics $T'_{(1)} \leq T'_{(2)} \leq \ldots \leq T'_{(n)}$ of i.i.d. random variables $T'_1, T'_2, \ldots, T'_n$, where the pdf of the common distribution of T'_j's is given by $v(x)/m(t),\ 0 \leq x \leq t$, we have

$$\begin{aligned} & E\left[\prod_{i=1}^{N(t)} \left(q(T_i) + p(T_i)\,\overline{G}(T_i,\ t - T_i)\right) \mid N(t) = n\right] \\ &= E\left[\prod_{i=1}^{n} \left(q(T'_{(i)}) + p(T'_{(i)})\,\overline{G}(T_{(i)},\ t - T'_{(i)})\right)\right] - E\left[\prod_{i=1}^{n} \left(q(T'_i) + p(T'_i)\,\overline{G}(T'_i,\ t\ \ T'_i)\right)\right] \\ &= \left(E\left[q(T'_i) + p(T'_i)\,\overline{G}(T'_i,\ t - T'_i)\right]\right)^n = \left(\frac{1}{m(t)} \int_0^t \left(q(x) + p(x)\,\overline{G}(x;\ t - x)\right) v(x)\, \mathrm{d}x\right)^n. \end{aligned} \tag{4.33}$$

From Eqs. (4.32) and (4.33),

$$P(T_S > t) = \sum_{n=0}^{\infty} \left(\frac{1}{m(t)} \int_0^t \left(q(x) + p(x)\,\overline{G}(x,\ t-x)\right) v(x)\,\mathrm{d}x \right)^n \cdot \frac{m(t)^n}{n!} e^{-m(t)}$$

$$= e^{-m(t)} \cdot \exp\left\{ \int_0^t \left(q(x) + p(x)\,\overline{G}(x,\ t-x)\right) v(x)\,\mathrm{d}x \right\}$$

$$= \exp\left\{ \int_0^t q(x)v(x)\,\mathrm{d}x + \int_0^t \overline{G}(x,\ t-x)p(x)\,v(x)\,\mathrm{d}x - \int_0^t v(x)\,\mathrm{d}x \right\}$$

$$= \exp\left\{ -\int_0^t G(x,\ t-x)\,p(x)\,v(x)\,\mathrm{d}x \right\}.$$

Therefore, by Leibnitz rule, the failure rate function of the system, $\lambda_S(t)$, is given in the following meaningful and rather simple form:

$$\lambda_S(t) = \int_0^t g(x; t-x)p(x)\,v(x)\,\mathrm{d}x + G(t,\ 0)p(t)\,v(t),\ t \geq 0. \tag{4.34}$$

□

Formally, the split of effects to effective and ineffective shocks does not add any mathematical complexity because of the NHPP nature of the arrival process. This means that the result would be the same if we had only one type of effects and the NHPP with the rate function $p(t)\,v(t)$. However, from the practical point of view and keeping in mind that we are generalizing here the classical extreme shock model with two types of effects, this splitting seems to be reasonable. Furthermore, we can consider the case of the multitype delayed consequences of shocks ($n > 1$), where the shock that occurs at time t causes the delayed (with distribution $G_i(t,x)$) effect of type i with probability $p_i(t)$, whereas the probability of 'no effect' is $1 - \sum_{i=1}^n p_i(t)$. Obviously, this model is the same as the single-type model with $G(t,x) = \sum_{i=1}^n p_i^*(t)\,G_i(t,x)$ and $p(t) = \sum_{i=1}^n p_i(t)$, where $p_i^*(t) = p_i(t)/\sum_{i=1}^n p_i(t)$. Therefore, similar to Theorem 4.6,

$$P(T_S \geq t) = \exp\left\{ -\int_0^t \left(\sum_{i=1}^n p_i(x)\,G_i(x,\ t-x) \right) v(x)\,\mathrm{d}x \right\},\ t \geq 0$$

and

$$\lambda_S(t) = \int_0^t \left(\sum_{i=1}^n p_i(x)\,g_i(x,\ t-x) \right) v(x)\,\mathrm{d}x + \left(\sum_{i=1}^n p_i(t)G_i(t,\ 0) \right) v(t).$$

4.5 Cumulative Shock Model with Initiated Wear Processes

Consider now a cumulative model for the IEs, where the accumulated wear can result in a system's failure when it reaches the given boundary. Our setting that follows is different from the conventional one. In the conventional setting, the wear caused by a shock is incurred at the moment of the corresponding shock (see Sect. 4.1). In our model, however, the wear process, triggered by a shock, is activated at the moment of a shock's occurrence and continuously increases with time.

Denote by $W(t,u)$ the random wear incurred in u units of time after a single shock (IE) that has occurred at time t. Let $W(t,\ 0)\equiv 0$, for all $t\geq 0$. Assume that $W(t,u)$ is stochastically increasing (see Sect. 2.8) in t and u, that is,

$$W(t_1,\ u)\leq_{st}W(t_2,\ u) \text{ for all } t_2>t_1>0 \text{ and for all } u>0;$$

and

$$W(t_1,\ u)\leq_{st}W(t,u) \text{ for all } u_2>u>0 \text{ and for all } t>0.$$

An example for this type of $W(t,\ u)$ is the gamma process, with the pdf for $W(t,\ u)$ given by

$$f(w,\ t,\ u)=\frac{\beta^{\alpha(t,u)}\cdot w^{\alpha(t,u)-1}\exp\{-\beta w\}}{\Gamma(\alpha(t,\ u))},\quad w\geq 0,$$

where $\alpha(t,\ 0)=0$, for all $t\geq 0$, and $\alpha(t,\ u)$ is strictly increasing in both t and u.

If all shocks from the initial process trigger wear, then the accumulated wear from all shocks in $[0,\ t)$ is

$$W(t)=\sum_{i=0}^{N(t)}W(T_i,t\ -\ T_i),$$

which can be considered as a general form of a shot noise process (see Sect. 4.3). Assume that each shock with probability $p(t)$ results in an immediate failure (termination), otherwise, with probability $q(t)$ it triggers the wear process in the way described above. The failure also occurs when the accumulated wear reaches the random boundary R and we are interested in obtaining the distribution of the time to failure, T_S.

The corresponding conditional survival probability for this model can be written as [7]

$$\begin{aligned}&P(T_S>t|N(s),0\leq s\leq t;\ W(T_i,\ t-T_i),\ i=1,2,\ldots,N(t);\ R)\\&=\prod_{i=0}^{N(t)}q(T_i)\cdot I\left(\sum_{i=0}^{N(t)}W(T_i,\ t-T_i)\leq R\right).\end{aligned}$$

For obtaining the explicit expression for the unconditional survival probability in this case assume additionally that R is the exponentially distributed (with parameter λ) random variable.

Theorem 4.7 *Let the shock process be the NHPP with* rate $v(t)$ *and suppose that* $m^{-1}(t)$ *exists (for* $t > 0$*). Then*

$$P(T_S \geq t) = \exp\left\{ -\int_0^t v(x)\,\mathrm{d}x + \int_0^t M_{W(x,\,t-x)}(-\lambda) \cdot q(x)v(x)\,\mathrm{d}x \right\},\ t \geq 0,$$

and the corresponding failure rate function is

$$\lambda_S(t) = p(t)\,v(t) \ - \int_0^t \frac{\mathrm{d}}{\mathrm{d}t}\left(M_{W(x,\,t-x)}(-\lambda)\right) \cdot q(x)\,v(x)\,\mathrm{d}x,\ t \geq 0,$$

where $M_{W(t,\,u)}(\cdot)$ *is the mgf of* $W(t,\ u)$ *(for fixed* t *and* u*).*

Proof Given the assumptions, we can directly 'integrate out' the variable R and define the corresponding probability in the following way:

$$\begin{aligned} &P(T_S > t|N(s),\ 0 \leq s \leq t;\ W(T_i, t \ - \ T_i),\ i = 1, 2, \ldots, N(t)) \\ &= \left(\prod_{i=0}^{N(t)} q(T_i)\right) \cdot \exp\left\{ - \int_0^{\sum_{i=0}^{N(t)} W(T_i, t-T_i)} \lambda \mathrm{d}u \right\} \\ &= \exp\left\{ -\lambda \sum_{i=1}^{N(t)} W(T_i,\ t - T_i) + \sum_{i=1}^{N(t)} \ln q(T_i) \right\}. \end{aligned}$$

Thus, the survival function can be obtained as

$$P(T_S > t) = E\left[E\left[\exp\left\{ -\lambda \sum_{i=1}^{N(t)} W(T_i, t \ - \ T_i) \ + \ \sum_{i=1}^{N(t)} \ln q(T_i) \right\} |N(t)\right]\right].$$

Following the same procedure described in the Proof of Theorem 4.6,

$$\begin{aligned} &E\left[\exp\left\{ -\lambda \sum_{i=1}^{N(t)} W(T_i,\ t - T_i) + \sum_{i=1}^{N(t)} \ln q(T_i) \right\} |N(t) = n\right] \\ &= \left(E\left[\exp\{ -\lambda\, W(T_1',\ t - T_1') + \ln q(T_1') \}\right]\right)^n. \end{aligned}$$

Observe that,

$$E\left[\exp\{-\lambda W(T_1',\ t-T_1')+\ln q(T_1')\}\right] = \frac{1}{m(t)}\int_0^t \left(q(x)\,M_{W(x,\,t-x)}(-\lambda)\right)\nu(x)\,\mathrm{d}x.$$

Hence,

$$E\left[\exp\left\{-\lambda\sum_{i=1}^{N(t)} W(T_i,\ t-T_i)+\sum_{i=1}^{N(t)}\ln q(T_i)\right\}|N(t)=n\right]$$
$$=\left(\frac{1}{m(t)}\int_0^t \left(q(x)\,M_{W(x,\,t-x)}(-\lambda)\right)\nu(x)\,\mathrm{d}x\right)^n.$$

Finally,

$$P(T_S>t)=\exp\left\{-\int_0^t \nu(x)\,\mathrm{d}x + \int_0^t M_{W(x,\,t-x)}(-\lambda)\cdot q(x)\,\nu(x)\,\mathrm{d}x\right\}.$$

Therefore, by Leibnitz rule, the failure rate function of the system, $\lambda_S(t)$, is

$$\lambda_S(t)=\left(1 - M_{W(t,\,0)}(-\lambda)\cdot q(t)\right)\nu(t)-\int_0^t \frac{\mathrm{d}}{\mathrm{d}t}\left(M_{W(x,\,t-x)}(-\lambda)\right)\cdot q(x)\,\nu(x)\,\mathrm{d}x$$

$$=p(t)\nu(t)-\int_0^t \frac{\mathrm{d}}{\mathrm{d}t}\left(M_{W(x,\,t-x)}(-\lambda)\right)\cdot q(x)\,\nu(x)\,\mathrm{d}x. \qquad \square$$

Let, for simplicity, $\lim_{t\to\infty}\nu(t)\equiv\nu(\infty)\equiv\nu_0<\infty,\ \nu_0>0;\ p(t)\equiv p,\ q(t)\equiv q$. It is clear from general considerations that $\lim_{t\to\infty}\lambda_S(t)=\lim_{t\to\infty}\nu(t)=\nu_0$ monotonically approaching the limit from below. Indeed, consider a system that had survived in $[0,t)$, which means that the next interval $[t,\ t+\mathrm{d}t)$ starts with the same 'resource' R, as the boundary is exponentially distributed. Due to the fact that all previous nonfatal shocks accumulate wear and all triggered wear processes are increasing, as t increases ($W(t)\to\infty$ as $t\to\infty$), the resource R is 'consumed more intensively' with time. This obviously means that the probability of failure in $[t,\ t+\mathrm{d}t)$ is increasing in t and, therefore, $\lambda_S(t)$ is increasing. Eventually, when $t\to\infty$, each triggering shock becomes fatal in the limit, which means that

$$\lim_{t\to\infty} \lambda_S(t) = \lim_{t\to\infty} v(t) = v_0.$$

The following example illustrates these considerations.

Example 4.5 Suppose that $W(t, u)$ follows the gamma process, that is, the pdf of $W(t, u)$ is

$$f(w; t, u) = \frac{\beta^{\alpha(t, u)} \cdot w^{\alpha(t, u)-1} \exp\{-\beta w\}}{\Gamma(\alpha(t, u))}, \quad w \geq 0,$$

where $\alpha(t, 0) = 0$ for all $t \geq 0$, and $\alpha(t, u)$ is strictly increasing in both t and u. Then

$$M_{W(x, t-x)}(-\lambda) = \left(\frac{\beta}{\beta+\lambda}\right)^{\alpha(x, t-x)},$$

and

$$\frac{\mathrm{d}}{\mathrm{d}t}\left(M_{W(x,t-x)}(-\lambda)\right) = \frac{\mathrm{d}}{\mathrm{d}t}(\alpha(x,t - x)) \ln\left(\frac{\beta}{\beta+\lambda}\right) \cdot \left(\frac{\beta}{\beta+\lambda}\right)^{\alpha(x,t-x)}.$$

Let $v(t) = v,\ q(t) = q,\ t \geq 0,\ \alpha(t, u) = \alpha u,\ t, u \geq 0$. Then

$$\begin{aligned}\int_0^t \frac{\mathrm{d}}{\mathrm{d}t}\left(M_{W(x,t-x)}(-\lambda)\right) \cdot q(x)\, v(x)\, \mathrm{d}x &= \int_0^t \alpha \cdot \ln\left(\frac{\beta}{\beta+\lambda}\right) \cdot \left(\frac{\beta}{\beta+\lambda}\right)^{\alpha(t-x)} \cdot q\, v \mathrm{d}x \\ &= \int_0^{\alpha t} \ln\left(\frac{\beta}{\beta+\lambda}\right) \cdot \left(\frac{\beta}{\beta+\lambda}\right)^{x} \cdot q\, v \mathrm{d}x \\ &= qv\left(\left(\frac{\beta}{\beta+\lambda}\right)^{\alpha t} - 1\right).\end{aligned}$$

Therefore, we have

$$\lambda_S(t) = pv + qv\left(1 - \left(\frac{\beta}{\beta+\lambda}\right)^{\alpha t}\right), \quad t \geq 0$$

and

$$\lim_{t\to\infty} \lambda_S(t) \equiv v,$$

which illustrates the fact that every triggering shock in the limit becomes fatal.

4.6 'Curable' Shock Processes

In this section, we generalize the setting of Sect. 4.4 to the case when each failure that was initiated (and delayed), has a chance to be repaired or cured as well. Therefore, as previously, consider a system subject to the NHPP of IEs $\{N(t),\ t \geq 0\}$ to be called shocks. Let the rate of this process be $\nu(t)$ and the corresponding arrival times be denoted as $T_1 < T_2 < T_3 \ldots$. Assume that the ith shock triggers the failure process of the system which can result in its failure after a random time $D(T_i)$, $i = 1, 2, \ldots$, where for each fixed $t \geq 0$, the delay $D(t)$ is a non-negative, continuous random variable. Let $G(t,\ x) \equiv P(D(t) \leq x)$, $\bar{G}(t,\ x) \equiv 1 - G(t,\ x)$, and $g(t,\ x)$ be the Cdf, the survival function, and the pdf of $D(t)$, respectively. Assume now that with probability $q(t,\ x) = 1 - p(t,\ x)$, where t is the time of a shock's occurrence and x is the corresponding delay, each failure can be instantaneously *cured (repaired)*, as if this shock did not trigger the failure process at all. For instance, it can be an instantaneous overhaul of an operating system by the new one that was not exposed to shocks before. It should be noted that this operation is executed *at time* $t + x$ *and not at time* t, as in the classical extreme shock model without delay. Different cure models have been considered mostly in the biostatistical literature (see Aalen et al. [1] and references therein). Usually, these models deal with a population that contains a subpopulation that is not susceptible to, e.g., a disease (i.e., 'cured') after some treatment. This setting is often described by the multiplicative frailty model with the frailty parameter having a mass at 0. It means that there exists a nonsusceptible (cured) subpopulation with the hazard rate equal to 0. In our case, however, the interpretation is different, but the mathematical description is also based on considering the corresponding improper distributions [9].

For simplicity of notation, consider the t-independent case, when $D(t) \equiv D$, $G(t,\ x) \equiv G(x)$, $g(t,\ x) \equiv g(x)$ and $p(t,\ x) \equiv p(x)$. The results can be easily modified to the t-dependent setting. Having in mind that D denotes the time of delay, let D_C be the time from the occurrence of an IE to the system failure caused by this IE. Note that D_C is an improper random variable, as $D_C \equiv \infty$ (with a non-zero probability) when the corresponding IE does not result in an ultimate system failure due to cure. Then the improper survival function that describes D_C is:

$$\bar{G}_C(x) \equiv 1 - \int_0^x p(u)g(u)\,\mathrm{d}u \tag{4.35}$$

with the corresponding density:

$$g_C(x) = p(x)\,g(x). \tag{4.36}$$

Thus, the EE that has occurred in $[x,\ x + \mathrm{d}x)$ is fatal with probability $p(x)$ and is cured with probability $q(x)$. For the specific case, $p(x) \equiv p$, we can say that the proportion p of events of interest results in failure, whereas 'the proportion $1 - p$ is cured'

Another setting, which yields a similar description, is as follows: let each IE along with the failure development mechanism ignites a repair mechanism described by the repair time R with the Cdf $K(t)$. If $R > D$, then the EE is fatal, otherwise it will be repaired before the failure ($R \leq D$) and therefore, can formally be considered as cured. Thus, probability $p(x)$ in (4.36) has a specific, meaningful form in this case

$$p(x) = 1 - K(x).$$

After describing the setting, we are ready now to derive the formal result. The proof is relatively straightforward and similar to the proofs of the previous sections of this chapter; however the explicit result to be obtained is really meaningful. We are interested in describing the lifetime of our system T_S (time to the first fatal EE). The corresponding conditional survival function is given by

$$\begin{aligned} P(T_S > t|N(s),\ 0 \leq s \leq t;\ D_{C1}, D_{C2}, \ldots, D_{CN(t)}) \\ = \prod_{i=1}^{N(t)} (I(D_{Ci} > t - T_i)), \end{aligned} \tag{4.37}$$

where the indicators are defined as

$$I(D_{Ci} > t - T_i) = \begin{cases} 1, & \text{if } D_{Ci} > t - T_i \\ 0, & \text{otherwise} \end{cases}.$$

Let

$$J_i = \begin{cases} 1, & \text{if the ith cure process is successful,} \\ 0, & \text{otherwise.} \end{cases}$$

We assume that given the shock process, (i) $J_i,\ i = 1, 2, \ldots$, are mutually independent; (ii) $D_i, i = 1, 2, \ldots$, are mutually independent; (iii) $\{J_i,\ i = 1, 2, \ldots\}$, $\{D_i,\ i = 1, 2, \ldots\}$ are mutually independent. Therefore, D_{Ci} $i = 1, 2, \ldots$, are also mutually independent.

Integrating out all conditional random quantities in (4.37) under the basic assumptions described above, we arrive at the following theorem, which modifies Theorem 4.6 [11]:

Theorem 4.8 *Let $m^{-1}(t)$ exist for $t > 0$. Then*

$$P(T_S \geq t) = \exp\left\{ -\int_0^t G_C(t - u)\nu(u)\, du \right\},\ t \geq 0, \tag{4.38}$$

and the failure rate function of the system is

$$\lambda_S(t) = \int_0^t p(t-u)\, g(t-u)\, v(u)\, du, \; t \geq 0. \tag{4.39}$$

Proof From (4.37),

$$P(T_S > t | N(t), T_1, T_2, \ldots, T_{N(t)}; D_{C1}, D_{C2}, \ldots, D_{CN(t)}) \\ = \prod_{i=1}^{N(t)} (I(D_{Ci} > t - T_i)).$$

Due to the conditional independence assumption described above, we can 'integrate out' D_{Ci}'s separately and define the corresponding probability in the following way:

$$P(T_S > t | N(t), T_1, T_2, \ldots, T_n) = \prod_{i=1}^{N(t)} (\bar{G}_C(t - T_i)).$$

Therefore,

$$P(T_S > t) = E\left[\prod_{i=1}^{N(t)} (\overline{G}_C(t - T_i))\right] = E\left[E\left[\prod_{i=1}^{N(t)} (\overline{G}_C(t - T_i)) \mid N(t)\right]\right]. \tag{4.40}$$

The joint distribution of $T_1, T_2, \ldots, T_n$ given $N(t) = n$ is the same as the joint distribution of order statistics $T'_{(1)} \leq T'_{(2)} \leq \ldots \leq T'_{(n)}$ of i.i.d. random variables $T'_1, T'_2, \ldots, T'_n$, where the p.d.f. of the common distribution of T'_j's is given by $v(x)/m(t)$, $0 \leq x \leq t$:

$$(T_1, T_2, \ldots, T_n | N(t) = n) =^d (T'_{(1)}, T'_{(2)}, \ldots, T'_{(n)}).$$

Then

$$\begin{aligned} & E\left[\prod_{i=1}^{N(t)} (\overline{G}_C(t - T_i)) | N(t) = n\right] \\ & = E\left[\prod_{i=1}^{n} \left(\overline{G}_C(t - T'_{(i)})\right)\right] \\ & = E\left[\prod_{i=1}^{n} (\overline{G}_C(t - T'_i))\right] \\ & = \left(E\left[\overline{G}_C(t - T'_i)\right]\right)^n \\ & = \left(\frac{1}{m(t)} \int_0^t (\overline{G}(t - u))\, v(u)\, du\right)^n. \end{aligned} \tag{4.41}$$

From Eqs. (4.40) and (4.41),

$$
\begin{aligned}
P(T_S > t) &= \sum_{n=0}^{\infty} \left(\frac{1}{m(t)} \int_0^t \left(\overline{G_C}(t - u)\right) v(u)\, \mathrm{d}u \right)^n \cdot \frac{m(t)^n}{n!} e^{-m(t)} \\
&= e^{-m(t)} \cdot \exp\left\{ \int_0^t \left(\overline{G}_C(t - u)\right) v(u)\, \mathrm{d}u \right\} \\
&= \exp\left\{ \int_0^t \overline{G}_C(t - u)\, v(u)\, \mathrm{d}x - \int_0^t v(u)\, \mathrm{d}u \right\} \\
&= \exp\left\{ - \int_0^t G_C(t - u)\, v(u)\, \mathrm{d}u \right\},
\end{aligned}
$$

where $G_C(t-u)$ is defined by (4.35). Therefore, using Leibnitz rule and Eq. (4.36), $\lambda_S(t)$ can be obtained in the following meaningful and a rather simple form:

$$
\lambda_S(t) = \int_0^t g_C(t - u)\, v(u)\, \mathrm{d}u = \int_0^t p(t - u)\, g(t - u)\, v(u)\, \mathrm{d}u. \tag{4.42}
$$

□

We will show now that under certain assumptions the $p(t) \Leftrightarrow q(t)$ model (4.1) and the current one are asymptotically equivalent. Indeed, assume that $\lim_{t\to\infty} v(t) \equiv v < \infty$. Without loss of generality, let $p(t)$ and $v(t)$ be the continuous functions with $p(t) > 0$, for all $t \geq 0$. Then the failure rate (4.42) tends to a constant as $t \to \infty$, i.e.,

$$
\begin{aligned}
\lim_{t\to\infty} \lambda_S(t) &= \lim_{t\to\infty} \int_0^t p(t-u)\, g(t-u)\, v(u)\, \mathrm{d}u \\
&= v \int_0^{\infty} p(u) g(u)\, \mathrm{d}u.
\end{aligned}
$$

The latter integral obviously is finite as $g(t)$ is the pdf and $p(t) < 1$ for all $t > 0$. Specifically, when $\lim_{t\to\infty} p(t) = p$,

$$
\lim_{t\to\infty} \lambda_S(t) = vp.
$$

Thus, under the given assumptions, the failure rate (4.42), 'asymptotically converges' (as $t \to \infty$) to that of the classical extreme shock model (4.1).

4.7 Stress–Strength Model with Delay and Cure

Consider now a more specific and practical model with delay and possible cure that can be applied, e.g., in reliability modeling of materials and mechanical structures. Let, as previously, $v(t)$ be the rate of the NHPP process of shocks (IEs) affecting our system and S_i denote the magnitude of the ith shock (stress). Assume that $S_i, i = 1, 2, \ldots$ are i.i.d. random variables with the common Cdf $F_S(s)$ $(\overline{F_S}(s) \equiv 1 - F_S(s))$ and the corresponding pdf $f_S(s)$. The system is characterized by its strength to resist stresses. Let first, the strength of the system Y be a constant, i.e., $Y = y$. Assume that for each $i = 1, 2, ..,$ the operable system immediately fails if $S_i > y$ (fatal immediate failure) and the EE is triggered with the delay time and possible cure (as in the previous section) if $S_i \leq y$. It is clear that due to the described operation of thinning, the initial NHPP splits into two NHPP processes with rates $\bar{F}_S(y)\, v(t)$ and $F_S(y)\, v(t)$. Therefore, combining results of the previous section with the *classical extreme shock* model (4.1), Eqs. (4.38) and (4.39) can be generalized to

$$P(T_S > t|Y = y) = \exp\left\{-\bar{F}_S(y)\int_0^t v(u)\,du\right\}\exp\left\{-F_S(y)\int_0^t G_C(t - u)\,v(u)\,du\right\},\ t \geq 0, \tag{4.43}$$

$$\lambda_S(t|Y = y) = \bar{F}_S(y)\,v(t) + F_S(y)\int_0^t p(t - u)\,g(t - u)\,v(u)\,du,\ t \geq 0, \tag{4.44}$$

accordingly.

In practice, due to various reasons, the strength of a system Y can be considered as a random variable. Let its support be, e.g., $[0, \infty)$. Denote by $H_Y(y)$ $(\overline{H}_Y(y) \equiv 1 - H_Y(y))$ and by $h_Y(y)$, the corresponding Cdf and the pdf, respectively. The first guess in generalizing (4.43) and (4.44) to the case of a random Y would be just to replace $F_S(u)$ and $\overline{F_S}(u)$ in these equations by the expectations

$$\int_0^\infty F_S(y)\,h_Y(y)\,dy \text{ and } \int_0^\infty \bar{F}_S(y)\,h_Y(y)\,dy, \tag{4.45}$$

accordingly. However, it is not true, as the proper conditioning should be imposed (on condition that the previous shocks have been survived). This operation is similar to the Bayesian update of information. It can be easily seen from (4.43) and (4.44) that the model can be considered now as a mixture, or equivalently as a frailty model with the frailty parameter Y (see the next Chapter). Therefore, the mixture (observed) survival function for the lifetime T_S is obtained directly from (4.43) as the corresponding expectation:

$$
\begin{aligned}
P(T_S > t) &= \int\limits_0^\infty P(T_S \geq t | Y = y)\, h_Y(y)\, \mathrm{d}y \\
&= \int\limits_0^\infty \exp\left\{ - \int\limits_0^t (\bar{F}_S(y)\, v(u)\, \mathrm{d}u \;+\; F_S(y)\, G_C(t \;-\; u)\, v(u))\, \mathrm{d}u \right\} h_Y(y)\, \mathrm{d}y,
\end{aligned} \tag{4.46}
$$

whereas the failure rate is the following *conditional expectation*:

$$
\lambda_S(t) = \int\limits_0^\infty \lambda_S(t | Y = y)\, h_Y(y | T_S > t)\, \mathrm{d}y, \tag{4.47}
$$

where $h_Y(y|T_S > t)$ is the pdf of the random variable $Y|T_S > t$, or equivalently, $\lambda_S(t)$, in accordance with the definition, is

$$
\lambda_S(t) \;=\; -\frac{P'(T_S > t)}{P(T_S > t)}.
$$

From (4.43), $h_Y(y|T_S > t)$ can be obtained as

$$
\begin{aligned}
h_Y(y|T_S > t) \;&= \exp\left\{ -\bar{F}_S(y) \int\limits_0^t v(u)\, \mathrm{d}u \right\} \exp\left\{ -F_S(y) \int\limits_0^t G_C(t \;-\; u)\, v(u)\, \mathrm{d}u \right\} h_Y(y) \\
&\times \left(\int\limits_0^\infty \exp\left\{ - \int\limits_0^t (\bar{F}_S(x)\, v(u)\, \mathrm{d}u \;+\; F_S(x)\, G_C(t \;-\; u)\, v(u))\, \mathrm{d}u \right\} h_Y(x)\, \mathrm{d}x \right)^{-1}
\end{aligned} \tag{4.48}
$$

Equations (4.44), (4.47) and (4.48) show that the explicit form of $\lambda_S(t)$ is rather cumbersome and numerical methods should be used for calculating it in practice. However, our goal here is to emphasize the relevant methodological issues.

Specifically, when there is only a fatal immediate failure (i.e., without delays), Eq. (4.46) simplifies to

$$
P(T_S > t) = \int\limits_0^\infty \exp\left\{ -\overline{F}_S(y) \int\limits_0^t v(u)\, \mathrm{d}u \right\} h_Y(y)\, \mathrm{d}y \tag{4.49}
$$

and after the change in the order of integration, the corresponding failure rate becomes

$$
\lambda_S(t) = \frac{\int_0^\infty \int_0^s \exp\left\{ -\overline{F}_S(y) \int\limits_0^t v(u)\, \mathrm{d}u \right\} h_Y(y)\, \mathrm{d}y f_S(s)\, \mathrm{d}s}{\int\limits_0^\infty \exp\left\{ -\overline{F}_S(y) \int\limits_0^t v(u)\, \mathrm{d}u \right\} h_Y(y)\, \mathrm{d}y} v(t). \tag{4.50}
$$

The right-hand side of Eq. (4.50) is still much more complex than the corresponding failure rate for the fixed strength model, which is the simple product, $\bar{F}_S(y)\,\nu(t)$. The price for this simplicity is in neglecting the random nature of the strength of a system.

4.8 Survival of Systems with Protection Subject to Two Types of External Attacks

Consider a large system (LS) that, because of its importance and (or) large economic value, should be protected from possible harmful attacks or intrusions. At many instances, this protective function is performed by a specially designed defence system (DS). Therefore, the attacker wants to destroy the DS partially or completely and then to attack the LS [12].

Let the maximum level of performance of the DS be described by the value of the *initial* defence capacity, D_M—to be interpreted as, e.g., the total number of defence units, service points, firewalls, etc. For instance, we may imagine a system that executes defence against aircraft or missile strikes on some important object (as, e.g., a power station or a marine port during combat). Another more 'peaceful example' is the computer network that should be protected from hack-attacks aimed at disabling firewalls.

The attacker executes two types of attacks—those that target the DS and those that target the system itself. We will model these actions by two different stochastic point processes to be called for convenience, the A1 and the A2 shock processes, respectively. The shocks from the A1 process damage, i.e., destroy certain parts of the DS. We assume that the DS is repairable and, therefore, this effect is temporal. Given the stochastic nature of the setting, the actual *defence capacity* at time t can be modeled by a stochastic process $\{D(t),\, t \geq 0\}$. For example, it may be maximal for long periods of time, i.e., $D(t) = D_M$, or severely hampered when $D(t) << D_M$. Thus, distinct from the conventional shock models with accumulated damage, our model describes a nonmonotonic damage process, which accounts for, e.g., the corresponding repair actions.

The DS defends the nonrepairable LS from the A2 process of shocks that are aimed to destroy the LS or, in other words, to completely terminate its operation. In accordance with reliability terminology, we will call this event a *failure*. Assume that, similar to the classical extreme shock models, each shock from the A2 process results in the LS failure with probability $p(t)$ or it is 'perfectly' survived with the complementary probability $q(t) = 1 - p(t)$. The latter means in our case that the DS has neutralized the attack. It is natural to assume that these probabilities are the functions of the defence capacity in the following sense: for each realization of $D(t) = d(t)$, the failure probability $p(t)$ is a decreasing function of the actual defence capacity, i.e., $p(t) = p^*(d(t))$, where $p^*(\cdot)$ is strictly decreasing in its argument. As the simplest and meaningful scenario, one may define a proportion-type function:

$$p^*(d(t)) = (D_M - d(t))/D_M.$$

The failure of the LS occurs when the attack on it is not neutralized by the DS. We are interested in the survival probability of the LS in $[0, t)$. An obvious specific case is when instead of the A2 shock process, only one attack at time instant $t' \in [0, t)$ is executed with the corresponding survival probability $p(t') = p^*(d(t'))$. The foregoing setting indicates that the description of the stochastic process $\{D(t),\ t \geq 0\}$ is the crucial part of our approach. In order to obtain the mathematically tractable solution, the relatively simple stochastic point processes need to be adopted as the corresponding models for the A1 and the A2 shock processes.

For a formal description, denote

(i) $\{N(t),\ t \geq 0\}$ the NHPP process of the A1 shocks with rate $v(t)$ and (ordered) arrival times $R_i,\ i = 0,\ 1,\ 2, \ldots,\ R_1 < R_2 < R_3, \ldots,$ where $i = 0$ *formally* means that there were no events in $[0,\ t)$.

(ii) $\{Q(t),\ t \geq 0\}$—the NHPP process of the A2 shocks with rate $w(t)$ and ordered arrival times $B_i,\ i = 1,\ 2, \ldots,\ B_1 < B_2 < B_3, \ldots,$ where $i = 0$ *formally* means that there were no events in $[0,\ t)$. The specific case of the only one A2 event in $[0,\ t)$ will be also considered.

Assume that, when $D(t) = D$, the A2 shock at time t directly destroys the operating LS with probability

$$p(t \mid D(t) = D) = 1 - \alpha \frac{D}{D_M}$$

and is survived with the complementary probability

$$q(t \mid D(t) = D) \equiv 1 - p(t \mid D(t) = D) = \alpha \frac{D}{D_M}, \tag{4.51}$$

where $\{D(t),\ t \geq 0\}$ is a stochastic process that models the defence capacity of the DS, $D_M = D(0)$ is its fixed initial maximal value and α $(0 < \alpha \leq 1)$ is a constant. The coefficient α shows the protection coverage of the LS by the DS. Specifically, when $\alpha = 1$ and $D(t) = D_M$, the DS executes the 100 % protection of the LS from the A2 shock at time t. In what follows, for simplicity of notation, we will assume that $\alpha = 1$, whereas the general case is obtained by a trivial modification. It should be noted that Eq. (4.51) means that the survival probability for the A2 shock is proportional to the normalized defence capacity $D(t)/D_M$.

We must set now the model for the process $\{D(t),\ t \geq 0\}$, which is the major challenge in this setting. Let the ith A1 shock causes the damage $W_i,\ i = 1,\ 2, \ldots$ to the DS. We assume that this effect 'expires' in a random time τ_i (e.g., the repair facility is restoring the DS from the consequences of this shock). As the damages are accumulated,

$$D(t) = D_M - \sum_{i=1}^{N(t)} W_i 1(t - R_i < \tau_i), \tag{4.52}$$

where $1(\cdot)$ is the corresponding indicator. Obviously, the stochastic process $\{D(t), t \geq 0\}$ should not be negative and we will discuss it for the specific models to follow.

The number of A1 shocks that contribute toward the total damage at time t can be obviously defined as the following stochastic process

$$X(t) = \sum_{i=1}^{N(t)} 1(t - R_i \leq \tau_i), \tag{4.53}$$

In other words, $X(t)$ counts the number of A1 shocks with 'active' damage (not eliminated or vanished) at time t. Assume further that

(iii) τ_i, $i = 1, 2, 3, \ldots$ are i.i.d. random variables with the Cdf $G(t)$ and mean $\bar{\tau}_G$.
(iv) W_i, $i = 1, 2, 3, \ldots$ are i.i.d. random variables with finite expectation $E[W_i] = d_w$ (for Model 1 to follow).
(v) $\{N(t), t \geq 0\}$, $\{Q(t), t \geq 0\}$, W_i, $i = 1, 2, \ldots$ and τ_i, $i = 1, 2, \ldots$ are independent of each other.

We will consider two models for damage accumulation and the resulting probabilities of interest.
Model 1. In accordance with (4.51) ($\alpha = 1$),

$$q_1(t|W_i = w_i, i = 1, 2, \ldots, X(t) = r) = \frac{D_M - \sum_{i=1}^{r} w_{ji}}{D_M}, \tag{4.54}$$

where, $j_1 < j_2 < \ldots < j_r$ are the subscripts of W_i for which $\{t - R_i < \tau_i\}$ is satisfied and the subscript "1" in q_1 stands for the first model. Assume initially that there is only one A2 shock, whereas the case of the process of A2 shocks will be considered further. The unconditional probability of survival under a single A2 shock at time t is the corresponding expectation that, in accordance with Wald's equality, can be written as

$$\begin{aligned} q_1(t) &= E[q_1(t|W_i i = 1, 2, \ldots, X(t))] \\ &= \frac{D_M - E\left[\sum_{i=1}^{X(t)} W_{j_i}\right]}{D_M} = 1 - \frac{E[X(t)]\, d_w}{D_M}. \end{aligned} \tag{4.55}$$

In this model, we implicitly assume that *damages are relatively small compared with the full size* D_M, i.e.,$d_w \ll D_M$ and the rate of the A1 process is not too large, in order (4.52) to be positive (i.e., the probability that it is formally negative is negligible). These assumptions in a broader context will be discussed later.
Model 2. Model 1 traditionally describes accumulation of damage via the i.i.d. increments. However, in view of our two shock processes setting, it can be interesting and appealing to consider a different *new* scenario when each shock *decreases proportionally* the defence capacity [12]. The damage in this case

depends on the value of the defence capacity: the larger $D(t)$ corresponds to the larger damage from a shock. This assumption seems to be often more realistic than the i.i.d. one, as at many instances, the size of the damage depends on the size of the attacked system. Suppose that a single A2 shock has occurred at time t. Then our assumption can be formalized as

$$D(t) = kD(t-), \tag{4.56}$$

where the proportionality factor $k(0<k<1)$ describes the efficiency of attacks for each shock from the A1 process and "$t-$" denotes the time instant just prior to t.

As the defence system starts at $t = 0$ at 'full size', its capacity at time t is given by the following random variable (for each fixed t), or equivalently, by the stochastic process $\{D(t),\ t \geq 0\}$:

$$D(t) = D_M k^{X(t)}, \tag{4.57}$$

as the effect of all other damages caused by the process $N(t),\ t \geq 0$ (not counted by (4.53)), was eliminated (repaired). In contrast to Model 1, $D(t)$ is always positive and no additional assumption for that is needed. In accordance with (4.51):

$$q_2(t|X(t) = r) = k^r. \tag{4.58}$$

The unconditional probability of survival under a shock at time t is the corresponding expectation with respect to $X(t)$:

$$q_2(t) = E[q_2(t|X(t))] = E[k^{X(t)}]. \tag{4.59}$$

In practice, k is usually close to 1 meaning that only a small portion of the defence capability is lost on each A1 shock.

Denote, as previously, by T_S the time to failure of the LS. Now we are ready for obtaining the survival probability, $\Pr(T_S > t)$. As follows from (4.55) and (4.59), in order to describe the process $\{D(t),\ t \geq 0\}$ and to derive $\Pr(T_S > t)$ *for both models,* we need to obtain the discrete distribution of $X(t)$ given by Eq. (4.53). The proof of the following theorem is rather straightforward and similar to the proofs of the previous sections and, therefore, it is omitted. However, this result will be basic for our further derivations in this section.

Theorem 4.9 *Let* $m_\nu(t) \equiv E(N(t)) = \int_0^t \nu(x)\,\mathrm{d}x$ *denotes the cumulative rate of the A1 process of shocks and suppose that* $m_\nu^{-1}(t),\ t > 0$ *exists. Then, the distribution of* $X(t)$ *for each fixed* t *is given by the following formula:*

$$\Pr(X(t) = r) = \frac{\left(\int_0^t \nu(x)\,\bar{G}(t-x)\,\mathrm{d}x\right)^r \exp\left\{-\int_0^t \nu(x)\bar{G}(t-x)\mathrm{d}x\right\}}{r!}, \tag{4.60}$$

where $\bar{G}(t) \equiv 1 - G(t)$ *is the survival probability for* $\tau_i,\ i = 1, 2, 3, \ldots$

Consider first, the probability of survival under a single A2 shock at time t, which can be already of a practical interest in applications. In fact, this is our $q(t)$ defined for both models by expectations (4.55) and (4.59), respectively. The following theorem gives the corresponding expressions.

Theorem 4.10 *The probability of survival of the operating LS under a single A2 shock at time t is*

$$q_1(t) = 1 - \frac{\left[\int_0^t \nu(x)\,\bar{G}(t - x)\,\mathrm{d}x\right] d_w}{D_M} \tag{4.61}$$

for Model 1 and

$$q_2(t) = \exp\left\{-(1 - k)\int_0^t \nu(x)\,\bar{G}(t - x)\,\mathrm{d}x\right\} \tag{4.62}$$

for Model 2.

Proof It immediately follows from Eq. (4.60) that

$$E[X(t)] = \int_0^t \nu(x)\,\bar{G}(t - x)\,\mathrm{d}x$$

and, therefore, (4.61) holds.

Similarly, for Model 2,

$$\begin{aligned} q_2(t) &= E\left[k^{X(t)}\right] \\ &= \sum_{r=0}^{\infty} k^r \frac{\left(\int_0^t \nu(x)\,\bar{G}(t - x)\,\mathrm{d}x\right)^r \exp\left\{-\int_0^t \nu(x)\,\bar{G}(t - x)\,\mathrm{d}x\right\}}{r!} \\ &= \exp\left\{-(1 - k)\int_0^t \nu(x)\,\bar{G}(t - x)\,\mathrm{d}x\right\}. \end{aligned}$$

□

Theorem 4.11 *Let* $\nu(t) = \nu$, $t \in [0, \infty)$ *or* $\lim_{t \to \infty} \nu(t) = \nu$. *Then the stationary values of* $q_i(t)$, i.e., $\lim_{t \to \infty} q_i(t) = q_i, i = 1, 2$ *are given by*

$$q_1 = 1 - \frac{\bar{\tau}_G d_w}{\bar{\tau}_N D_M}, \tag{4.63}$$

$$q_2 = \exp\left\{-(1 - k)\frac{\bar{\tau}_G}{\bar{\tau}_N}\right\}, \tag{4.64}$$

where $\bar{\tau}_G = \int_0^\infty \bar{G}(x)\,\mathrm{d}x$ *is the mean time which corresponds to random variables* τ_i, $i = 1, 2, \ldots$ *and* $\bar{\tau}_N = 1/v$ *is the mean time (exactly or asymptotically as* $t \to \infty$*) between successive A1 shocks.*

Theorem 4.11 is intuitively obvious and can be proved in a straightforward way by using the variable substitution $y = t - x$ for the integrals in (4.61) and (4.62) and by applying the Lebesgue's Dominated Convergence Theorem afterward. When $\bar{\tau}_G/\bar{\tau}_N << 1$, which means a very quick repair of damage with respect to the time between successive A1 shocks, Model 2 reduces to a very simple (and usually not practically justified) setting when the repair periods after different A1 shocks do not overlap. In this case, the probability of failure that corresponds to (4.64) is just $p_2 = 1 - q_2 \approx (1 - k)\,\bar{\tau}_G/\bar{\tau}_N$.

It follows from the above reasoning that the stationary variant of (4.60) (i.e., for t sufficiently large and $v(t) = v$, $t \in [0, \infty)$ or $\lim_{t\to\infty} v(t) = v$) can be of interest. Denote, $\bar{\tau}_G/\bar{\tau}_N \equiv \eta$. Then the stationary distribution for (4.60) is the Poisson random variable with this parameter:

$$\Pr(X_S = r) = \frac{\eta^r \exp\{-\eta\}}{r!}. \tag{4.65}$$

Theorem 4.10 provides a simple way of obtaining the probability of failure of the LS under a single attack at time t.

We are ready now to consider the A2 process of shocks and to derive the corresponding probability of system's survival, $P(T_S > t)$ under the attacks of two types. However, it turns out that this problem is much more complex than it looks from the first sight and, therefore, additional assumptions should be imposed in order to simplify it and to obtain results that potentially can have practical value. First of all, we must answer the question: are the probabilities $q_i(t)$ ($p_i(t)$) obtained in Theorem 4.10 suitable for using in the classical $p(t) \Leftrightarrow q(t)$ model? Recall that in this extreme shock model, each event from the Poisson process of shocks with rate $w(t)$ is survived with probability $q(t)$ and 'kills' a system with the complementary probability $p(t) = 1 - q(t)$ independently of all previous history. In this case, the system's survival probability in $[0, t)$ is given by the following exponential representation (see also Eq. (4.1):

$$P(T_S > t) \equiv \bar{F}_S(t) = \exp\left(-\int_0^t p(u)\,w(u)\,\mathrm{d}u\right), \tag{4.66}$$

and, therefore, the corresponding failure rate function $\lambda_S(t)$ is

$$\lambda_S(t) = p(t)w(t),\ t \geq 0. \tag{4.67}$$

From the first glance, it looks that we have already everything in place for using (4.61) and (4.62) in Eq. (4.66). However, it can be shown, that certain dependence on history prevents from that and the only way to deal with this complexity for obtaining some practically meaningful results is to consider additional assumptions that allow for additional simplification of the model.

Let both A1 and A2 be the *homogeneous* Poisson shock processes with rates v and w, respectively. Let the A2 shocks be sufficiently rare when compared with the dynamics of the $X(t)$ process

$$\bar{\tau}_Q \equiv \frac{1}{w} \gg \frac{1}{v} \equiv \bar{\tau}_N; \ \bar{\tau}_G \ll \bar{\tau}_Q, \tag{4.68}$$

which makes sense in practice, as the intensity of attacks on the LS could be considered as much smaller than that on the DS. The second inequality in (4.68) implies that the mean time of repair of the DS is much smaller than the mean inter-arrival time of the potentially terminal A2 shocks, which is also a reasonable assumption in practice. Inequalities (4.68) can be considered as the analogue to the *fast repair* conditions (see e.g., Ushakov and Harrison [28]). Finkelstein and Zarudnij [20] have used the similar assumptions for approximating the multiple availability on stochastic demand (i.e., the repairable system should be available at all demands that occur in accordance with the homogeneous Poisson process in $[0, t)$). Assumptions (4.68) 'can help to forget the history' of the process $X(t)$ and, therefore, a simple $p(t) \Leftrightarrow q(t)$ model (4.66)–(4.67) holds. Indeed, under these assumptions the correlation between values of the process $X(t)$ at instants of occurrence of the A2 shocks is negligible as the time between successive A2 shocks is sufficiently large. Therefore, the probabilities of survival under each A2 shock for both models are given approximately by Eq. (4.66), whereas the following result holds asymptotically:

Theorem 4.12 *Let $v(t) = v$, $w(t) = w$; $w/v \to 0$, $\bar{\tau}_G/\bar{\tau}_Q \to 0$ and t is sufficiently large: $t \gg \bar{\tau}_Q$. Then the probabilities of survival for two models, in accordance with Theorem 4.11, are*

$$P_1(T_S > t) = \exp\left\{-w\left[\eta\frac{d_w}{D_M}\right]t\right\}(1 + o(1)), \tag{4.69}$$

$$P_2(T_S > t) = \exp\{-w[1 - \exp\{-(1 - k)\eta\}]t\}(1 + o(1)), \tag{4.70}$$

where $\eta \equiv \bar{\tau}_G/\bar{\tau}_N$.

It should be noted that for the sufficiently small t, when $t \ll \bar{\tau}_Q$, we can approximately consider the case of only one A2 shock that is arriving in accordance with the distribution $F(t) = 1 - \exp\left\{-\int_0^t w(u)\,du\right\}$. Then

$$P_i(T_S > t) = \int_0^t q_i(u)f(u)\,du + \exp\left\{-\int_0^t w(u)\,du\right\},$$

where $q_i(u)$, $i = 1, 2$ are given by Eqs. (4.61) and (4.62) and $f(u) = F'(t)$. Obviously, as in this case the A2 process can be approximately regarded as one first event, we do not need any other assumptions on the A1 process. Dealing with the A2 process of shocks, however, creates more mathematical difficulties and, therefore, a number of assumptions and simplifications have been made to arrive at approximations (4.69) and (4.70).

4.9 Geometric Process of Shocks

The nonhomogeneous Poisson process (NHPP), due to its relative probabilistic simplicity, is definitely the most popular counting (point) process in applications and, specifically, in shock modeling. It often allows for rather simple and compact expressions for the probabilities of interest for the basic and generalized settings as was shown in the Sect. 4.8. However, in practice, the point events do not necessarily possess the property of independent increments and the number of events in the fixed interval of time does not necessarily follow the Poisson distribution. Therefore, other *distribution-based* counting processes should also be considered and, therefore, in this section, we will suggest another distribution-based class of counting processes (with dependent increments) that still allows for compact, explicit relationships for some applications [10].

The counting (point) processes that describe 'events' in the real world should share certain natural properties that can be formulated in the following way:

(i) two or more events cannot occur 'at the same time' (i.e., the process is *orderly*),
(ii) the mean number of occurrences in $(0, t]$ as a function of t, i.e., $\Lambda(t) \equiv E[N(t)]$, is sufficiently 'smooth', so that its derivative that is called the *rate* or *intensity*, exists at every t, i.e., $\Lambda'(t) = \lambda(t), t \geq 0$, or $\Lambda(t) = \int_0^t \lambda(u)\,du$.

It is well-known that these statements (for the sufficiently small Δt) can be formalized as

(a) $N(0) = 0$.
(b) $P(N(t + \Delta t) - N(t) = 1) = \lambda(t)\,\Delta t + o(\Delta t)$.
(c) $P(N(t + \Delta t) - N(t) \geq 2) = o(\Delta t)$.

For the sake of notation, let us denote the general class of point processes, which satisfy (a), (b), and (c) by ***G***. Clearly, if we adopt *additionally*

(d) $\{N(t),\ t \geq 0\}$ has independent increments,

then we arrive at the NHPP. It is also well-known that assumptions (a)–(d) result in the Poisson distribution of the number of events in $(t_1,\ t_2]$. Thus, in what follows, in accordance with our intention stated above, we will 'depart' from the governing *Poisson* distribution.

Definition 4.1 The counting process $\{N(t),\ t \geq 0\}$ belongs to the Class of Geometric Counting Processes (CGCP), i.e., $\{N(t),\ t \geq 0\} \in \Gamma$, if

(a) $N(0) = 0$.

(b)

$$P(N(t_2) - N(t_1) = k) = \left(\frac{1}{1 + \Lambda(t_2) - \Lambda(t_1)}\right)\left(\frac{\Lambda(t_2) - \Lambda(t_1)}{1 + \Lambda(t_2) - \Lambda(t_1)}\right)^k, \quad k = 0,\ 1,\ 2,\ldots \tag{4.71}$$

It is easy to see that properties (b) and (c) of the general class $\boldsymbol{G}$ can be derived from (4.71):

(b) $P(N(t+\Delta t) - N(t) = 1)$
$= \lambda(t)\,\Delta t + \left\{-\lambda(t)\,\Delta t + \left(\frac{1}{1+\Lambda(t+\Delta t)-\Lambda(t)}\right)\left(\frac{\Lambda(t+\Delta t)-\Lambda(t)}{1+\Lambda(t+\Delta t)-\Lambda(t)}\right)\right\}$,

where the second term in the right-hand side is clearly $o(\Delta t)$;

(c) $P(N(t+\Delta t) - N(t) \geq 2) = \left(\frac{\Lambda(t+\Delta t)-\Lambda(t)}{1+\Lambda(t+\Delta t)-\Lambda(t)}\right)^2$, which is obviously $o(\Delta t)$.

Therefore, the CGCP becomes a subclass of $\boldsymbol{G}$.

Observe that the counting distribution in (4.71) is obtained from the *time-dependent* reparametrization of the geometric distribution:

$$P(N = k) = d(1-d)^k,\ k = 0,\ 1,\ 2,\ldots,$$

where $0 < d < 1$.

In accordance with (4.71), the mean number of events in $(t_1, t_2]$ is

$$E[N(t_2) - N(t_1)] = \Lambda(t_2) - \Lambda(t_1) = \int_{t_1}^{t_2} \lambda(u)\,du.$$

Specifically,

$$P(N(t) = k) = \left(\frac{1}{1+\Lambda(t)}\right)\left(\frac{\Lambda(t)}{1+\Lambda(t)}\right)^k,\ k = 0,\ 1,\ 2,\ldots, \tag{4.72}$$

where $E[N(t)] = \Lambda(t) = \int_0^t \lambda(u)du$.

Thus NHPP and $\{N(t),\ t \geq 0\} \in \Gamma$ can have the same rate, but the crucial difference is that the members of the latter class, as intended, *do not possess the property of independent increments*, which can be easily seen from the following considerations.

Definition 4.2 The orderly counting process $\{N(t),\ t \geq 0\}$ with $N(0) = 0$ possesses the weak positive (negative) dependence, if

$$Cov(I(\{N(s+t) - N(s) = 0\}),\ I(\{N(s) = 0\})) > 0\ (<0), \tag{4.73}$$

where $I(\cdot)$ is the indicator function for the corresponding event.

The intuitive meaning of (4.73) for the positive (negative) dependence case is that the two events $\{N(s)=0\}$ and $\{N(s+t)-N(s)=0\}$ have the 'tendency' to occur simultaneously (not to occur simultaneously). We will also interpret this definition in the other equivalent form after the following simple theorem.

Theorem 4.13 *The counting process* $\{N(t),\ t\geq 0\}\in\Gamma$, *possesses the weak positive dependence property.*

Proof Observe that, from (4.71),

$$\begin{aligned}
&Cov(I(\{N(s+t)-N(s)=0\}),\ I(\{N(s)=0\}))\\
&=E[I(\{N(s+t)-N(s)=0\},\ \{N(s)=0\})]-E[I(\{N(s+t)-N(s)=0\})]\,E[I(\{N(s)=0\})]\\
&=P(N(s+t)-N(s)=0,\ N(s)=0)-P(N(s+t)-N(s)=0)\,P(N(s)=0)\\
&=P(N(s+t)=0)-P(N(s+t)-N(s)=0)\,P(N(s)=0)\\
&=\frac{[1+\Lambda(s)][1+\Lambda(s+t)-\Lambda(s)]-[1+\Lambda(s+t)]}{[1+\Lambda(s+t)][1+\Lambda(s+t)-\Lambda(s)][1+\Lambda(s)]}>0.
\end{aligned}$$

□

It follows from the proof that, as $P(N(s)=0)>0$, inequality (4.73) (for positive dependence) is equivalent to

$$P(N(s+t)-N(s)=0|N(s)=0)>P(N(s+t)-N(s)=0) \tag{4.74}$$

or to

$$P(N(s+t)-N(s)\geq 1|N(s)=0)<P(N(s+t)-N(s)\geq 1).$$

The latter means that the absence of events in $(0,\ s]$ decreases the probability of events in $(s,\ s+t]$. This seems to be a more natural interpretation of a (weak) positive dependence.

In order to consider the rate and the corresponding conditional characteristic, we replace t in (4.74) by the infinitesimal dt. Then

$$\begin{aligned}
&P(N(s+\mathrm{d}t)-N(s)=0|N(s)=0)-P(N(s+\mathrm{d}t)-N(s)=0)\\
&=\frac{\int_0^s\lambda(u)\,\mathrm{d}u\int_s^{s+\mathrm{d}t}\lambda(u)\,\mathrm{d}u}{\left(1+\int_0^{s+\mathrm{d}t}\lambda(u)\,\mathrm{d}u\right)\left(1+\int_s^{s+\mathrm{d}t}\lambda(u)\,\mathrm{d}u\right)}\\
&=\frac{\lambda(s)\int_0^t\lambda(u)\,\mathrm{d}u}{(1+\int_0^s\lambda(u)\,\mathrm{d}u+\lambda(s)\,\mathrm{d}t)(1+\lambda(s)\,\mathrm{d}t)}(1+o(1))\,\mathrm{d}t\\
&=\frac{\lambda(s)\int_0^s\lambda(u)\,\mathrm{d}u}{(1+\int_0^s\lambda(u)\,\mathrm{d}u)}(1+o(1))\,\mathrm{d}t=\frac{\lambda(s)\Lambda(s)}{(1+\Lambda(s))}(1+o(1))\,\mathrm{d}t,
\end{aligned}$$

which is obviously positive. However, we can say now more about the corresponding dependence properties. As $o(1)$ can be made as small as we wish, it is sufficient to consider $\lambda(s)\Lambda(s)/(1+\Lambda(s))$. This expression (for $\lambda'(s)<\infty$) is increasing in s when

$$(\lambda'(s)\Lambda(s) + \lambda^2(s))\,(1 + \Lambda(s)) - \lambda(s)^2\Lambda(s) = \lambda'(s)\Lambda(s)\,(1 + \Lambda(s)) + \lambda^2(s) > 0, \tag{4.75}$$

which holds, for instance, for increasing $\lambda(s)$. Specifically, when $\lambda(s) \equiv \lambda$, the left-hand side of (4.75) is equal to λ^2. Thus, the dependence of the defined type is 'getting stronger' with s increasing.

Taking into account that $\{N(t),\ t \geq 0\} \in \Gamma$ is orderly, i.e.,

$$\begin{aligned} &P(N(s + \mathrm{d}t) - N(s) = 0|N(s) = 0) - P(N(s + \mathrm{d}t) - N(s) = 0) \\ &\quad = -(P(N(s + \mathrm{d}t) - N(s) = 1|N(s) = 0) - P(N(s + \mathrm{d}t) - N(s) = 1)) + o(\mathrm{d}t), \end{aligned}$$

the difference between the conditional rate of $\{N(t),\ t \geq 0\} \in \Gamma$ (the intensity function) on condition that there were no events in $(0,\ s]$ and its unconditional rate, is obviously also increasing in s when (4.75) holds.

As previously, we will consider shocks as events of point processes. The described weak dependence means now that the absence of shocks in $(0,\ s]$ decreases the probability of a shock in $(s,\ s + \mathrm{d}t]$, which can be natural for certain types of shock processes. For instance, the probability of an earthquake is usually larger when the previous earthquake occurred recently, compared with the case when it occurred earlier. A similar argument can be true for heart attacks. For another example, suppose that the 'realization' of a shock process is the homogeneous Poisson process (HPP) with a constant rate, but the rate is determined randomly at $t = 0$ (i.e., the conditional Poisson process). It is well-known [27], that the conditional Poisson process has dependent increments. It can be easily shown that it possesses our weak positive dependence property, i.e., the absence of a shock in $(0,\ s]$ decreases the probability of a shock in $(s,\ s + \mathrm{d}t]$.

The NHPP has another important limitation in terms of the mean and variance relationship for the counting random variable $\mathrm{Var}[N(t)] = E[N(t)]$, for all $t \geq 0$. However, for $\{N(t),\ t \geq 0\} \in \Gamma$,

$$\mathrm{Var}[N(t)] = \Lambda(t)\,(1 + \Lambda(t)) > E[N(t)], \tag{4.76}$$

which can describe many other cases that are not covered by the NHPP.

Thus, in our formulation, the rates of the NHPP and the members of the CGCP, $\{N(t),\ t \geq 0\} \in \Gamma$ can be the same, but because of the dependence of increments, the corresponding probabilistic properties are different. Different members of this class can possess different dependence structures sharing some common features (e.g., the positive dependence of the described type).

Usually for the corresponding stochastic modeling, we need a sufficiently complete description of a relevant stochastic process. However, there are settings when probabilistic reasoning and explicit results do not depend on certain properties of the processes. The shock models to be considered in the following examples are the perfect examples of that. It turns out that the results to be derived are valid *for any member* $\{N(t),\ t \geq 0\} \in \Gamma$ and therefore, they *do not depend on the specific dependence structure* of this process [10]. Therefore, in practice, in order to apply the proposed CGCP, it is sufficient to check the validity of (4.71).

Example 4.6 **Extreme Shock model**. Consider an extreme shock model (see 4.1) for the specific case $p(t) = p$ and let the shock process be from the CGCP, i.e., $\{N(t),\ t \geq 0\} \in \Gamma$, with rate $\lambda(t)$ and arrival times $T_i,\ i = 1,\ 2, \ldots$. Then, due to the assumption of independence,

$$P(T_S > t | N(t) = n) = q^n,$$

and

$$\begin{aligned} P(T_S > t) &= E[P(T_S > t | N(t))] = E[q^{N(t)}] \\ &= \sum_{n=0}^{\infty} q^n \left(\frac{1}{1+\Lambda(t)}\right)\left(\frac{\Lambda(t)}{1+\Lambda(t)}\right)^n = \frac{1}{1+\Lambda(t)p}. \end{aligned}$$

The corresponding failure rate function is

$$\lambda_S(t) = -\frac{d \ln P(T_S > t)}{dt} = \frac{\lambda(t)p}{1+\Lambda(t)p}.$$

Thus, the survival probability and the failure rate are obtained *without specifying the dependence structure of the shock process*. It should be noted that when the process of shocks is NHPP,

$$\lambda_S(t) = p\lambda(t),\ \forall t \geq 0$$

and the *shape* of $\lambda_S(t)$ coincides with that of $\lambda(t)$. However, in the considered case, the result can be dramatically different. Assume that $\lambda(t)$ is differentiable, then

$$\lambda'_S(t) = \frac{\lambda'(t)p - (\lambda(t)p)^2}{(1+\Lambda(t)p)^2},$$

and thus, $\lambda_S(t)$ is increasing (decreasing) in (t_1, t_2) iff

$$\lambda'(t) \geq p(\lambda(t))^2\ (\lambda'(t) \leq p(\lambda(t))^2)$$

in $(t_1,\ t_2)$.

Let, specifically, $\lambda(t) = \lambda,\ \forall t \geq 0$, and therefore, the failure rate, $\lambda_S(t)$ is constant when shocks follow the HPP pattern. However, if it is the process, $\{N(t),\ t \geq 0\} \in \Gamma$ with the same rate λ, then the system failure rate, $\lambda_S(t) = p\lambda/(1+p\lambda t)$ is strictly decreasing with time. This can be loosely interpreted in the following way: equation $P(T_S > t) = E[q^{N(t)}]$, which defines the survival probability for the extreme shock model with an arbitrary point process $\{N(t),\ t \geq 0)$ means that the larger t for the survived system results in the 'sparser' shocks in time. The latter, due to the independent increments property of the Poisson process, does not change the probability of a system's failure in the infinitesimal interval of time $[t,\ t+dt)$. However, for$\{N(t),\ t \geq 0\} \in \Gamma$, as prompted by (4.74), it *decreases* the chance of shocks in the next interval, which eventually results in the decreasing failure rate.

Example 4.7 **Cumulative Shock Model**. Let, as previously, a system be subject to the process $\{N(t),\ t \geq 0\} \in \Gamma$ of shocks with arrival times $T_i,\ i = 1,\ 2, \ldots$. Assume that the ith shock increases the wear of a system by a random increment $W_i \geq 0$. In accordance with this setting, a random accumulated wear of a system at time t is

$$W(t) = \sum_{i=0}^{N(t)} W_i.$$

As previously, assume that the system fails when the accumulated wear exceeds a random boundary R, i.e., $W(t) > R$. The corresponding survival function in this case is given by

$$P(T_S > t) = P(W(t) \leq R). \tag{4.77}$$

Explicit derivations in (4.77) can be performed in specific, mathematically tractable cases.

Case 1. Suppose that $W_i,\ i = 1,\ 2, \ldots$ are i.i.d. and exponential with mean θ. Denote, for the sake of notation, the random variable with this distribution by W. Let $f_R(r)$ be the pdf of the random boundary R. First of all, the mgf of $W(t)$, $M_{W(t)}(z)$, can be expressed as

$$\begin{aligned} M_{W(t)}(z) &= E[\exp\{zW(t)\}] = \sum_{n=0}^{\infty} E[\exp\{zW\}]^n \left(\frac{1}{1+\Lambda(t)}\right)\left(\frac{\Lambda(t)}{1+\Lambda(t)}\right)^n \\ &= \frac{1}{1+\Lambda(t)[1-(1-\theta z)^{-1}]} = \frac{1}{1+\Lambda(t)} \cdot M_0(z) + \frac{\Lambda(t)}{1+\Lambda(t)} \cdot M_{\exp[\theta(1+\Lambda(t))]}(z), \end{aligned} \tag{4.78}$$

where $M_0(z) \equiv 1$ corresponds to the mgf of the degenerate distribution with probability 1 at 0 and

$$M_{\exp[\theta(1+\Lambda(t))]}(z) \equiv \left(\frac{1}{1-\theta(1+\Lambda(t))z}\right)$$

corresponds to the mgf of an exponential distribution with mean $\theta(1+\Lambda(t))$. It follows from (4.78) that the mgf of $W(t)$ is given by the weighted average of the mgf's of two random variables, which implies that the distribution of $W(t)$ is the mixture of the corresponding distributions. Therefore, $W(t)$ has the point mass at 0 (no shocks had occurred in $[0, t]$),

$$P(W(t) = 0) = \frac{1}{1+\Lambda(t)},$$

and, for $x > 0$, $W(t)$ has the pdf

$$f_{W(t)}(x) = \frac{\Lambda(t)}{\theta(1+\Lambda(t))^2}\exp\left\{-\frac{x}{\theta(1+\Lambda(t))}\right\},\ x\geq 0.$$

Then the Cdf of $W(t)$ is given by

$$F_{W(t)}(x) = 1 - \frac{\Lambda(t)}{1+\Lambda(t)}\exp\left\{-\frac{x}{\theta(1+\Lambda(t))}\right\},\ x\geq 0.$$

Finally, the survival function of a system can now be defined as

$$\begin{aligned}P(T_S > t) &= \int_0^\infty F_{W(t)}(r) f_R(r)\,\mathrm{d}r,\ t\geq 0\\ &= 1 - \frac{\Lambda(t)}{1+\Lambda(t)}\int_0^\infty \exp\left\{-\frac{r}{\theta(1+\Lambda(t))}\right\} f_R(r)\,\mathrm{d}r,\ t\geq 0.\end{aligned}$$

Case 2. Suppose that the distribution of the random boundary R is now exponential with mean θ. Let $M_W(z)$ be the mgf of an arbitrary distributed random variable W (W_i are i.i.d)).

Observe that, as the distribution of the random boundary R is exponential, the accumulated wear until time t, $W(t) = \sum_{i=0}^{N(t)} W_i$ does not affect the failure process of the system after time t. That is, on the next shock, the probability of a system's failure due to the accumulated wear is just $P(R \leq W_{N(t)+1})$, and does not depend on the wear accumulation history, i.e.,

$$\begin{aligned}&P(R\geq W_1 + W_2 + \ldots + W_n | R \geq W_1 + W_2 + \ldots + W_{n-1})\\ &= P(R\geq W_n),\ \text{for all } n = 1, 2, \ldots, W_1, W_2, \ldots,\end{aligned}$$

where $W_1 + W_2 + \ldots + W_{n-1} \equiv 0$ when $n = 1$. Then, finally, each shock results in the immediate failure of a system with probability $P(R < W)$ and it does not cause any change in the system with probability $P(R \geq W)$. This interpretation of the model implies that the cumulative shock model in this setting corresponds to the extreme shock model considered previously and

$$p = P(R < W) = 1 - P(R\geq W) = 1 - M_W(-\theta).$$

Therefore,

$$P(T_S > t) = \frac{1}{1 + \Lambda(t)\,(1 - M_W(-\theta))},\ t\geq 0,$$

and the corresponding failure rate is

$$\lambda_S(t) = \frac{\lambda(t)(1 - M_W(-\theta))}{1 + \Lambda(t)(1 - M_W(-\theta))},\ t\geq 0.$$

Finally, the combined shock model (see also Sect. 4.1 for a more general setting) can be also considered. Assume that the ith shock, as in the extreme shock model, causes immediate system's failure with probability p, but in contrast to this model, with probability q it increases the wear of a system by a random increment $W_i \geq 0$. The failure occurs when a critical shock (that destroys a system with probability p) occurs or the random accumulated wear $W(t)$ reaches the random boundary R. Therefore,

$$P(T_S > t|N(s),\ 0 \leq s \leq t;\ W_1, W_2, \ldots, W_{N(t)};\ R) = q^{N(t)} I\left(\sum_{i=0}^{N(t)} W_i \leq R\right)$$

and the survival function of a system is

$$P(T_S > t) = E[q^{N(t)} I(W(t) \leq R)].$$

As previously, for simplicity, let the distribution of a random boundary R be exponential with mean θ. In a similar way, it can be shown that

$$P(T_S > t|N(t) = n) = E\left[\prod_{i=1}^{n} q \exp\{-\theta W_i\}\right] = (qM_W(-\theta))^n.$$

Finally,

$$P(T_S > t) = \frac{1}{1 + \Lambda(t)(1 - qM_W(-\theta))}.$$

And the failure rate function is

$$\lambda_S(t) = -\frac{d \ln P(T_S > t)}{dt} = \frac{\lambda(t)(1 - qM_W(-\theta))}{1 + \Lambda(t)(1 - qM_W(-\theta))}.$$

Thus, we have shown that survival probabilities for some shock models can be effectively obtained for any process that belongs to the CGCP without specifying its dependence structure [10].

4.10 Information-Based Thinning of Shock Processes

4.10.1 General Setting

In this section, we consider some of the settings of the previous sections from a more general viewpoint that employs the operation of thinning of point processes [15]. Thinning of point processes is often applied in stochastic modeling when different types of point events (in terms of their impact, e.g., on a system) occur. In the previous sections, we were mostly interested in the corresponding survival probabilities and, therefore, there was a sequence of 'survival events' and one final event of failure. Now we will be interested in two sequences of events and will use this characterization for further discussion of the strength–stress model of Sect. 4.7.

When the initial point process is the NHPP, the thinned processes are also NHPP independent of each other [15]. The crucial assumption in obtaining this well-known result is that the classification of occurring point events is independent of all other events, including the history of the process. However, in practice, this classification is often dependent on the history. In this section, we define and describe the thinned processes for the *history-dependent* case using different levels of available information and apply our general results to the strength–stress type shock model, which is meaningful in reliability applications. For each considered level of information, we construct the corresponding conditional intensity function and interpret the obtained results.

Let us define the setting in formal terms. Suppose that each event from the NHPP, $\{N(t),\ t \geq 0\}$ with rate (intensity function) $v(t)$ is classified as the Type I event with probability $p(t)$ or as the Type II event with the complementary probability $1 - p(t)$. It is well-known (see, e.g., [4], [5]) that the corresponding stochastic processes $\{N_1(t),\ t \geq 0\}$ and $\{N_2(t),\ t \geq 0\}$ are NHPPs with rates $p(t)v(t)$ and $(1 - p(t))v(t)$, respectively, and they are stochastically independent. This operation for $p(t) \equiv p$ is usually called in the literature 'the thinning of the point process' [15]. As stated above, in reality, classification of events is often history-dependent and the point process is not necessarily Poisson. Therefore, considering history-dependent thinning appears to be an interesting and important problem both from theoretical and practical points of view. The following setting considered in Sect. 4.7 can be helpful as a relevant example.

Suppose that an object (e.g., a system or an organism) is characterized by an *unobserved random quantity* U(e.g., strength or vitality). The object is 'exposed' to a marked NHPP with rate $v(t)$, arrival times $T_1 < T_2 < T_3 \ldots$ and random marks $S_i,\ i = 1, 2, \ldots$, that can be interpreted as some stresses or demands. If $S_i > U$, then the Type I event occurs; if $S_i \leq U$ then the Type II event occurs. We are interested in probabilistic description of the processes of Type I and Type II events. It should be noted that probabilities $P(S_i > U),\ i = 2, 3, \ldots$ already depend on the history, as the distribution of U is *updated by the previous information*, as was mentioned in Sect. 4.7 [8].

First, we will characterize the 'conditional properties' of $\{N_1(t), t \geq 0\}$ and $\{N_2(t),\ t \geq 0\}$, $(N(t) = N_1(t) + N_2(t))$. In various practical problems, we are often interested in the conditional intensity of one of the processes, as only this process 'impacts' our system. The conditional intensity or the intensity process and Eq. (2.12) e.g., for the thinned process, $\{N_1(t),\ t \geq 0\}$ is defined as

$$\begin{aligned}\lambda_1(t|H_{1t-}) &= \lim_{\Delta t \to 0} \frac{E[N_1((t + \Delta t)-) - N_1(t-)|H_{1t-}]}{\Delta t} \\ &= \lim_{\Delta t \to 0} \frac{P[N_1((t + \Delta t)-) - N_1(t-) = 1|H_{1t-}]}{\Delta t},\end{aligned} \tag{4.79}$$

where $H_{1t-} = \{N_1(t-), T_{11}, T_{12}, \ldots, T_{1N_1(t-)}\}$ is the history of the Type I process before time t and T_{1i}, $i = 1, 2, \ldots$ are the corresponding sequential arrival times. In practice, we often observe the process $\{N_1(t),\ t \geq 0\}$, e.g., as the process of some 'effective events' that can cause certain 'detectable changes' (or consequences) in the system. On the other hand, $\{N_2(t),\ t \geq 0\}$ can be the process of 'ineffective events' that have no impact on the system at all. Therefore, the 'observed history' H_{1t-} is our 'available information' that is used for describing $\{N_1(t),\ t \geq 0\}$ via the corresponding conditional intensity, whereas the ineffective events are often (but not necessarily) not observed and thus information on $\{N_2(t),\ t \geq 0\}$ is not available.

As the conditional intensity fully describes the underlying point process, it can obviously be used for defining the corresponding conditional failure rates, which describe the times to events of interest. For example, assume that our system *fails* at the kth Type I event (e.g., due to accumulation of some damage), whereas Type II events, as previously, are ineffective. Then, given $N_1(t-) = k - 1$, the conditional intensity $\lambda_1(t|H_{1t-})$ in (4.79) can be viewed as the conditional failure rate (given the history). Specifically, when our system fails at the first Type I event, the history of our interest becomes $H_{1t-} = \{N_1(t-) = 0\}$. Alternatively, let the system fail on the kth Type I event with probability $p(k)$ and survives with probability $1 - p(k)$ independent of all other events. Then, given $N_1(t-) = k - 1$, the conditional failure rate (on condition that the history H_{1t-} is given) at time t is $\lambda_1(t|H_{1t-})\,p(k)$. Thus, the Type 1 event could terminate the process, which is important for different reliability settings.

As illustrated in the above examples, different conditions can be defined that characterize 'fatal events'. However, we are primarily interested in a general description of the process $\{N_1(t),\ t \geq 0\}$ via its conditional intensity $\lambda_1(t|H_{1t-})$ (without termination). Thus, we will focus first on the conditional intensity (4.79) for a general history $H_{1t-} = \{N_1(t-),\ T_{11},\ T_{12}, \ldots,\ T_{1N_1(t-)}\}$. For convenience, at some instances, the notation H_{1t-} for denoting the corresponding realization $\{N_1(t-) = n_1,\ T_{11} = t_{11},\ T_{12} = t_{12}, \ldots, T_{1N_1(t-)} = t_{1n_1}\}$ will be used as well. Furthermore, the case when the given history is partial, i.e., $\lambda_1(t|H_{1t-}^P)$, where H_{1t-}^P is the partial history of H_{1t-}, will also be investigated. For example, there can be situations when the arrival times are not observed/recorded but only the number of Type I events is observed/recorded. In this case, the 'available information' at hand is only $N_1(t-)$.

Coming back to the specific stress–strength example, note that, when $\{N(t),\ t \geq 0\}$ is the NHPP, U is deterministic, $U = u$ and $S_i, i = 1, 2, \ldots$ are i.i.d. with the common Cdf $F_S(s)$, the processes $\{N_1(t),\ t \geq 0\}$ and $\{N_2(t),\ t \geq 0\}$ are NHPPes. Moreover, they are stochastically independent with rates $p(t)\nu(t)$ and $(1 - p(t))\,\nu(t)$, respectively, where $p(t) = P(S_i > u)$. Thus, obviously,

$$\lambda_1(t|H_{1t-}) = \lim_{\Delta t \to 0} \frac{E[N_1((t + \Delta t)-) - N_1(t-)|H_{1t-}]}{\Delta t} = P(S_i > u)\,\nu(t),$$

as the process $\{N_1(t),\ t \geq 0\}$ possesses the property of independent increments.

We will come back to discussing the case when U is random after a general formulation of the operation of thinning [8].

4.10.2 Formal Description of the Information-Dependent Thinning

Let $\{N(t),\ t \geq 0\}$ denote an orderly point process of events with arrival times $T_i,\ i = 1, 2, \ldots$. We assume that this process is *external* for the system in the sense that it may influence its performance but is not influenced by it [21]. On each event from $\{N(t),\ t \geq 0\}$, depending on the history of the processes $\{N(t), t \geq 0\}$, $\{N_1(t),\ t \geq 0\}$ (note that, $N(t) = N_1(t) + N_2(t)$ and see the corresponding description in the previous subsection) and also on some other random history process up to t, Φ_{t-}, the event is classified as belonging to either the Type I or to the Type II category. Specifically, $\Phi_{t-} \equiv \Phi$ can be just a random variable as, e.g., the random quantity U in the previous example. The conditional probability of the Type I event in the infinitesimal interval of time can be formally written as

$$\begin{aligned}
&P[N_1((t+\mathrm{d}t)-) - N_1(t-) = 1|H_{1t-}, H_{t-}, \Phi_{(t+\mathrm{d}t)-}] \\
&\quad = P[N_1((t+\mathrm{d}t)-) - N_1(t-) = 1|H_{1t-}, H_{t-}, \Phi_{(t+\mathrm{d}t)-}, N((t+\mathrm{d}t)-) - N(t-) = 1] \\
&\quad \times P[N((t+\mathrm{d}t)-) - N(t-) = 1|H_{1t-}, H_{t-}, \Phi_{(t+\mathrm{d}t)-}] \\
&\quad + P[N_1((t+\mathrm{d}t)-) - N_1(t-) = 1|H_{1t-}, H_{t-}, \Phi_{(t+\mathrm{d}t)-}, N((t+\mathrm{d}t)-) - N(t-) = 0] \\
&\quad \times P[N((t+\mathrm{d}t)-) - N(t-) = 0|H_{1t-}, H_{t-}, \Phi_{(t+\mathrm{d}t)-}] \\
&\quad = P[N_1((t+\mathrm{d}t)-) - N_1(t-) = 1|H_{1t-}, H_{t-}, \Phi_{(t+\mathrm{d}t)-}, N((t+\mathrm{d}t)-) - N(t-) = 1] \\
&\quad \times P[N((t+\mathrm{d}t)-) - N(t-) = 1|H_{t-}],
\end{aligned} \tag{4.80}$$

where

$$P[N((t+\mathrm{d}t)-) - N(t-) = 1|H_{1t-},\ H_{t-},\ \Phi_{(t+\mathrm{d}t)-}]$$

reduces to

$$P[N((t+\mathrm{d}t)-) - N(t-) = 1|H_{t-}],$$

as the initial point process is defined as external. It should be noted that H_{t-} is the history of the initial process $\{N(t),\ t \geq 0\}$ and it does not contain the information on the type of events and on the corresponding arrival times of events. In other words, mathematically, H_{t-} 'does not define' H_{1t-} and we need both of them for conditioning. Accordingly, from (4.80),

$$\begin{aligned}
&P[N_1((t+\mathrm{d}t)-) - N_1(t-) = 1|H_{1t-}, H_{t-}, \Phi_{(t+\mathrm{d}t)-}] \\
&= P[N_1((t+\mathrm{d}t)-) - N_1(t-) = 1|H_{1t-}, H_{t-}, \Phi_{(t+\mathrm{d}t)-}, N((t+\mathrm{d}t)-) - N(t-) \\
&= 1] \cdot \nu(t|H_{t-})\,\mathrm{d}t,
\end{aligned}$$

where $v(t|H_{t-})$ is the conditional intensity for $N(t), t \geq 0$

$$v(t|H_{t-}) \equiv \lim_{\Delta t \to 0} \frac{P[N((t+\Delta t)-) - N(t-) = 1|H_{t-}]}{\Delta t}.$$

Therefore, we arrive at the following result ([8] for the conditional intensity for a general history-dependent thinned process:

Theorem 4.14 *Under the given assumptions, the conditional intensity* $\lambda_1(t|H_{1t-})$ *is defined by the following expression:*

$$\begin{aligned}\lambda_1(t|H_{1t-}) &= E[P[N_1((t+\mathrm{d}t)-) - N_1(t-) \\ &= 1|H_{1t-}, H_{t-}, \Phi_{(t+\mathrm{d}t)-}, N((t+\mathrm{d}t)-) - N(t-) = 1] \cdot v(t|H_{t-})],\end{aligned} \tag{4.81}$$

where the expectation is with respect to the joint conditional distribution $(H_{t-},\ \Phi_{(t+\mathrm{d}t)-}|H_{1t-})$.

Theorem 4.14 holds for general orderly point processes. Furthermore, when we observe only the partial history H_{1t-}^P, the conditional intensity $\lambda_1(t|H_{1t-}^P)$ can be obtained from (4.81) by replacing H_{1t-} by H_{1t-}^P and by applying an appropriately modified conditional distribution $(H_{t-},\ \Phi_{(t+\mathrm{d}t)-}|H_{1t-}^P)$.

In what follows, we will simplify the setting and consider the case when the dependence on the history in the second multiplier in (4.81) is eliminated, whereas it is preserved for the first multiplier. Therefore, $v(t|H_{t-})$ is substituted by the rate of the corresponding NHPP, $v(t)$. This assumption enables to derive the closed-form results of the following subsection.

4.10.3 Stress–Strength Type Classification Model

Consider first, the case when only the *partial information* $H_{1t-}^P = \{N_1(t-)\}$ is observed, which means that the corresponding arrival times are not observed. Thus, only the number of Type 1 events is available. Then, formally,

$$\begin{aligned}&\lambda_1(t|H_{1t-}^P) \\ &= E[P[N_1((t+\mathrm{d}t)-) - N_1(t-) = 1|H_{1t-}^P,\ H_{t-},\ \Phi_{(t+\mathrm{d}t)-},\ N((t+\mathrm{d}t)-) \\ &\quad - N(t-) = 1]] \cdot v(t),\end{aligned} \tag{4.82}$$

where the expectation is with respect to the joint conditional distribution $(H_{t-},\ \Phi_{(t+\mathrm{d}t)-}|H_{1t-}^P)$. Denote the pdf and the Cdf of a random quantity (strength) U by $g_U(u)$ and $G_U(u)$, respectively. In this case, $\Phi_{(t+\mathrm{d}t)-} = \{S_1, S_2, \ldots, S_{N((t+\mathrm{d}t)-)}; U\}$ and

$$P[N_1((t+\mathrm{d}t)-)-N_1(t-)=1|H^P_{1t-},\,H_{t-},\,\Phi_{(t+\mathrm{d}t)-},\,N((t+\mathrm{d}t)-)-N(t-)=1]$$
$$=I(S_{N(t-)+1}>U),$$

where the conditional distribution of $U|H^P_{1t-}$ does depend on the history H^P_{1t-} and, as previously, S_i denotes the value of stress on the ith event. Therefore, in accordance with Theorem 4.14, $\lambda_1(t|H^P_{1t-})$ can be obtained as

$$\lambda_1(t|H^P_{1t-})=P(S_{N(t-)+1}>U|H^P_{1t-})\cdot v(t).$$

As the distribution of $S_{N(t-)+1}$ does not depend on the history $H^P_{1t-}=\{N_1(t-)\}$, it is sufficient to derive the distribution for $U|H^P_{1t-}$. Given $U=u$, the process $\{N_1(t),\ t\geq 0\}$ is the NHPP with intensity $\overline{F_S}(u)\,v(t)$ and thus the conditional distribution of $N_1(t-)|U$ is

$$P(N_1(t-)=n_1|U=u)=\frac{\left(\overline{F_S}(u)\int_0^t v(x)\,\mathrm{d}x\right)^{n_1}}{n_1!}\exp\left\{-\overline{F_S}(u)\int_0^t v(x)\,\mathrm{d}x\right\}.$$

Therefore, the conditional distribution of $U|N_1(t-)$ is

$$\frac{\frac{\left(\overline{F_S}(u)\int_0^t v(x)\,\mathrm{d}x\right)^{n_1}}{n_1!}\exp\{-\overline{F_S}(u)\int_0^t v(x)\,\mathrm{d}x\}\cdot g_U(u)}{\int_0^\infty\frac{\left(\overline{F_S}(w)\int_0^t v(x)\,\mathrm{d}x\right)^{n_1}}{n_1!}\exp\{-\overline{F_S}(w)\int_0^t v(x)\,\mathrm{d}x\}\cdot g_U(w)\,dw}.$$

Finally, from (4.82),

$$\lambda_1(t|H^P_{1t-})=\frac{\int_0^\infty\overline{F_S}(u)\cdot\frac{\left(\overline{F_S}(u)\int_0^t v(x)\,\mathrm{d}x\right)^{n_1}}{n_1!}\exp\{-\overline{F_S}(u)\int_0^t v(x)\,\mathrm{d}x\}\cdot g_U(u)\,\mathrm{d}u}{\int_0^\infty\frac{\left(\overline{F_S}(w)\int_0^t v(x)\,\mathrm{d}x\right)^{n_1}}{n_1!}\exp\{-\overline{F_S}(w)\int_0^t v(x)\,\mathrm{d}x\}\cdot g_U(w)\,dw}\cdot v(t). \tag{4.83}$$

For the specific case when $H^P_{1t-}=\{N_1(t-)=0\}$, i.e., $n_1=0$, the conditional intensity $\lambda_1(t|H^P_{1t-})$ in (4.83) reduces to

$$\lambda_S(t)=\frac{\int_0^\infty\int_0^s\exp\{-\overline{F}_S(r)\int_0^t v(x)\,\mathrm{d}x\}\cdot g_U(r)\,\mathrm{d}r f_S(s)\,\mathrm{d}s}{\int_0^\infty\exp\{-\overline{F}_S(r)\int_0^t v(x)\,\mathrm{d}x\}g_U(r)\,\mathrm{d}r}v(t),$$

which is, obviously the same as Eq. (4.50).

Consider now the case when the *full history*

$$H_{1t-}=\{N_1(t-)=n_1,\ T_{11}=t_{11},\ T_{12}=t_{12},\ldots,\ T_{1N_1(t-)}=t_{1n_1}\}$$

is observed and, therefore, is available. The crucial step in deriving the conditional intensity in the previous case was to obtain the conditional distribution of $U|H^P_{1t-}$. Intuitively, as the distribution of U depends only on 'the number of successes' up to t, but not on the arrival times of events, it seems that the full history H_{1t-}can be reduced to the partial history H^P_{1t-} 'without loss of relevant information' (i.e., the full history H_{1t-} is redundant). Thus it would be meaningful to see whether this statement is true or not. To show this, consider, as before,

$$P[N_1((t+\mathrm{d}t)-) - N_1(t-) = 1|H_{1t-},\, H_{t-},\, \Phi_{(t+\mathrm{d}t)-},\, N((t+\mathrm{d}t)-) - N(t-) = 1] \\ = I(S_{N(t-)+1} > U).$$

In accordance with Theorem 4.14, $\lambda_1(t|H_{1t-})$ can be obtained as

$$\lambda_1(t|H_{1t-}) = P(S_{N(t-)+1} > U|H_{1t-}) \cdot v(t).$$

It is sufficient to derive the distribution for $U|H_{1t-}$. Note that the joint conditional distribution of $(N_1(t-), T_{11}, T_{12}, \ldots, T_{1N_1(t-)}|U)$ is given by

$$\exp\left\{\int_0^{t_{11}} \overline{F_S}(u)\, v(x)\,\mathrm{d}x\right\}\overline{F_S}(u)\, v(t_{11})\, \exp\left\{-\int_{t_{11}}^{t_{12}} \overline{F_S}(u)\, v(x)\,\mathrm{d}x\right\}\overline{F_S}(u)\, v(t_2)\ldots$$
$$\times \exp\left\{-\int_{t_{1(n_1-1)}}^{t_{1n_1}} \overline{F_S}(u)\, v(x)\,\mathrm{d}x\right\}\overline{F_S}(u)\, v(t_{1n_1})\, \exp\left\{-\int_{t_{1n_1}}^{t} \overline{F_S}(u)\, v(x)\,\mathrm{d}x\right\}$$
$$= \left(\overline{F_S}(u)\right)^{n_1} v(t_{11})v(t_{12})\ldots v(t_{1n})\exp\left\{-\overline{F_S}(u)\int_0^t v(x)\,\mathrm{d}x\right\}.$$

Therefore, the conditional distribution of $(U|N_1(t-), T_{11}, T_{12}, \ldots, T_{1N_1(t-)})$ is

$$\frac{\left(\overline{F_S}(u)\right)^{n_1} v(t_{11})v(t_{12})\ldots v(t_{1n_1})\exp\left\{-\overline{F_S}(u)\int_0^t v(x)\,\mathrm{d}x\right\} \cdot g_U(u)}{\int_0^\infty \left(\overline{F_S}(w)\right)^{n_1} v(t_{11})v(t_{12})\ldots v(t_{1n_1})\exp\left\{-\overline{F_S}(w)\int_0^t v(x)\,\mathrm{d}x\right\} \cdot g_U(w)\,dw}.$$

Finally, from (4.81)

$$\lambda_1(t|H_{1t-}) = \frac{\int_0^\infty \left(\overline{F_S}(u)\right)^{n_1+1} v(t_{11})v(t_{12})\ldots v(t_{1n_1})\exp\left\{-\overline{F_S}(u)\int_0^t v(x)\,\mathrm{d}x\right\} \cdot g_U(u)\,\mathrm{d}u}{\int_0^\infty \left(\overline{F_S}(w)\right)^{n_1} v(t_{11})v(t_{12})\ldots v(t_{1n_1})\exp\left\{-\overline{F_S}(w)\int_0^t v(x)\,\mathrm{d}x\right\} \cdot g_U(w)\,dw} \cdot v(t). \tag{4.84}$$

It can be seen that $\lambda_1(t|H_{1t-})$ in Eq. (4.84) and that in Eq. (4.83) are identical and, therefore, H_{1t-} can be reduced to the partial history H^P_{1t-} "without loss of relevant information" as our initial intuition prompted us.

Note that, as the external point process is the NHPP, $\lambda(t|H_{t-}) = v(t)$. Then, using $\lambda(t|H_{1t-}) = \lambda_1(t|H_{1t-}) + \lambda_2(t|H_{1t-})$, the following relationship holds:

$$\lambda_2(t|H_{1t-}) \equiv \lim_{\Delta t \to 0} \frac{P[N_2((t+\Delta t)-) - N_2(t-) = 1|H_{1t-}]}{\Delta t} = v(t) - \lambda_1(t|H_{1t-}).$$

It is clear that the conditional probability that the event that happened at time t belongs to $\{N_1(t),\ t \geq 0\}$ is

$$\frac{\lambda_1(t|H_{1t-})}{\lambda_1(t|H_{1t-}) + \lambda_2(t|H_{1t-})}.$$

Obviously, both processes $\{N_1(t),\ t \geq 0\}$ and $\{N_2(t),\ t \geq 0\}$ are not NHPPs now.

The case when we observe the full history of $\{N_1(t),\ t \geq 0\}$ and $\{N_2(t),\ t \geq 0\}$, can be considered in a similar way [8].

References

1. Aalen OO, Borgan O, Gjessing HK (2008) Survival and event history analysis. A process point of view. Springer, New York
2. Anderson PK, Borgan O, Gill RD, Keiding N (1993) Statistical models based on counting processes. Springer-Verlag, New York
3. Beichelt FE, Fischer K (1980) General failure model applied to preventive maintenance policies. IEEE Trans Reliab 29:39–41
4. Block HW, Borges WS, Savits TH (1985) Age-dependent minimal repair. J Appl Probab 22:370–386
5. Cha JH, Finkelstein M (2009) On a terminating shock process with independent wear increments. J Appl Probab 46:353–362
6. Cha JH, Finkelstein M (2011) On new classes of extreme shock models and some generalizations. J Appl Probab 48:258–270
7. Cha JH, Finkelstein M (2012a) Stochastic survival models with events triggered by external shocks. Probab Eng Inf Sci 26:183–195
8. Cha JH, Finkelstein M (2012b) Information-based thinning of point processes and its application to shock models. J Stat Plan Inference 142:2345–2350
9. Cha JH, Finkelstein M (2012c) A note on the curable shock processes. J Stat Plan Inference 142:3146–3151
10. Cha JH, Finkelstein M (2013a) A note on the class of geometric point processes. Proba Eng Inf Sci 27:177–186
11. Cha JH, Finkelstein M (2013b) On generalized shock models for deteriorating systems. Appl Stoch Models Bus Ind 29. doi: 10.1002/asmb.1933
12. Cha JH, Finkelstein M, Marais F (2013). Survival of systems with protection subject to two types of external attacks. Ann Oper Res. doi:10.1007/s10479-013-1315-6
13. Cha JH, Mi J (2011) On a stochastic survival model for a system under randomly variable environment. Methodol Comput Appl Probab 13:549–561
14. Cinlar E (1975) Introduction to stochastic processes. Prentice Hall, Englewood Cliffs, New Jersey
15. Cox DR, Isham V (1980) Point processes. University Press, Cambridge
16. Finkelstein M (2007) On statistical and information-based virtual age of degrading systems. Reliab Eng Syst Saf 92:676–682
17. Finkelstein M (2008) Failure rate modelling for reliability and risk. Springer, London

18. Finkelstein M (1999) Wearing-out components in variable environment. Reliab Eng Sys Saf 66:235–242
19. Finkelstein M, Marais F (2010) On terminating Poisson processes in some shock models. Reliab Eng Syst Saf 95:874–879
20. Finkelstein M, Zarudnij VI (2002) Laplace transform methods and fast repair approximations for multiple availability and its generalizations. IEEE Trans Reliab 51:168–177
21. Fleming TR, Harrington DP (1991) Counting processes and survival analysis. Wiley, New York
22. Kebir Y (1991) On hazard rate processes. Nav Res Logist 38:865–877
23. Lemoine AJ, Wenocur ML (1985) On failure modeling. Nav Res Logist 32:497–508
24. Lemoine AJ, Wenocur ML (1986) A note on shot-noise and reliability modeling. Oper Res 1986(34):320–323
25. Nachlas JA (2005) Reliability engineering: probabilistic models and maintenance methods. Taylor & Francis, Boca Raton
26. Rice J (1977) On generalized shot noise. Adv Appl Probab 9:553–565
27. Ross SM (1996) Stochastic processes, 2nd edn. Wiley, New York
28. Ushakov IA, Harrison RA (1994) Handbook of reliability engineering. Wiley, New York

Chapter 5
Heterogeneous Populations

Homogeneity of objects is a unique property that is very rare in nature and in industry. It can be created in the laboratory, but not outside it. Therefore, one can hardly find homogeneous populations in real life; however, most of reliability modeling deals with homogeneous cases. Due to instability of production processes, environmental and other factors, most populations of manufactured items in real life are heterogeneous. Similar considerations are obviously true for biological items (organisms). Neglecting heterogeneity can lead to serious errors in reliability assessment of items and, as a consequence, to crucial economic losses. Stochastic analysis of heterogeneous populations presents a significant challenge to developing mathematical descriptions of the corresponding reliability indices. On the other hand, everything depends on the definition, on what we understand by homogeneous and heterogeneous populations. From the statistical point of view, these terms mean the following.

In homogeneous populations, the lifetimes of items form a sequence of independent and identically distributed random variables (i.i.d.) with the common Cdf $F(t)$ pdf $f(t)$, and the failure rate, $\lambda(t)$. However, due to instability of production processes, environmental and other factors, most populations of manufactured items in real life (and biological organisms in nature as well) are heterogeneous. This means that these populations can be often considered as a *finite or non-finite collection of homogeneous subpopulations* [which are frequently ordered in some suitable stochastic sense, e.g., in the sense of the hazard rate ordering (2.70)].

As an illustrative discrete example, we can think about the collection of $n = 2$ subpopulations of statistically identical items produced at different facilities and mixed together in one population. Assume for simplicity, that each subpopulation consists of a sufficiently large (infinite) number of items. Let the first subpopulation be described by the failure rate $\lambda(t)$ (baseline failure rate), whereas the second subpopulation, due to the better production quality has a smaller failure rate $k\lambda(t)$, where k is a fixed constant such that $0 < k < 1$. Let the proportions of both subpopulations in the population be π_1 and π_2, $\pi_1 + \pi_2 = 1$. An item is selected at random from the described heterogeneous population and therefore, we do not know to which subpopulation it belongs (although the proportions can be

M. Finkelstein and J. H. Cha, *Stochastic Modeling for Reliability*,
Springer Series in Reliability Engineering, DOI: 10.1007/978-1-4471-5028-2_5,

known at some instances). This choice can be described by the discrete random variable Z (unobserved) with the possible values "1" and "k" and the corresponding probability masses $\pi(1) = \pi_1$, $\pi(k) = \pi_2$. Based on the description of Z, the failure rates of the subpopulation with $Z = z$ can be now specified as $\lambda(t, z)$: $\lambda(t, 1) = \lambda(t)$ and $\lambda(t, k) = k\lambda(t)$. In the literature, the random variable Z is often called "frailty". Frailty describes the susceptibility to failures of items from different ordered subpopulations. Various frailty models have been studied in numerous statistical publications. However, as most of the settings that were considered in reliability theory and practice are homogeneous, the concept of frailty has not been sufficiently elaborated in the reliability literature so far.

Instability of production processes, environmental and other factors can obviously result in more than $n = 2$ 'quality levels' and in the continuous frailty model as well. Let, as previously, $\lambda(t)$ denote now the failure rate of some baseline subpopulation. For illustration of the continuous frailty concept, consider the multiplicative (proportional) frailty model. In this model, the failure rates of all other subpopulations are defined as $\lambda(t, z) \equiv z\lambda(t)$, where z is the realization of Z with support, e.g., in $[0, \infty)$. Thus, the failure rate is larger (smaller) for larger (smaller) values of z and we see here the explicit ordering of the corresponding subpopulations in the sense of the hazard rate ordering (2.70). The frailty Z is now the continuous random variable. The term "frailty" was introduced in Vaupel et al. [63] for the gamma-distributed frailty Z. It is worth noting, however, that this specific case of the gamma-frailty model was, in fact, first considered by the British actuary Robert Beard [7, 8].

Mixtures of distributions usually present an effective mathematical tool for modeling heterogeneity, especially when we are interested in the failure rate, which is the conditional characteristic. The introductory Sect. 2.3 was devoted to the shape of the failure rate in the homogeneous setting, which is really important in many applications (reliability, demography, risk analysis, etc.). In heterogeneous populations, the analysis of the shape of the mixture (population) failure rate starts to be even more meaningful. It is well known, e.g., that mixtures of decreasing failure rate (DFR) distributions are always DFR [6]. On the other hand, mixtures of increasing failure rate (IFR) distributions can decrease, at least in some intervals of time. Note that the IFR distributions are often used to model lifetimes governed by the aging processes. Therefore, the operation of mixing can dramatically change the pattern of population aging, e.g., from positive aging (IFR) to negative aging (DFR).

In Sects. 5.1–5.6, on the basis of Finkelstein [28, 29], we will present a brief survey of results relevant for our further discussion in this and in the subsequent chapters. In the rest of this chapter, some new applications of the mixture failure rate modeling will be considered.

5.1 Failure Rate of Mixture of Two Distributions

Suppose, for instance, that a population of some manufactured items consists of items with and without manufacturing defects. The time to failure of an item picked up at random from this population can be obviously described in terms of mixtures. We start with a mixture of two lifetime distributions $F_1(t)$ and $F_2(t)$ with the pdfs $f_1(t)$ and $f_2(t)$ and failure rates $\lambda_1(t)$ and $\lambda_2(t)$, respectively, whereas the Cdf, pdf, and the failure rate of the mixture itself are denoted by $F_m(t), f_m(t)$ and $\lambda_m(t)$, accordingly.

Let the masses π and $1-\pi$ define the discrete mixture distribution. The mixture survival function and the mixture pdf are

$$\begin{aligned} \bar{F}_m(t) &= \pi\bar{F}_1(t) + (1-\pi)\bar{F}_2(t), \\ f_m(t) &= \pi f_1(t) + (1-\pi)f_2(t), \end{aligned} \tag{5.1}$$

respectively. In accordance with the definition of the failure rate (2.4), the mixture failure rate in this case is

$$\lambda_m(t) = \frac{\pi f_1(t) + (1-\pi)f_2(t)}{\pi\bar{F}_1(t) + (1-\pi)\bar{F}_2(t)}.$$

As $\lambda_i(t) = f_i(t)/\bar{F}_i(t),\ i = 1,2$, this can be transformed into

$$\lambda_m(t) = \pi(t)\lambda_1(t) + (1-\pi(t))\lambda_2(t), \tag{5.2}$$

where the time-dependent probabilities are

$$\pi(t) = \frac{\pi\bar{F}_1(t)}{\pi\bar{F}_1(t) + (1-\pi)\bar{F}_2(t)}, \quad 1-\pi(t) = \frac{(1-\pi)\bar{F}_2(t)}{\pi\bar{F}_1(t) + (1-\pi)\bar{F}_2(t)}, \tag{5.3}$$

It follows from Eq. (5.2) that $\lambda_m(t)$ is contained between $\min\{\lambda_1(t), \lambda_2(t)\}$ and $\max\{\lambda_1(t), \lambda_2(t)\}$. Specifically, if the failure rates are ordered as $\lambda_1(t) \le \lambda_2(t)$, then

$$\lambda_1(t) \le \lambda_m(t) \le \lambda_2(t).$$

Differentiating (5.1) results in [51]:

$$\lambda'_m(t) = \pi(t)\lambda'_1(t) + (1-\pi(t))\lambda'_2(t) - \pi(t))(1-\pi(t)(\lambda_1(t) - \lambda_2(t))^2. \tag{5.4}$$

Assume that $\lambda_i(t)\ i - 1,2$ are DFR. Then the mixture failure rate is also decreasing, which is the well-known fact for general mixtures [6].

As $\bar{F}_i(0) = 1, i = 1,2$, the initial value of the mixture failure rate ($t = 0$) is just the 'ordinary' mixture of initial values of the two failure rates, i.e.,

$$\lambda_m(0) = \pi\lambda_1(0) + (1-\pi)\lambda_2(0).$$

When $t > 0$, the conditional probabilities $\pi(t)$ and $1 - \pi(t)$ are obviously not equal to π and $1 - \pi$, respectively. Assume that $\lambda_1(t) \leq \lambda_2(t)$. Dividing the numerator and the denominator in the first equation in (5.3) by $\bar{F}_1(t)$ it is easy to see that the proportion of the survived up to t items in the mixed population, i.e., $\pi(t)$ is increasing ($(1 - \pi(t))$ is decreasing). This effect can be meaningfully interpreted in the following way: *the weakest items are dying out first*. Therefore,

$$\lambda_m(t) < \pi\lambda_1(t) + (1 - \pi)\lambda_2(t), \ \ t > 0. \tag{5.5}$$

Thus, $\lambda_m(t)$ is always smaller than the expectation $\pi\lambda_1(t) + (1 - \pi)\lambda_2(t)$.

Assume now that both $\lambda_1(t)$ and $\lambda_2(t)$ are increasing for $t \geq 0$. Can the mixture failure rate initially (at, least, for small t) decrease? Equation (5.4) helps us to give the positive answer to this question. The corresponding sufficient condition is

$$\pi\lambda_1'(t) + (1 - \pi)\lambda_2'(t) - \pi(1 - \pi)(\lambda_1(0) - \lambda_2(0))^2 < 0, \tag{5.6}$$

where the derivatives are obtained at $t = 0$. Inequality (5.6), e.g., means that if $|\lambda_1(0) - \lambda_2(0)|$ is sufficiently large, then the mixture failure rate is initially decreasing no matter how fast the failure rates $\lambda_1(t)$ and $\lambda_2(t)$ are increasing in the neighborhood of 0, which is a remarkable fact, indeed. Let, for instance,

$$\lambda_1(t) = c_1 t + a_1, \ \lambda_2(t) = c_2 t + a_2, \ 0 < c_1 < c_2, \ 0 < a_1 < a_2,$$

Then, if

$$a_2 - a_1 > \left(\frac{\pi c_1 + (1 - \pi_1)c_2}{\pi(1 - \pi)} \right)^{1/2},$$

$\lambda_m(t)$ is initially decreasing.

What about the asymptotic (for large t) behavior of $\lambda_m(t)$? Due to the *weakest populations are dying first* principle the intuitive guess would be: the mixture failure rate tends (in some suitable sense) to the failure rate of the strongest population as $t \to \infty$. Block and Joe [13] give some general conditions for this convergence. We will just consider here an important specific case of proportional failure rates that allows formulating these conditions explicitly:

$$\lambda_1(t) \equiv \lambda(t, z_1) = z_1\lambda(t), \ \lambda_2(t) \equiv \lambda(t, z_2) = z_2\lambda(t), \ z_2 > z_1,$$

where $\lambda(t)$ is some baseline failure rate. We will distinguish between the convergence

$$\lambda_m(t) - \lambda(t, z_1) \to 0 \text{ as } t \to \infty \tag{5.7}$$

and the asymptotic equivalence

$$\lambda_m(t) = \lambda(t, z_1)(1 + o(1)) \text{ as } t \to \infty, \tag{5.8}$$

which will mostly be used in the following alternative notation: $\lambda_m(t) \sim \lambda(t, z_1)$ as $t \to \infty$.

When $\lambda(t)$ has a finite limit as $t \to \infty$, these relationships coincide. The following theorem [32] specifies the corresponding conditions:

Theorem 5.1 *Consider the mixture model (5.1)–(5.3), where*

$$\lambda(t, z_1) = z_1\lambda(t),\ \lambda(t, z_2) = z_2\lambda(t);\ z_2 > z_1 > 0,$$

and $\lambda(t) \to \infty$ *as* $t \to \infty$*.Then*

- *Relationship (5.8) holds;*
- *Relationship (5.7) holds if*

$$\lambda(t)\exp\{-(z_2 - z_1)\int_0^t \lambda(u)\mathrm{d}u\} \to 0 \text{ as } t \to \infty. \tag{5.9}$$

The proof is straightforward and is based on considering the quotient $\lambda_m(t)/\lambda(t, z_1)$ as in Block and Joe [13].

□

Condition (5.9) is a rather weak one. In essence, it states that the pdf of a distribution with an ultimately increasing failure rate tends to 0 as $t \to \infty$. All distributions that are typically used in lifetime data analysis meet this requirement.

Similar reasoning can be used for describing the shape of the failure rate for the mixture of $n > 2$ distributions [13, 28].

We have described some approaches to analyze the general pattern of the shape of the mixture failure rate for two distributions focusing on initial and tail behavior. The concrete shapes can be versatile. We will just present here a few examples. More information on specific shapes of the mixture failure rate of two distributions can be found in Gurland and Sethuraman [40], Gupta and Waren [39], Block et al. [14, 18], Lai and Xie [43], Navarro and Hernandez [51], Finkelstein [28], and Block et al. [16]. Note that the different shapes of the mixture mortality rate were analyzed in various demographic applications.

- As follows from Gupta and Waren [39], the mixture of two gamma distributions with increasing failure rates (with the same scale parameter) can result either in the increasing mixture failure rate or in the modified bathtub (MBT) mixture failure rate (it first increases and then behaves like a bathtub (BT) failure rate). This shape agrees with our general reasoning of this section, as it can be easily verified that condition (5.6) does not hold in this case and therefore the initial decreasing is not possible.
- Similar shapes occur for the mixtures of two Weibull distributions with increasing failure rates. Note that in this case, MBT shape results when p in Eq. (5.1) is less than some ξ, $0 < \xi < 1$ and the mixture failure rate increases for $p \geq \xi$.

- Navarro and Hernandez [51] state that the mixture failure rate of two truncated normal distributions (we are dealing with lifetime random variables), depending on parameters involved, can also be increasing, BT-shaped or MBT-shaped. The BT shape obtained via the generalized mixtures (when p is a real number and not necessarily $p \in [0, 1]$) where studied in Navarro and Hernandez [52].
- Block et al. [18] give explicit conditions which describe the possible shapes of the mixture failure rate for two increasing linear failure rates. Again the possible shapes in this case are IFR, BT, and MBT (for the non-crossing linear failure rates).
- Block et al. [16] present an interesting generalization when one of the distributions is itself a continuous mixture of exponentials (and therefore, decreasing) and the other is a gamma distribution. It is shown that for the specific values of parameters involved the mixture failure rate has a BT shape. In essence, these authors are 'constructing' the BT shape using the specifically decreasing in $(0, \infty)$ to $\zeta > \lambda_0 > 0$ failure rate of the first distribution and the increasing to λ_0 failure rate of the second distribution. Note that, as follows from (5.3), $\lambda_m(t)$ is contained between these two failure rates. Block et al. [16] also prove that mixtures of DFR gamma distributions with an IFR gamma distribution are bathtub-shaped and mixtures of modified Weibull distributions (the failure rate is decreasing not to 0, as for 'ordinary' Weibull distribution, but to ζ) with an IFR gamma distribution have also the bathtub-shaped failure rate.

5.2 Continuous Mixtures

Let Z be now a continuous mixing random variable (frailty) with support in $[0, \infty)$ and the pdf $\pi(z)$. Other intervals of support can be also considered. Similar to the previous section, the mixture survival function and the mixture pdf are defined as the following expectations:

$$\begin{aligned} \bar{F}_m(t) &= \int_0^\infty \bar{F}(t, z)\pi(z)\mathrm{d}z, \\ f_m(t) &= \int_0^\infty f(t, z)\pi(z)\mathrm{d}z, \end{aligned} \tag{5.10}$$

respectively, where the notation for conditional functions $\bar{F}(t|Z = z) = \bar{F}(t, z)$ and $f(t|Z = z) = f(t, z)$ means that a lifetime distribution is indexed by parameter z. The corresponding conditional failure rate is denoted by $\lambda(t, z)$, whereas the mixture (observed) failure rate is

$$\lambda_m(t) = \frac{\int_0^\infty f(t, z)\pi(z)\mathrm{d}z}{\int_0^\infty \bar{F}(t, z)\pi(z)\mathrm{d}z}. \tag{5.11}$$

Equation (5.11) can be transformed to [47]:

$$\lambda_m(t) = \int_0^\infty \lambda(t,z)\pi(z|t)\mathrm{d}z, \; \pi(z|t) = \frac{\pi(z)\bar{F}(t,z)}{\int_0^\infty \bar{F}(t,z)\pi(z)\mathrm{d}z}, \tag{5.12}$$

where $\pi(z|t)$ denotes the conditional pdf of Z on condition that $T > t$, i.e., an item described by a lifetime T with the Cdf $F_m(t)$ had survived in $[0,t]$. Denote this random variable by $Z|t$. Obviously the masses $\pi(t)$ and $1-\pi(t)$ in (5.1) correspond to $\pi(z|t)$ in the continuous case.

Under the mild assumptions (see Theorem 5.2), a property that is similar to the discrete case (5.5) holds for the continuous case as well, i.e.,

$$\lambda_m(t) < \lambda_P(t) \equiv \int_0^\infty \lambda(t,z)\pi(z)\mathrm{d}z, \quad t > 0; \; \lambda_m(0) = \lambda_P(t) \tag{5.13}$$

meaning that the mixture failure rate is always smaller than the 'ordinary' expectation. Thus, owing to conditioning, the mixture failure rate is smaller than the unconditional one for each $t > 0$, which, as in the discrete case, can be interpreted via *the weakest populations are dying out first* principle. As time increases, those subpopulations that have larger failure rates have larger chances of dying and, therefore, the proportion of subpopulations with a smaller failure rate increases.

The following theorem [33] states also the condition for $\lambda_P(t) - \lambda_m(t)$ to increase:

Theorem 5.2 *Let the failure rate $\lambda(t,z)$ be differentiable with respect to both arguments and be ordered as*

$$\lambda(t,z_1) < \lambda(t,z_2), \quad z_1 < z_2, \forall z_1, z_2 \in [a,b], \; t \geq 0. \tag{5.14}$$

Then

- *Inequality (5.13) holds;*
- *If, additionally, $\partial\lambda(t,z)/\partial z$is increasing in t, then $\lambda_P(t) - \lambda_m(t)$ is increasing.*

We will consider now two important applications specific in cases of model (5.12). Let $\lambda(t,z)$ be indexed by parameter z in the following *additive* way:

$$\lambda(t,z) = \lambda(t) + z, \tag{5.15}$$

where $\lambda(t)$ is a deterministic, continuous, and positive function for $t > 0$. It can be viewed as some baseline failure rate. Equation (5.15) defines for $z \in [0,\infty)$ a family of 'horizontally parallel' functions. We will be interested in an increasing $\lambda(t)$. Applying (5.12) to this model results in

$$\lambda_m(t) = \lambda(t) + \frac{\int_0^\infty z\bar{F}(t,z)\pi(z)\mathrm{d}z}{\int_0^\infty \bar{F}(t,z)\pi(z)\mathrm{d}\theta} = \lambda(t) + E[Z|t], \tag{5.16}$$

where, in accordance with (5.12), $E[Z|t]$ denotes the expectation of the random variable $Z|t$. It can be easily shown by direct derivation that $E'[Z|t] = -\mathrm{Var}(Z|t) < 0$. Differentiating (5.16) and using this property, we obtain the following result [32, 47].

Theorem 5.3 *Let $\lambda(t)$ be an increasing, convex function in $[0,\infty)$. Assume that* $\mathrm{Var}(Z|t)$ *is decreasing in $t \in [0,\infty)$ and*

$$\mathrm{Var}(Z|0) > \lambda'(0).$$

Then $\lambda_m(t)$ decreases in $[0,c)$ and increases in $[c,\infty)$, where c can be uniquely defined from the following equation:

$$\mathrm{Var}(Z|t) = \lambda'(t).$$

It follows from this theorem that the corresponding model of mixing results in the bathtub shape of the mixture failure rate: it first decreases and then increases, converging to the failure rate of the strongest population, which is $\lambda(t)$ in our case. It seems that the conditional variance $\mathrm{Var}(Z|t)$ should decrease, as the "weak populations are dying out first" when t increases. It turns out, however, that this intuitive reasoning is not true for the general case and some specific distributions can result in initially increasing $\mathrm{Var}(Z|t)$. The corresponding counter-example can be found in Finkelstein and Esaulova [32]. It is also shown that $\mathrm{Var}(Z|t)$ is always decreasing in $[0,\infty)$ when Z is gamma-distributed.

The most popular and elaborated applications model of mixing is the multiplicative one:

$$\lambda(t,z) = z\lambda(t), \tag{5.17}$$

where, as previously, the baseline $\lambda(t)$ is a deterministic, continuous, and positive function for $t > 0$. In survival analysis, Eq. (5.17) is usually called a multiplicative frailty model (proportional hazards). The mixture failure rate in this case is

$$\lambda_m(t) = \int_0^\infty \lambda(t,z)\pi(z|t)\mathrm{d}z = \lambda(t)E[Z|t]. \tag{5.18}$$

Differentiating both sides gives

$$\lambda_m'(t) = \lambda'(t)E[Z|t] + \lambda(t)E'[Z|t]. \tag{5.19}$$

Thus, when $\lambda(0) = 0$, the failure rate $\lambda_m(t)$ increases in the neighborhood of $t = 0$. Further behavior of this function depends on the other parameters involved. Similar to the additive case, $E'[Z|t] = -\lambda(t)\mathrm{Var}(Z|t) < 0$, which means that $E[Z|t]$ is decreasing in t [38]. Therefore, it follows from Eq. (5.18) that the function

$\lambda_m(t)/\lambda(t)$ is a decreasing one, which imply that $\lambda(t)$ and $\lambda_m(t)$ cross at most at only one point. It immediately follows from Eq. (5.19) that when $\lambda(t)$ is decreasing, $\lambda_m(t)$ is also decreasing (another proof of this well-known property). When $\lambda(0) \neq 0$ and

$$\frac{\lambda'(0)}{\lambda^2(0)} \leq \frac{\text{Var}(Z)}{E[Z]},$$

the mixture failure rate is decreasing in $[0, \varepsilon)$, $\varepsilon > 0$ meaning, e.g., that for the fixed $E[Z]$ the variance of Z should be sufficiently large.

Asymptotic behavior of $\lambda_m(t)$ as $t \to \infty$ for this and other (more general models will be discussed in Sect. 5.4). Note that, the accelerated life model (ALM) to be studied in this section does not allow the foregoing reasoning based on considering expectation $E[Z|t]$.

5.3 Examples

5.3.1 Weibull and Gompertz Distributions

Consider multiplicative frailty model (5.17). Let Z be a gamma-distributed random variable with shape parameter α and scale parameter β and let $\lambda(t) = \gamma t^{\gamma-1}$, $\gamma > 1$ be the increasing failure rate of the Weibull distribution, $\lim_{t\to\infty} \lambda(t) = \infty$. The mixture failure rate $\lambda_m(t)$ in this case, can be obtained by the direct integration, as in Finkelstein [28] (see also [38]):

$$\lambda_m(t) = \frac{\alpha\beta\gamma t^{\gamma-1}}{1 + \beta t^{\gamma}}. \tag{5.20}$$

The shape of the mixture failure rate differs dramatically from the shape of the increasing baseline failure rate $\lambda(t)$. Thus $\lambda_m(t)$ is equal to 0 at $t = 0$, increases to a maximum at

$$t_{\max} = \left(\frac{\gamma - 1}{\beta}\right)^{\frac{1}{\gamma}}$$

and then decreases to 0 as $t \to \infty$ (Fig. 5.1).

Weibull distribution with $\gamma > 1$ is often used for modeling aging processes as its failure rate is increasing. Therefore the mixture model results in the dramatically different shape (the upside-down bathtub shape). This phenomenon should certainly be taken in account in reliability practice.

The described shape of the mixture failure rate was observed for a heterogeneous sample of miniature light bulbs [28]. The failure rate of the homogeneous population of these light bulbs, however, follows the Weibull law. Therefore the observed shape complies with the predicted one.

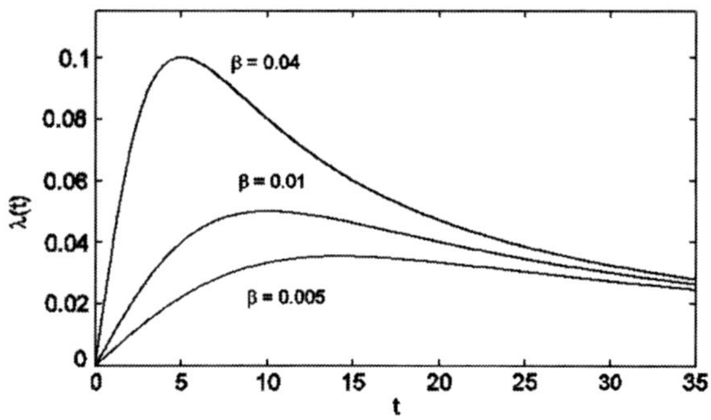

Fig. 5.1 The mixture failure rate for the Weibull baseline distribution, $\gamma = 2$, $\alpha = 1$

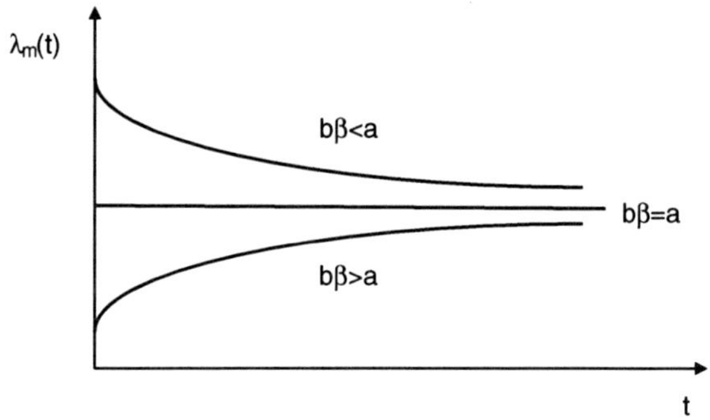

Fig. 5.2 Gamma-Gompertz mixture failure rate

Let again the mixing distribution be the gamma distribution with shape parameter c and scale parameter β, whereas the baseline distribution be the Gompertz distribution with the failure rate $\lambda(t) = a\exp\{bt\}$, $a, b > 0$. Owing to its computational simplicity, the gamma-frailty model is practically the only one widely used in applications so far. Direct computation in accordance with Eq. (5.12) for this baseline failure rate results in

$$\lambda_m(t) = \frac{bc\exp\{bt\}}{\exp\{bt\} + \left(\frac{b\beta}{a} - 1\right)}. \tag{5.21}$$

If $b\beta = a$, then $\lambda_m(t) \equiv bc$. However, if $b\beta > a$, then $\lambda_m(t)$ increases to bc and if $b\beta < a$, it decreases to bc (Fig. 5.2).

Thus, we are mixing exponentially increasing failure rates and as a result obtaining a slowly increasing (decreasing) mixture failure rate, which converges to a constant value.

5.3.2 Reliability Theory of Aging

Consider now a discrete frailty parameter, $Z = N$ with the Cdf $F_0(n) \equiv P(N \leq n)$. We will be interested in the following meaningful reliability interpretation.

Let N be a random number of initially (at $t = 0$) operating independent and identically distributed components with constant failure rates λ. Assume that these components form a parallel system, which, according to Gavrilov and Gavrilova [36], models the lifetime of an organism (generalization to the series-parallel structure is straightforward). These authors also provide a biological justification of the model. In each realization $N = n,\ n \geq 1$, the degradation process of pure death can be defined as just the number of failed components. When this number reaches n, the death of an organism occurs. Denote by $\lambda_n(t)$ the mortality (failure) rate, which describes T_n—the time to death for the fixed $N = n,\ n = 1, 2, \ldots$ ($n = 0$ is excluded, as there should be at least one operating component at $t = 0$). It is shown in Gavrilov and Gavrilova [36] that as $t \to 0$, this mortality rate tends to an increasing power function (the Weibull law), which is a remarkable fact. On the other hand, for random N, similar to (5.2), (5.3) and (5.11, 5.12), the observed (mixture) mortality rate is given as the following conditional expectation with respect to N:

$$\lambda_m(t) = E[\lambda_N(t)|T > t], \tag{5.22}$$

where T, as usual, denotes the lifetime of interest. Therefore, as previously, $\lambda_m(t)$ is a conditional expectation (on condition that the system is operable at t) of a random mortality rate $\lambda_N(t)$. Note that, for small t, this operation can approximately result in the unconditional expectation

$$\lambda_m(t) \approx E[\lambda_N(t)] = \sum_{n=1}^{\infty} P_n \lambda_n(t), \tag{5.23}$$

where $P_n \equiv \Pr[N = n]$, but the limiting transition, as $t \to 0$, should be performed carefully in this case. As $t \to \infty$, we observe the following mortality plateau [34]:

$$\lambda_m(t) \to \lambda. \tag{5.24}$$

This is due to the fact that the conditional probability that only one component with the failure rate λ is operating tends to 1 as $t \to \infty$ (on condition that the system is operating).

Assume now that N is Poisson distributed with parameter η (on condition that the system is operable at $t = 0$). Therefore

$$P_n = \frac{\exp\{-\eta\}\eta^n}{n!(1 - \exp\{-\eta\})}, \quad n = 1, 2, \ldots.$$

It can be shown via direct integration that the time to death in our simplified model has the following Cdf [55]:

$$F(t) = \Pr[T \leq t] = \frac{1 - \exp\{-\eta \exp\{-\lambda t\}\}}{1 - \exp\{-\eta\}}. \tag{5.25}$$

The corresponding mixture mortality rate is

$$\lambda_m(t) = \frac{F'(t)}{1 - F(t)} = \frac{\eta\lambda\exp\{-\lambda t\}}{\exp\{\eta\exp\{-\lambda t\}\} - 1}. \tag{5.26}$$

Performing, as $t \to \infty$, the limiting transition in (5.26), we also arrive at the mortality plateau (5.5).

In fact, the mortality rate given by Eq. (5.26) is far from the exponentially increasing Gompertz law. The Gompertz law can erroneously follow (as in Gavrilov and Gavrilova [36]) from (5.23) if this approximation is used formally, without considering a proper conditioning in (5.23). However, for some specific values of parameters and sufficiently small t, exponential approximation can still hold. The relevant discussion can be found in Steinsaltz and Evans [55].

5.4 Mixture Failure Rate for Large *t*

The failure (mortality) rate behavior for large t, is important for objects at the last phase of their useful life (e.g., the above mentioned mortality plateaus). Among the first to consider the limiting behavior of mixture failure rates for the continuous mixtures were Clarotti and Spizzichino [23]. They showed that the mixture failure rate for a family of exponential distributions with parameter $\alpha \in [a, \infty)$ converges to the failure rate of the strongest population, which is a in this case. Block et al. [17], Block et al. [14], and Li [44] extended this to a general case (see also [15]). As the approach (and obtained important mathematical results) of these authors is very general and some assumptions are rather restrictive, it does not provide specific asymptotic relationship that can be used in practical analysis for mixed populations. In order to be able to perform this analysis, Finkelstein and Esaulova [33] developed an approach that was applied to reasonably general survival model that allows for explicit asymptotic relationships and covers (as specific cases) three most popular in survival analysis frailty models: additive, proportional, and accelerated life. The main results that were obtained using this approach are discussed below. The corresponding proofs that are quite technical can be found in this paper.

Let $T \geq 0$ be a lifetime with the cdf $F(t)$, pdf $f(t)$, and the failure rate $\lambda(t)$. Let, as previously, these functions be indexed by the realization of the frailty parameter $Z = z$, i.e., $F(t,z)$, $f(t,z)$, $\lambda(t,z)$, respectively. Consider the following general survival model:

$$\Lambda(t,z) = A(z\phi(t)) + \psi(t), \tag{5.27}$$

where $\Lambda(t,z) \equiv \int_0^t \lambda(t,z)$ denotes the corresponding cumulative failure rate and $A(\cdot)$, $\psi(\cdot)$ and $\phi(\cdot)$ are increasing differentiable functions of their arguments. The meaning of relationship (5.27): we perform a scale transformation $\phi(t)$ in the argument of the cumulated failure rate $\Lambda(t)$ and 'insert' a frailty parameter. An important feature of the model is that parameter z is a multiplier.

This model includes a number of well-known survival analysis and reliability specific cases, i.e.,
Additive Model: Let

$$A(u) \equiv u,\ \phi(t) = t,\ \ \psi(0) = 0.$$

Then

$$\lambda(t,z) = z + \psi'(t), \quad \Lambda(t,z) = zt + \psi(t). \tag{5.28}$$

PH (multiplicative) Model: Let

$$A(u) \equiv u,\ \phi(t) = \Lambda(t).$$

Then

$$\begin{aligned} \lambda(t,z) &= z\lambda(t), \\ \Lambda(t,z) = z\Lambda(t) &= z\int_0^t \lambda(u)\mathrm{d}u. \end{aligned} \tag{5.29}$$

Accelerated Life Model: Let

$$A(u) \equiv \Lambda(u),\ \phi(t) = t.$$

Then

$$\Lambda(t,z) = \int_0^{zt} \lambda(u)\mathrm{d}u = \Lambda(zt), \tag{5.30}$$

$$\lambda(t,z) = z\lambda(zt). \tag{5.31}$$

We are interested in asymptotic behavior (as $t \to \infty$) of $\lambda_m(t)$. For simplicity of notation (and, in fact, not loosing the generality), we will assume further that $\psi(t) = 0$.

Theorem 5.4 *Let the cumulative failure rate* $\Lambda(t,z)$ *be given by Eq.* (5.27) ($\psi(t) = 0$) *and let the mixing pdf* $\pi(z),\ z \in [0,\infty)$ *be defined as*

$$\pi(z) = z^{\alpha}\pi_1(z), \tag{5.32}$$

where $\alpha > -1$ *and* $\pi_1(z),\ \pi_1(0) \neq 0$ *is a function bounded in* $[0,\infty)$ *and continuous at* $z = 0$. *Assume also that* $\phi(t) \to \infty$ *as* $t \to \infty$ *and that* $A(s)$ *satisfies*

$$\int_0^{\infty} \exp\{-A(s)\}s^{\alpha}\mathrm{d}s < \infty. \tag{5.33}$$

Then

$$\lambda_m(t) \sim (\alpha + 1)\frac{\phi'(t)}{\phi(t)}, \tag{5.34}$$

where, as usual, asymptotic notation $a(t) \sim b(t)$ *as* $t \to \infty$ *means that* $\lim_{t\to\infty} a(t)/b(t) = 1$. *As we had mentioned, another possible notation for* (5.34) *is* $\lambda_m(t) = (\alpha + 1)\phi'(t)/\phi(t)(1 + o(1))$.

The *proof* of this result is cumbersome and is based on Abelian-type theorems for the corresponding asymptotic integrals. That is why the multiplicative form in $A(z\phi(t))$ is so important.

□

The specific case of this theorem for the multiplicative model (5.31) was independently considered by Steinsaltz and Wachter [56]. Assumption (5.32) just states the 'form' of the admissible mixing distribution and holds for the main lifetime distributions, such as Weibull, gamma, truncated normal, etc. However, it does not hold for a lognormal distribution, as the corresponding asymptote is proportional to $1/z$ when $z \to 0$. Assumption (5.33) is a very weak one (weaker than just having a finite expectation for a lifetime) and can be omitted in practical analysis.

A crucial feature of this result is that the asymptotic behavior of the mixture failure rate *depends only on the behavior of the mixing distribution in the neighborhood of* 0 and on the derivative of the logarithm of the scale function $\phi(t)$, i.e.,

$$(\log \phi(t))' = \phi'(t)/\phi(t).$$

When $\pi(0) \neq 0$ and $\pi(z)$ is bounded in $[0, \infty)$, the result does not depend on the mixing distribution at all, as $\alpha = 0$ in this case. Intuitively, the qualitative meaning is quite clear: as $t \to \infty$, only the most robust survivors are left and in, accordance with (5.27), this corresponds to the small values of z (weak populations are dying out first).

It is easy to see that for the multiplicative model (5.29), Eq. (5.34) reduces to

$$\lambda_m(t) \sim \frac{(\alpha + 1)\lambda(t)}{\int_0^t \lambda(u)\mathrm{d}u}. \tag{5.35}$$

and to

$$\lambda_m(t) \sim \frac{\alpha + 1}{t} \tag{5.36}$$

for the ALM (5.30), (5.31).

It should be noted that (5.36) is a really surprising result, as the shape of the mixture failure rate for large t does not depend on the baseline distribution $F(t)$. It is also dramatically different from the multiplicative case (5.35). This means that the 'nature' of the ALM is such that it ignores' the baseline distribution for large t.

Comparing (5.35) and (5.36), we see that the latter never results in the asymptotically flat observed failure rate (the mortality plateau in human mortality studies), whereas the multiplicative model can have this possibility, as in the case of the gamma-frailty model for the Gompertz distribution (see Eq. 5.21).

Note that, by direct integration, Eq. (5.21) can be generalized to the case of an arbitrary (absolutely continuous) baseline distribution characterized by the failure rate $\lambda(t)$:

$$\lambda_m(t) = \frac{c\lambda(t)}{\beta + \Lambda(t)} = \frac{c\lambda(t)}{\beta + \int_0^t \lambda(u)\mathrm{d}u}. \tag{5.37}$$

It is clear that $c = \alpha + 1$ for the gamma pdf and this formula perfectly comply with the general asymptotic result (5.34) and a classical result by Vaupel et al. [63].

Let, for instance, $\pi(z)$ be the uniform density in $[0, 1]$ and let also $\lambda(t) = \exp\{t\}(a, b = 1$ for simplicity of notation). Then $\lambda(t, z) = z\exp\{t\}$ and

$$\int_0^\infty \bar{F}(t,z)\pi(z)\mathrm{d}z = \frac{1}{\omega}(1 - \exp\{-\omega\}),$$

$$\int_0^\infty f(t,z)\pi(z)\mathrm{d}z = (\omega + 1)\left[-\frac{\exp\{-\omega\}}{\omega} + \frac{1}{\omega^2}(1 - \exp\{-\omega\})\right],$$

where $\omega = \exp\{t\} - 1$ and $\omega \to \infty$ as $t \to \infty$. Therefore, in accordance with Eq. (5.11),

$$\lim_{t\to\infty} \lambda_m(t) = 1.$$

The same limit holds for $\lambda_m(t)$ in (5.37) for the considered specific values of parameters. This example illustrates the fact that the asymptotic value of the mixture failure rate does not depend on a mixing distribution if $\pi(0) \neq 0$.

Theorem 5.4 deals with the case when the support of a mixing distribution includes 0, i.e., $z \in [0, \infty)$. In this case, the strongest population cannot usually be properly defined. If, however, the support is separated from 0, the mixture failure rate can tend to the failure rate of the strongest population as $t \to \infty$. The following theorem [33] states reasonable conditions for this convergence (we assume, for simplicity, as previously, that $\psi(t) = 0$):

Theorem 5.5 *Let, as in Theorem 5.4, the class c by Eq. (5.27), where $\phi(t) \to \infty$, $\psi(t) = 0$ and let $A(s)$ be twice differentiable.Assume that, as $s \to \infty$*

$$\frac{A''(s)}{(A'(s))^2} \to 0 \tag{5.38}$$

and

$$sA'(s) \to \infty. \tag{5.39}$$

Also assume that for all $b, c > a,\ b < c$, the quotient $A'(bs)/A'(cs)$ is bounded as $s \to \infty$. Finally, let the mixing pdf $\pi(z)$ be defined in $[a, \infty),\ a > 0$, bounded in this interval and continuous at $z = a$ and $\pi(a) \neq 0$. Then

$$\lambda_m(t) \sim a\phi'(t)A'(a\phi(t)). \tag{5.40}$$

The assumptions of this theorem are rather natural and hold at least for the specific models under consideration and for the main lifetime distributions. Assume additionally that the family of failure rates $\lambda(t, z)$ is ordered in z (as for additive or multiplicative models), i.e.,

$$\lambda(t, z_1) < \lambda(t, z_2), \quad z_1 < z_2, \forall z_1, z_2 \in [a, \infty],\ a > 0. \tag{5.41}$$

The right-hand side of (5.40) can be interpreted in this case as the failure rate of the strongest population. Specifically, for the multiplicative model:

$$\lambda_m(t) \sim a\lambda(t). \tag{5.42}$$

Thus, as intuition suggests, the mixture failure rate asymptotically does not depend on a mixing distribution. A similar result holds also for the case when there is a singularity in the pdf of the mixing distribution of the form:

$$\pi(z) = (z - a)^{\alpha}\pi_1(z - a), \tag{5.43}$$

where $\alpha > -1$ and $\pi_1(z - a)$ is bounded, $\pi_1(0) \neq 0$.

Missov and Finkelstein [49] have generalized these results to the wider class of mixing distributions. It turned out that the mixing pdf (5.32) in Theorem 5.4 can be of a more general form

$$\pi(z) = z^{\alpha}G(z)\pi_1(z),$$

where $G(z)$ is a regularly varying function. Recall (Bingham et al. [11]) that a positive function $G(t)$ defined on $(0.\infty)$ is slowly varying at 0 if for every $k > 0$,

$$\lim_{t \to 0} \frac{G(kt)}{G(t)} = 1.$$

Moreover, a positive function $R(t)$ defined on $(0.\infty)$ is regularly varying at 0 with power $-\infty < p < \infty$, if

$$\lim_{t \to 0} \frac{R(t)}{t^p G(t)} = 1,$$

where the function $G(t)$ is slowly varying at 0.

5.5 Mortality Plateaus

As it was already mentioned, demographers had recently observed the deceleration in human mortality at advanced ages which eventually results in human mortality plateaus [58]. The most reasonable explanation of this fact is via the concept of heterogeneity of human population which obviously takes place. The following refers to the interpretation of our results for this application.

- As follows from Eq. (5.36), the ALM (5.31) never results in the asymptotically flat failure rate. Moreover, it asymptotically tends to 0 and does not depend on a baseline distribution, which is Gompertz for the case under consideration
- The only function $g(t)$, for which $g(t)/\int_o^t g(u)\mathrm{d}u$ tends to a constant as $t \to \infty$, is the exponential function. Therefore, as follows from Relationship (5.35), the asymptotically flat rate in the multiplicative model (5.29) can result via mixing of a random lifetime distributed *only* in accordance with the Gompertz distribution or in accordance with a distribution with the failure rate that asymptotically converges to an exponential function.
- In accordance with Theorem 5.4, the admissible mixing distributions (i.e., the distributions that can lead to the asymptotically flat mortality rate) are those with behavior as $z^{\alpha}, \alpha > -1$ for $z \to 0$. The behavior outside the neighborhood of 0 does not contribute to asymptotic properties of the failure rate. Therefore, the power law (Weibull distribution), the gamma distribution, and some other distributions are admissible. Note that, when the mixing pdf is such that $\pi(0) \neq 0$ has a finite limit when $z \to 0$ (as, e.g., for the exponential distribution), relationship (5.35) reduces to

$$\lambda_m(t) \sim \frac{\lambda(t)}{\int_0^t \lambda(u)\mathrm{d}u}$$

- And, therefore, the mixture mortality rate does not depend on the mixing distribution at all! The same result holds for, e.g., the mixing density that is $1/a,\ a > 0$ in $[0, a]$ and is 0 in (a, ∞) (uniform distribution).

In view of the foregoing discussion, the asymptotically flat rate (as for human populations) can be viewed as an indication of:

- that the mixing model is multiplicative,
- that the underlying distribution is definitely Gompertz or asymptotically converges to the Gompertz distribution,
- that the mixing pdf is proportional to $z^{\alpha}, z > -1$, when $z \to 0$, e.g., the gamma distribution. The form of this distribution outside neighborhood of 0 has no influence on the asymptotic behavior of $\lambda(t)$.

5.6 Inverse Problem

There can be different approaches to considering the inverse problem in mixing. In view of the results of Sect. 5.4, one can be interested in defining the class of mixing distributions that 'produce' the mixture failure rate of the form given by (5.34). The following theorem [49] solves this problem.

Theorem 5.6 *Let conditions of Theorem 5.4 hold and, therefore, Relation (5.34) takes place.Then the pdf $\pi(z)$ of the mixing (frailty) distribution satisfies for $z \to 0$*

$$\frac{\int_0^\infty \exp\{-A(z\phi(t))\}z\pi'(z)\mathrm{d}z}{\int_0^\infty \exp\{-A(z\phi(t))\}\pi(z)\mathrm{d}z} \sim \alpha. \tag{5.44}$$

Condition (5.44) is not easy to check. However, the following theorem [49] gives a simple sufficient condition.

Theorem 5.7 *Let $\pi(z)$ be a regularly varying function defined by $\pi(z) = z^\alpha G(z)$, where $\alpha > -1$ and $\pi'(z)$ be asymptotically monotone as $z \to 0$. Then Relationship (5.44) holds.*

A well-known fact from survival analysis states that the failure data alone do not uniquely define a mixing distribution and additional information (e.g., on covariates) should be taken into account (a problem of nonidentifiability, as, e.g., in Tsiatis [59] and Yashin and Manton [66]). On the other hand, the following *specific* inverse problem can be solved analytically, at least for additive and multiplicative models of mixing [28]:

Given the mixture failure rate $\lambda_m(t)$ and the mixing pdf $\pi(z)$, obtain the failure rate $\lambda(t)$ of the baseline distribution.

This means that under certain assumptions any shape of the mixture failure rate can be constructed by the proper choice of the baseline failure rate. To illustrate this statement, consider the additive model (5.28):

$$\bar{F}(t,z) = \exp\{-\Lambda(t) - zt\}, \quad f(t,z) = (\lambda(t) + z)\exp\{-\Lambda(t) - zt\}. \tag{5.45}$$

Therefore, the mixture survival function in (5.10) can be written via the Laplace transform as

$$\bar{F}_m(t) = \exp\{-\Lambda(t)\int_0^\infty \exp\{-zt\}\pi(z)\mathrm{d}z = \exp\{-\Lambda(t)\}\pi^*(t), \tag{5.46}$$

where, $\pi^*(t) = E[\exp\{-zt\}]$ is the Laplace transform of the mixing pdf $\pi(z)$. Therefore, Eq. (5.15) yields

$$\lambda_m(t) = \lambda(t) + \frac{\int_0^\infty z\exp\{-zt\}\pi(z)\mathrm{d}z}{\int_0^\infty \exp\{-zt\}\pi(z)\mathrm{d}z} = \lambda(t) - \frac{\mathrm{d}}{\mathrm{d}t}\log\pi^*(t) \tag{5.47}$$

and the solution of the inverse problem for this special case is given by the following relationship:

$$\lambda(t) = \lambda_m(t) + \frac{d}{dt}\log \pi^*(t) = \lambda_m(t) - E[Z|t]. \tag{5.48}$$

If the Laplace transform of the mixing distribution can be derived explicitly, then Eq. (5.48) gives a simple analytical solution for the inverse problem. Assume, e.g., that 'we want' the mixture failure rate to be constant, i.e., $\lambda_m(t) = c$. Then the baseline failure rate is obtained as

$$\lambda(t) = c - E[Z|t].$$

The corresponding survival function for the multiplicative model (5.17) is $\exp\{-z\Lambda(t)\}$ and the mixture survival function for this specific case is

$$\bar{F}_m(t) = \int_0^\infty \exp\{-z\Lambda(t)\}\pi(z)\mathrm{d}z = \pi^*(\Lambda(t)). \tag{5.49}$$

It is obtained in terms of the Laplace transform of the mixing distribution as a function of the cumulative baseline failure rate $\Lambda(t)$. Therefore,

$$\lambda_m(t) = -\frac{\mathrm{d}}{\mathrm{d}t}\log \pi^*(\Lambda(t)). \tag{5.50}$$

The general solution to the inverse problem in terms of the Laplace transform is also simple in this case. Note that,

$$\pi^*(\Lambda(t)) = \exp\{-\Lambda_m(t)\}, \tag{5.51}$$

where $\Lambda_m(t)$ denotes the cumulative mixture failure rate. Applying the inverse Laplace transform $L^{-1}(\cdot)$ to both sides of this equation finally results in

$$\lambda(t) = \Lambda'(t) = \frac{\mathrm{d}}{\mathrm{d}t}L^{-1}(\exp\{-\Lambda_m(t)\}). \tag{5.52}$$

The Laplace transform methodology in multiplicative and additive models is usually very effective. It constitutes a convenient tool for dealing with mixture failure rates when the Laplace transform of the mixing distribution can be obtained explicitly. The exponential family [41] presents a wide class of such distributions. The corresponding pdf is defined in this case as

$$\pi(z) = \frac{\exp\{-\theta z\}g(z)}{\eta(\theta)}, \tag{5.53}$$

where $g(z)$ and $\eta(z)$ are some positive functions and θ is a parameter. The function $\eta(\theta)$ plays the role of a normalizing constant ensuring that the pdf integrates to 1. The gamma, the inverse Gaussian, and the stable distributions are relevant examples. Note that, the Laplace transform of $\pi(z)$ depends only on the normalizing function $\eta(z)$ [41], i.e.,

$$\pi^*(s) \equiv \int_0^\infty \exp\{-sz\}\pi(z)\mathrm{d}z = \frac{\eta(\theta+s)}{\eta(\theta)}.$$

This means that under certain assumptions any shape of the mixture failure rate can be constructed by the proper choice of the baseline failure rate. Specifically, for the exponential family of mixing densities and the multiplicative model under consideration, the mixture failure rate is obtained as

$$\begin{aligned}\lambda_m(t) &= -\frac{\mathrm{d}}{\mathrm{d}t}\log\frac{\eta(\theta+\Lambda(t))}{\eta(\theta)} \\ &= -\lambda(t)\frac{\frac{d}{d(\theta+\Lambda(t))}\eta(\theta+\Lambda(t))}{\eta(\theta+\Lambda(t))}.\end{aligned} \tag{5.54}$$

Therefore, the solution to the inverse problem can be obtained in this case as the derivative of the following function:

$$\Lambda(t) = \eta^{-1}(\exp\{-\lambda_m(t)\}\eta(\theta)) - \theta. \tag{5.55}$$

It can be easily calculated [28] that when the mixing pdf is gamma with parameters α and β, the solution of the inverse problem is obtained as

$$\lambda(t) = \frac{\beta}{\alpha}\lambda_m(t)\exp\left\{\frac{\Lambda_m(t)}{\alpha}\right\}. \tag{5.56}$$

Assume that the mixture failure rate is constant, i.e., $\lambda_m(t) = c$. It follows from (5.56) that for obtaining a constant $\lambda_m(t)$ the baseline $\lambda(t)$ should be exponentially increasing, i.e.,

$$\lambda(t) = \frac{\beta}{\alpha}c\exp\left\{\frac{ct)}{\alpha}\right\}.$$

But this is what we would really expect. As we already mentioned, this result is really surprising: we are mixing the exponentially increasing family of failure rates and arriving at a constant mixture failure rate.

5.7 The Failure Rate Dynamics in Heterogeneous Populations

The mixture failure rate function and some other measures based on it (e.g., the reliability function, the mean residual life function, etc.) are conventionally considered as *measures of performance* (or quality) of items in heterogeneous populations. However, if we pick an operable item at random from this population, its individual failure rate at each instant of time can be considered as a random variable, whereas the mixture failure rate is defined as its expectation. As in the case of 'ordinary' random variables, other than expectation characteristics are also important. The obvious first choice is the corresponding variance.

As an example, consider a system that should perform an important mission. The quality of its performance can be described by the probability of operation without failures during a mission time. If a mission is important and its failure results, e.g., in substantial economic loss, then not only the population (mixture) failure rate of a system that defines the average value of this probability, but the deviations from this value due to heterogeneity of a population are of considerable interest. As the weakest items are dying out first, the composition of the ordered heterogeneous population is improving in the sense that proportions of stronger items are increasing. However, does it mean that the 'quality' (from a broader perspective) of the entire population is improving? Not necessarily, as this quality can depend also on the variability characteristics to be discussed in this section. Furthermore, when we are dealing with failures that may result in serious consequences, more attention should be paid to the items with a high risk of failure, i.e., the items with large failure rates. Therefore, the measures for quality of these items should be also defined.

We consider a heterogeneous population of items (components) that consists of different homogeneous subpopulations, that are modeled via the frailty Z. The numbers of items in populations are supposed to be sufficiently large and thus our problems can be statistically described in terms of infinite populations. As time progresses, the failed items are discarded and therefore, the composition of the population of survived items (which is, in fact, the conditional frailty $Z|T > t$) changes. Alternatively, an item is chosen at random from our heterogeneous population and if it did not fail in $[0, t)$, then our initial knowledge about its 'quality' which is described by the frailty Z is changing in accordance with $Z|T > t$ (see Eq. (5.12) and the discussion after it).

For illustrating the dynamics in variability characteristics, consider the case of $n = 2$ subpopulations that can be generalized to the arbitrary finite n. Denote the lifetime of a component from the strong subpopulation by T_S and its absolutely continuous Cdf, pdf, and the failure rate function by $F_1(t)$, $f_1(t)$ and $\lambda_1(t)$, respectively. Similarly, the lifetime, the Cdf, the pdf, and the failure rate function of a weak component are T_W, $F_2(t)$, $f_2(t)$ and $\lambda_2(t)$, accordingly. Formal definitions of the strong and weak subpopulations will be given after presenting the necessary notation. The initial ($t = 0$) composition of our mixed population is as follows: the proportion of strong items is π, whereas the proportion of weak items is $1 - \pi$, which means that the distribution of the discrete frailty Z with realizations z_1 and z_2 in this case is

$$\pi(z) = \begin{cases} \pi, & z = z_1 \\ 1 - \pi, & z = z_2 \end{cases}$$

and z_1, z_2 ($z_1 < z_2$), correspond to the strong and the weak subpopulations, respectively. In accordance with Eqs. (5.1)–(5.3):

The mixture (population) survival function is

$$\bar{F}_m(t) = \pi\bar{F}_1(t) + (1-\pi)\bar{F}_2(t).$$

The mixture (observed) failure rate is

$$\lambda_m(t) = \frac{\pi f_1(t) + (1-\pi)f_2(t)}{\pi\bar{F}_1(t) + (1-\pi)\bar{F}_2(t)} = \pi(t)\lambda_1(t) + (1-\pi(t))\lambda_2(t), \tag{5.57}$$

where the time-dependent probabilities are

$$\pi(t) = \frac{\pi\bar{F}_1(t)}{\pi\bar{F}_1(t) + (1-\pi)\bar{F}_2(t)}, \quad 1-\pi(t) = \frac{(1-\pi)\bar{F}_2(t)}{\pi\bar{F}_1(t) + (1-\pi)\bar{F}_2(t)}.$$

Thus, the composition of our population is changing in time in accordance with the following distribution of $Z|t \equiv Z|T > t$:

$$\pi(z|t) = \begin{cases} \pi(t), & z = z_1 \\ 1-\pi(t), & z = z_2 \end{cases}.$$

Assume now that the populations are ordered (and therefore, the weak and the strong subpopulations are defined accordingly) in the sense of the failure rate ordering:

$$\lambda_2(t) \geq \lambda_1(t), \quad t \geq 0.$$

Then, it is easy to see that the proportion of strong items

$$\pi(t) = \frac{\pi}{\pi + (1-\pi)\bar{F}_2(t)/\bar{F}_1(t)},$$

is increasing as t is increasing. In the context of burn-in, e.g., it means that the quality of a population in the defined sense is improving as the time of burn-in is increasing.

Equation (5.57) defines the observed (mixture) failure rate, which is obviously an averaged characteristic. However, the above mixture setting implies that an operable item at time t can be described by a *random failure rate* $\lambda_R(t)$ with realizations $\lambda_1(t)$ and $\lambda_2(t)$:

$$\lambda_R(t) = \begin{cases} \lambda_1(t), & \text{with probability } \pi(t), \\ \lambda_2(t), & \text{with probability } 1-\pi(t). \end{cases} \tag{5.58}$$

Thus, we can also interpret (5.57) as the expectation of the random failure rate $\lambda_R(t)$

$$\lambda_m(t) = E[\lambda_R(t)].$$

Expectation is obviously an important characteristic, but, as in the case of 'ordinary random variables' we might be interested in moments and, first of all, in $\mathrm{Var}[\lambda_R(t)]$ as the variability measure of the population structure. This measure is important as we want to know (or control) the 'risks' (i.e., large deviations from the mean) that can occur in field usage. Therefore, $\lambda_m(t)$ and $\mathrm{Var}[\lambda_R(t)]$ can describe the quality of our heterogeneous population. It is reasonable to assume that the larger these characteristics are, the worse is the corresponding quality. Furthermore, at many instances, along with the absolute variability measure $\mathrm{Var}[\lambda_R(t)]$, the relative variability is of interest. Thus, in addition to $\mathrm{Var}[\lambda_R(t)]$, we will consider the measure for the 'relative deviation', i.e., the corresponding coefficient of variation:

$$CV[\lambda_R(t)] = \sqrt{\mathrm{Var}[\lambda_R(t)]}/E[\lambda_R(t)] = \sqrt{\mathrm{Var}[\lambda_R(t)]}/\lambda_m(t).$$

We will derive now general formulas for the measures of interest. In order to obtain $\mathrm{Var}[\lambda_R(t)]$, in accordance with (5.58), it is easier to consider the supplementary random variable $\lambda_{RC}(t)$, which is equal to $\lambda_1(t) - \lambda_2(t)$ with probability $\pi(t)$ and to 0 with probability $1 - \pi(t)$. Then

$$\mathrm{Var}[\lambda_R(t)] = \mathrm{Var}[\lambda_{RC}(t)] = (\lambda_1(t) - \lambda_2(t))^2 \pi(t)(1 - \pi(t)), \tag{5.59}$$

and

$$CV[\lambda_R(t)] = \sqrt{\mathrm{Var}[\lambda_R(t)]}/\lambda_m(t) = \frac{(\lambda_2(t) - \lambda_1(t))\sqrt{\pi(t)(1 - \pi(t))}}{\pi(t)\lambda_1(t) + (1 - \pi(t))\lambda_2(t)}. \tag{5.60}$$

As we know, the shape of the mixture failure rate is very important in describing heterogeneous populations. In accordance with the foregoing considerations, the shape of the functions $\mathrm{Var}[\lambda_R(t)]$ and $CV[\lambda_R(t)]$ is also of interest. For simplicity, we consider first the mixture of two exponential distributions. Let $\lambda_2(t) = \lambda_2 > \lambda_1(t) = \lambda_1$. Then, as a special case of Eq. (5.59),

$$\mathrm{Var}[\lambda_R(t)] = (\lambda_1 - \lambda_2)^2 \pi(t)(1 - \pi(t)),$$

and

$$\lambda_m'(t) = -(\lambda_1 - \lambda_2)^2 \pi(t)(1 - \pi(t)) = -\mathrm{Var}[\lambda_R(t)]. \tag{5.61}$$

Thus, the slope of the mixture failure rate in this case is equal to the variance of the random failure rate (with the negative sign). We can consider the following two cases:

(i) Let the initial proportion of strong components be larger than 0.5 ($\pi > 0.5$); then $\pi(t)(1 - \pi(t))$ strictly decreases in t from $\pi(0)(1 - \pi(0))$. Therefore, $\lambda_m(t)$ and $\mathrm{Var}[\lambda_R(t)]$ strictly decrease and, therefore, the population becomes 'better' (the failure rate is smaller) and more 'stable' (the variance is smaller). Observe that

$$CV'[\lambda_R(t)] = \frac{1}{2\sqrt{\pi(t)(1-\pi(t))}(\lambda_1\pi(t)+\lambda_2(1-\pi(t)))^2}$$
$$\times\ [(\lambda_2-\lambda_1)\pi'(t)\{1-2\pi(t)\}(\lambda_1\pi(t)+\lambda_2(1-\pi(t)))+2(\lambda_2-\lambda_1)^2\pi'(t)\pi(t)(1-\pi(t))]$$
$$= \frac{1}{2\sqrt{\pi(t)(1-\pi(t))}(\lambda_1\pi(t)+\lambda_2(1-\pi(t)))^2}(\lambda_2-\lambda_1)\pi'(t)\{\lambda_2(1-\pi(t))-\lambda_1\pi(t)\}.$$

Therefore, as $\pi'(t)$ is positive ($\pi(t)$ is increasing):

$$CV'[\lambda_R(t)] > 0 \Rightarrow \frac{\lambda_2}{\lambda_1} > \frac{\pi(t)}{1-\pi(t)}.$$

Obviously, $\pi(t)/(1-\pi(t))$ strictly increases to ∞ as t increases. Thus, when

$$\frac{\lambda_2}{\lambda_1} > \frac{\pi(0)}{1-\pi(0)}, \tag{5.62}$$

$CV[\lambda_R(t)]$ increases and then decreases with one change point t^* such that $\lambda_2/\lambda_1 = \pi(t^*)/(1-\pi(t^*))$. When

$$\frac{\lambda_2}{\lambda_1} < \frac{\pi(0)}{1-\pi(0)},$$

then $CV[\lambda_R(t)]$ monotonically decreases.

(ii) Let the initial proportion of strong components be smaller or equal to 0.5 ($\pi \le 0.5$). As it was stated, the proportion of remaining weak components $1-\pi(t)$ is always decreasing in time. Therefore, the first guess based on intuition would be that $\mathrm{Var}[\lambda_R(t)]$ (similar to (i)) is also decreasing. However, it is easy to see that at time t such that $\pi(t) = 0.5$, the function, $\mathrm{Var}[\lambda_R(t)]$ (and as follows from (5.61), $|\lambda'_m(t)|$ as well) has its maximum and only after this point it strictly decreases. In this case, Inequality (5.62) always holds and thus $CV[\lambda_R(t)]$ increases and then decreases with one change point t^* such that $\lambda_2/\lambda_1 = \pi(t^*)/(1-\pi(t^*))$.

Equation (5.59) can be used for analyzing the shape of $\mathrm{Var}[\lambda_R(t)]$ for time-dependent failure rates. Specifically, when $\lambda_2(t)-\lambda_1(t)$ is increasing and $\pi \le 0.5$, then $\pi(t)(1-\pi(t))$ first strictly increases and then decreases. Therefore, $\mathrm{Var}[\lambda_R(t)]$ *initially* strictly increases.

When $\lambda_1(t)-\lambda_2(t)$ is decreasing:

(i) If $\pi > 0.5$, then $\pi(t)(1-\pi(t))$ strictly decreases and $\mathrm{Var}[\lambda_R(t)]$ strictly decreases.
(ii) If $\pi \le 0.5$, then, $\pi(t)(1-\pi(t))$ strictly increases in $[0,t^*)$ and decreases in $[t^*,\infty)$, where t^* is the solution of the following equation: $\pi(t) = 0.5$. Thus $\mathrm{Var}[\lambda_R(t)]$ strictly decreases in $[t^*,\infty)$.

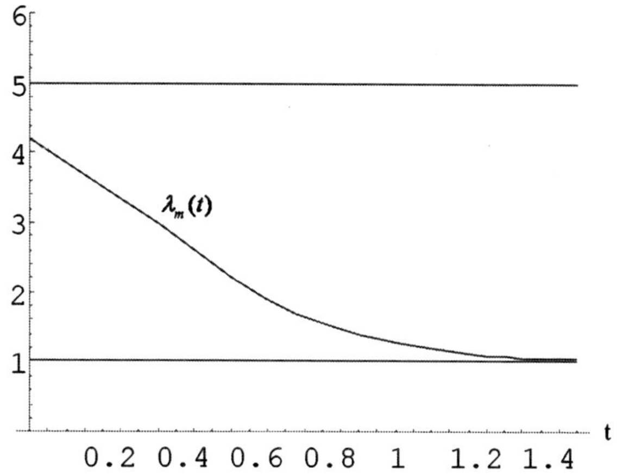

Fig. 5.3 Mixture Failure Rate $\lambda_m(t)$

Equation (5.60) can be used for analyzing the shape of $CV[\lambda_R(t)]$. For instance, if $\lambda_2(t) - \lambda_1(t)$ is decreasing and $\lambda_m(t)$ is increasing, then $CV[\lambda_R(t)]$ is strictly decreasing or it initially increases and then monotonically decreases.

Example 5.1 Let $\lambda_1(t) = 1$, $\lambda_2(t) = 5$ and $\pi = 0.2$. Then the mixture failure rate $\lambda_m(t)$ is given by Fig. 5.3.

Assume that an item has survived to age 0.4. As follows from the graph: $\lambda_m(0.4) \approx 3.0$. How much can we rely on this value? To answer this question, it is reasonable to consider Var$[\lambda_R(t)]$ given by Fig. 5.4.

We can see that Var$[\lambda_R(t)]$ has a maximum at $t \approx 0.4$ ($\pi(0.4) \approx 0.5$). This means that at $t = 0.4$, approximately 50 % of survived items have the failure rate with realization 5.0, and the other 50 % will have it 1.0, whereas the observed (mixture) failure rate $\lambda_m(t)$ is 3.0. However, as t increases from 0.4, we may more and more 'rely' on $\lambda_m(t)$ as variability decreases.

The above example is rather interesting: We may think that the population would become more and more 'stable' (monotonically) as $\lambda_m(t)$ (monotonically)

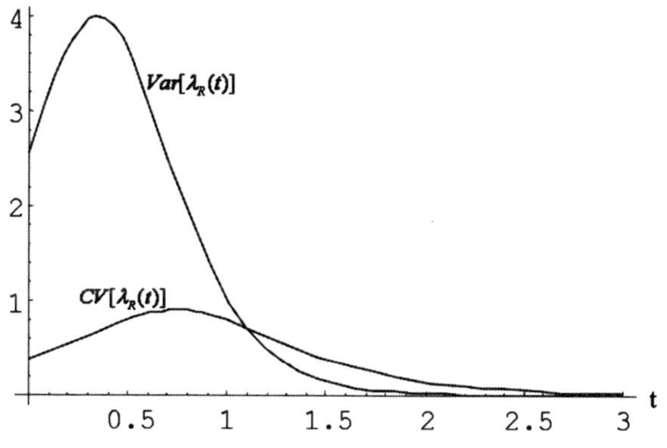

Fig. 5.4 $Var[\lambda_R(t)]$ and $CV[\lambda_R(t)]$

approaches the failure rate of the strongest subpopulation. However, it is not true, as the variance is not monotonic. The similar conclusion follows when considering $CV[\lambda_R(t)]$ (Fig. 5.4).

Similar consideration s can be applied to continuous mixtures defined by Eqs. (5.10)–(5.12). Let our subpopulations be ordered in the sense of the failure rate ordering:

$$\lambda(t, z_1) \leq \lambda(t, z_2), \quad z_1 < z_2, \forall z_1, z_2 \in [0, \infty), \ t \geq 0. \tag{5.63}$$

Denote the Cdfs of $\pi(z)$ and $\pi(z|t)$ by $\Pi(z)$ and $\Pi(z|t)$, respectively, and by $Z|t$ the conditional frailty (on condition that the item did not fail in $[0, t)$). The following simple result describes the important property of the family $\{Z|t\}_{t \geq 0}$.

Theorem 5.8 *Let our subpopulations be ordered in the sense of the failure rate ordering (5.64). Then the family of random variables $Z|t \equiv Z|T > t$ is DLR (decreasing in the sense of the likelihood ratio) in $t \in [0, \infty)$.*

Proof Recall that a random variable X (with the pdf $f(t)$) is smaller than a random variable Y (with the pdf $g(t)$) in the sense of the likelihood ratio ordering (LRO) if $f(t)/g(t)$ is decreasing in t (see also (2.71)).Therefore, the DLR property of the family $\{Z|t\}_{t \geq 0}$ means that for all $t_2 > t_1$, $Z|t_2$ is smaller than $Z|t_1$ in the sense of the LRO.

In accordance with the definition of the conditional mixing distribution (5.12) in the mixing model (5.11), the ratio of the corresponding densities for different instants of time is

$$L(z, t_1, t_2) = \frac{\pi(z|t_2)}{\pi(z|t_1)} = \frac{\bar{F}(t_2, z) \int_0^\infty \bar{F}(t_1, z)\pi(z)\mathrm{d}z}{\bar{F}(t_1, z) \int_0^\infty \bar{F}(t_2, z)\pi(z)\mathrm{d}z}.$$

Therefore, monotonicity in z of $L(z, t_1, t_2)$ is defined by the function

$$\frac{\bar{F}(t_2, z)}{\bar{F}(t_1, z)} = \exp\left\{ -\int_{t_1}^{t_2} \lambda(u, z)\mathrm{d}u \right\},$$

which, owing to ordering (5.63), is decreasing in z for all $t_2 > t_1$.

□

As the LRO ordering is stronger than the usual stochastic ordering, it means that $\Pi(z|t)$ is increasing in t for each $z > 0$. Therefore, in accordance with (5.63), the proportion of 'better' (with smaller failure rates) items is increasing.

For tractability, consider now the important specific case of the multiplicative model: $\lambda(t, z) = z\lambda(t)$. Therefore,

$$\lambda_R(t) = Z_t\lambda(t),$$

where $Z_t = Z|t$ and

$$\lambda_m(t) = E[\lambda_R(t)] = \lambda(t)\int_0^\infty z\pi(z|t)\mathrm{d}z = \lambda(t)E[Z|t].$$

Observe that

$$\mathrm{Var}[\lambda_R(t)] = (\lambda(t))^2\mathrm{Var}[Z_t] = (\lambda(t))^2\mathrm{Var}[Z|t],$$

and thus,

$$CV[\lambda_R(t)] = \frac{\sqrt{\mathrm{Var}[Z|t]}}{E[Z|t]} = CV[Z|t].$$

Furthermore, as $E'[Z_t] = E'[Z|t] = -\lambda(t)\mathrm{Var}[Z|t] < 0$,

$$\lambda'_m(t) = \lambda'(t)E[Z|t] - (\lambda(t))^2\mathrm{Var}[Z|t].$$

Specifically, when the population is a mixture of exponential distributions, we have

$$\lambda'_m(t) = -(\lambda(t))^2\mathrm{Var}[Z|t].$$

Example 5.2 Consider continuous mixture of exponentials. Let the conditional failure rate and the mixing distribution be $\lambda(t,z) = z$ and $\pi(z) = \theta\exp\{-\theta z\}$, respectively. Then

$$\lambda_m(t) = E[\lambda_R(t)] = E[Z|t] = 1/(\theta + t),$$

and

$$\mathrm{Var}[\lambda_R(t)] = \mathrm{Var}[Z|t] = 1/(\theta + t)^2.$$

Thus

$$CV[\lambda_R(t)] = 1.$$

Obviously, the quality of the population is defined only by $E[Z|t]$, which is decreasing in t. Therefore, the failure rates are 'improving' and the variance as well. However, the CV is constant, and this characteristic often more adequately describes variability especially when both the failure rate and its variance are decreasing in time.

5.8 Stochastic Intensity for Minimal Repairs in Heterogeneous Populations

In Sect. 2.5, we have defined and described the crucial for the reliability of repairable systems notion of minimal repair. This was done for items from homogeneous populations. It is really a challenge to define and study minimal repair in heterogeneous populations.

Consider a system with an absolutely continuous time to failure Cdf $F(t)$ and the failure rate $\lambda(t)$, which starts operating at $t = 0$. Assume that the repair action is performed instantaneously upon failure. Recall that the repair is usually qualified as *perfect* if the Cdf of the repaired object is $F(t)$ (as good as new) and as *minimal* at time x, if its Cdf is:

$$F(t|x) \equiv 1 - \frac{1 - F(t + x)}{1 - F(x)} \tag{5.64}$$

(as bad as old), which is equivalent to Eq. (2.26). Thus the minimal repair restores our system (in terms of the corresponding distribution) to the state it had prior to the failure.

Sometimes, upon failure, we can observe additional information about the state of an object (e.g., the structure of a system). This can allow us to define a more general type of repair, which is usually called the *information-based* (*or physical*) *minimal* repair. The information-based minimal repair brings our object back to the state (to be defined by the relevant information) it had just prior to the failure [4, 5, 10, 19, 26, 27, 50].

It is really challenging to generalize the notion of minimal repair to items from *heterogeneous populations.* The corresponding attempt was performed in Finkelstein [27] and further elaborated in Cha and Finkelstein [20]. Our presentation in this section will mostly follow the latter paper.

Let failures of repairable items be repaired instantaneously. Then the process of repairs can be described by a stochastic point process. A convenient way of mathematical description of these processes is using the concept of the stochastic intensity (the intensity process) $\lambda_t, t \geq 0$ defined by Relationship (2.12). A classical example of λ_t is the intensity process generated by the renewal process (perfect, instantaneous repairs):

$$\lambda_t = \sum_{n=0}^{\infty} \lambda(t - T_n) I(T_n \leq t < T_{n+1}), \quad T_0 = 0,$$

where $T_1 < T_2 < T_3 < \ldots$, are the random failure times. Another standard example is the 'deterministic stochastic intensity' $\lambda_t = \lambda(t)$ which defines the nonhomogeneous Poisson process (NHPP) of repairs with rate (intensity) $\lambda(t)$. It is well known that this example can also be interpreted as the process of *minimal* repairs.

As in the previous sections, we formally describe heterogeneous populations in the following way. Let $T \geq 0$ be a lifetime r.v. with the Cdf $F(t)$ ($\bar{F}(t) \equiv 1 - F(t)$). Assume that $F(t)$ is indexed by a r.v. Z, i.e.,

$$P(T \leq t|Z = z) \equiv P(T \leq t|z) \equiv F(t, z)$$

and that the pdf $f(t, z)$ exists. Then the corresponding failure rate $\lambda(t, z)$ is $f(t, z)/\bar{F}(t, z)$. Let Z be a frailty with support in $[a, b]$, $0 \leq a < b \leq \infty$, and the pdf $\pi(z)$. The above setting leads naturally to considering mixtures of distributions, which are useful for describing heterogeneity [see Eqs. (5.10–5.12)].

We can now define two types (scenarios) of minimal repair for heterogeneous populations, but in a more general context than in Finkelstein [27]. The first type of minimal repair does not employ any additional information and, therefore, the failed item is replaced by the statistically identical item. As the failure time distribution in this case is just the mixture (5.10), the stochastic intensity for the corresponding *process of minimal repairs* of this type is obviously equal to the mixture failure rate, i.e.,

$$\lambda_t = \lambda_m(t), \quad t \geq 0.$$

The second type of minimal repair (already information-based) restores an item to a statistically identical item with the same value of frailty Z. It can be realized in practice by performing the second 'operation' resulting in the 'classical' minimal repair when during the repair only a small part of a large system is replaced. It is natural to suggest that the state of an item is also defined by the corresponding realization of the frailty parameter (i.e., if $Z = z$ before the failure, it should be z after the failure). Thus (5.64) is modified to:

$$F(t, z|x) \equiv 1 - \frac{1 - F(t + x, z)}{1 - F(x, z)}.$$

Our main attention here focuses on this type of minimal repair, as it is the most 'interesting' from both a practical and a theoretical points of view.

Let us come back to the definition of the intensity process (2.12) and modify it with respect to the 'heterogeneous' case when the orderly point process is indexed by the frailty parameter Z. Observe that the stochastic intensity λ_t (unconditional with respect to frailty Z) can be specified now as:

$$\begin{aligned} \lambda_t &= \lim_{\Delta t \to 0} \frac{E[\Pr[N(t, t + \Delta t) = 1|H_t, Z]]}{\Delta t} \\ &= E\left[\lim_{\Delta t \to 0} \frac{\Pr[N(t, t + \Delta t) = 1|H_{t-}, Z]}{\Delta t}\right] \\ &= E[\lambda_{t,Z}], \end{aligned} \tag{5.65}$$

where the expectation is with respect to the conditional distribution $Z|H_t$ and

$$\lambda_{t,Z} \equiv \lim_{\Delta t \to 0} \frac{\Pr[N(t, t+\Delta t) = 1 | H_t, Z]}{\Delta t}. \tag{5.66}$$

Then $\lambda_{t,z}(Z = z)$ in (5.66) can be interpreted as the conditional (with respect to Z) stochastic intensity of the orderly point process, indexed by frailty Z.

We will specify now our point process. As before, let Z be the frailty of an item randomly selected at time $t = 0$ from our heterogeneous population. Upon each failure we perform the minimal repair of the second type. Note that, in this case, if $Z = z$ at time $t = 0$, then the corresponding realization is $\lambda_{t,z} = \lambda(t, z)$ for all $t \geq 0$ Z. Therefore, for the second type of minimal repair, $\lambda_{t,Z}$ in (5.66) is now given by

$$\lambda_{t,Z} = \lambda(t, Z), \quad t \geq 0,$$

and, in accordance with (5.65), the corresponding stochastic intensity λ_t is the expectation of $\lambda(t, Z)$ with respect to the distribution of $Z|H_t$. This operation means that, although the value of Z is chosen at $t = 0$ and is fixed, its distribution is updated with time as information about failures and survival times emerges (see the detailed procedure in what follows).

We see that stochastic modeling for the second type of minimal repair is dramatically different from that for the first type, as information about the operational history (failure times and survival times) updates the conditional frailty distribution $Z|H_t$.

In accordance with our considerations, it is clear that the stochastic intensity $\lambda_t = E[\lambda_{t,Z}]$ defined in (5.65) for $t \in [0, t_1)$, where t_1 is the realization of the failure time T_1, is just the mixture failure rate (5.12), i.e., $\lambda_m^1(t) = \lambda_m(t)$, as the information at hand is just the initial distribution $\pi(z)$ (and the fact that the item has survived in $[0, t)$).

Consider now the next interval $[t_1, t_2)$. Given the additional information (in addition to the initial distribution $\pi(z)$) that an item has failed at $t = t_1$, the pdf of frailty $Z = z$ (we repair an item to the state, defined by the same value of frailty) is

$$\pi_{02}(z) \equiv \frac{\lambda(t_1, z) \exp\{-\int_0^{t_1} \lambda(s, z) ds\} \cdot \pi(z)}{\int_a^b \lambda(t_1, z) \exp\{-\int_0^{t_1} \lambda(s, z) ds\} \cdot \pi(z) dz}. \tag{5.67}$$

Thus the 'initial frailty distribution' (at the start of the second cycle) just after the minimal repair is given by (5.67). Furthermore, the 'remaining survival function' at time $t = t_1$ is given by $[\bar{F}(t_1 + u, z)/\bar{F}(t_1, z)]$. Then, the conditional frailty distribution $Z|H_t$ in $[t_1, t_2)$ is

$$\frac{[\bar{F}(t, z)/\bar{F}(t_1, z)] \cdot \pi_{02}(z)}{\int_a^b [\bar{F}(t, z)/\bar{F}(t_1, z)] \cdot \pi_{02}(z) dz} = \frac{\lambda(t_1, z) \exp\{-\int_0^t \lambda(s, z) ds\} \cdot \pi(z)}{\int_a^b \lambda(t_1, z) \exp\{-\int_0^t \lambda(s, z) ds\} \cdot \pi(z) dz},$$

and the corresponding stochastic intensity is, in accordance with (5.65),

$$\lambda_m^2(t) = \int_a^b \lambda(t,z) \cdot \frac{\lambda(t_1,z)\exp\{-\int_0^t \lambda(s,z)\mathrm{d}s\} \cdot \pi(z)}{\int_a^b \lambda(t_1,z)\exp\{-\int_0^t \lambda(s,z)\mathrm{d}s\} \cdot \pi(z)\mathrm{d}z}\mathrm{d}z, \text{ in } [t_1,t_2). \quad (5.68)$$

Using another useful (Bayesian) interpretation, we can say that the item fails at time t_1 and, after repair, survives in $[t_1, t]$. Thus, the corresponding probability (conditional probability given $Z = z$ at $t = 0$) is

$$\lambda(t_1,z)\exp\{-\int_0^{t_1} \lambda(s,z)\mathrm{d}s\} \cdot \exp\left\{-\int_{t_1}^{t} \lambda(s,z)\mathrm{d}s\right\}dt_1$$
$$= \lambda(t_1,z)\exp\left\{-\int_0^{t} \lambda(s,z)\mathrm{d}s\right\}dt_1.$$

Given this information, the conditional frailty distribution $Z|H_t$ should be updated as

$$\frac{\lambda(t_1,z)\exp\{-\int_0^t \lambda(s,z)\mathrm{d}s\} \cdot \pi(z)}{\int_a^b \lambda(t_1,z)\exp\{-\int_0^t \lambda(s,z)\mathrm{d}s\} \cdot \pi(z)\mathrm{d}z},$$

which yields (5.68).

Consider now the intensity process in $[t_2, t_3)$. As we know that the item has failed at times t_1 and t_2 and after minimal repairs has survived to t_1-t_2, the corresponding probability (conditional probability given $Z = z$ at $t = 0$, divided by $dt_1 dt_2$) is

$$\lambda(t_1,z)\exp\left\{-\int_0^{t_1} \lambda(s,z)\mathrm{d}s\right\} \cdot \lambda(t_2,z)\exp\left\{-\int_{t_1}^{t_2} \lambda(s,z)\mathrm{d}s\right\} \cdot \exp\left\{-\int_{t_2}^{t} \lambda(s,z)\mathrm{d}s\right\}$$
$$= \lambda(t_1,z)\lambda(t_2,z)\exp\left\{-\int_0^{t} \lambda(s,z)\mathrm{d}s\right\}.$$

Given this information, the conditional frailty distribution $Z|H_t$ should be updated as

$$\frac{\lambda(t_1,z)\lambda(t_2,z)\exp\{-\int_0^t \lambda(s,z)\mathrm{d}s\} \cdot \pi(z)}{\int_a^b \lambda(t_1,z)\lambda(t_2,z)\exp\{-\int_0^t \lambda(s,z)\mathrm{d}s\} \cdot \pi(z)\mathrm{d}z}.$$

Thus, in $[t_2, t_3)$, as before,

$$\lambda_m^3(t) = \int_a^b \lambda(t,z) \cdot \frac{\lambda(t_1,z)\lambda(t_2,z)\exp\{-\int_0^t \lambda(s,z)\mathrm{d}s\} \cdot \pi(z)}{\int_a^b \lambda(t_1,z)\lambda(t_2,z)\exp\{-\int_0^t \lambda(s,z\mathrm{d}s\} \cdot \pi(z)dz}\mathrm{d}z, \text{ in } [t_2,t_3).$$

More generally, for $t \in [t_{n-1}, t_n)$, the conditional frailty distribution $Z|H_t$ is defined by

$$\pi^n(z|t_1,\ldots,t_{n-1}) \equiv \frac{\lambda(t_1,z)\cdots\lambda(t_{n-1},z)\exp\left\{-\int_0^t \lambda(s,z)\mathrm{d}s\right\}\cdot\pi(z)}{\int_a^b \lambda(t_1,z)\cdots\lambda(t_{n-1},z)\exp\left\{-\int_0^t \lambda(s,z)\mathrm{d}s\right\}\cdot\pi(z)\mathrm{d}z} \tag{5.69}$$

and, therefore,

$$\lambda_m^n(t) = \int_a^b \lambda(t,z)\pi^n(z|t_1,\ldots,t_{n-1})dz \text{ in } [t_{n-1}, t_n). \tag{5.70}$$

Based on (5.69) and (5.70), the corresponding stochastic intensity can now be defined as

$$\lambda_t = \sum_{n=1}^{\infty} \lambda_m^n(t) I(T_{n-1} \le t < T_n),\ T_0 \equiv 0. \tag{5.71}$$

The following result presents a useful ordering of stochastic intensities for minimal repairs of the first and the second types (Cha and Finkelstein [20]).

Theorem 5.9 *Let the values of $\lambda(t,z)$ be ordered with respect to z: for all $z_1, z_2 \in [a,b]$, $t \ge 0$*

$$\lambda(t,z_1) < \lambda(t,z_2),\ \text{if } z_1 < z_2.$$

Then

$$\lambda_m(t) \le \lambda_t,\ t \ge 0,$$

where λ_t is the stochastic intensity for the second type of minimal repair in (5.71).

Proof Note that if $X \le_{st} Y$ and $g(\cdot)$ is any increasing function, then $g(X) \le_{st} g(Y)$ and, accordingly, $E[g(X)] \le E[g(Y)]$. Observe that both $\lambda_m(t)$ and λ_t are expectations of $\lambda(t,Z)$ with respect to the mixing distributions

$$\pi(z|t) = \pi(z)\frac{\overline{F}(t,z)}{\int_a^b \overline{F}(t,z)\pi(z)\mathrm{d}z}$$

and

$$\pi^n(z|t_1,\ldots,t_{n-1}) \equiv \frac{\lambda(t_1,z)\cdots\lambda(t_{n-1},z)\exp\left\{-\int_0^t \lambda(s,z)\mathrm{d}s\right\}\cdot\pi(z)}{\int_a^b \lambda(t_1,z)\cdots\lambda(t_{n-1},z)\exp\left\{-\int_0^t \lambda(s,z)\mathrm{d}s\right\}\cdot\pi(z)\mathrm{d}z},$$

respectively. Then it is sufficient to show that

$$\Pi(v|t) \ge \Pi^n(v|t_1,\ldots,t_{n-1}), \tag{5.72}$$

for all $n \geq 1, 0 < t_1 < \ldots < t_{n-1} < t$, where $\Pi(z|t)$ and $\Pi^n(v|t_1, \ldots, t_{n-1})$ are the corresponding Cdfs. Observe that

$$\pi^n(z|t_1, \ldots, t_{n-1}) \equiv \frac{\lambda(t_1, z) \cdots \lambda(t_{n-1}, z)\overline{F}(t, z)\pi(z)}{\int_a^b \lambda(t_1, z) \cdots \lambda(t_{n-1}, z)\overline{F}(t, z)\pi(z)\mathrm{d}z}$$
$$= \frac{\lambda(t_1, z) \cdots \lambda(t_{n-1}, z) \cdot \pi(z|t)}{\int_a^b \lambda(t_1, z) \cdots \lambda(t_{n-1}, z)\pi(z|t)\mathrm{d}z}.$$

It is clear that there exist $a \leq z^*(a, v) \leq v$ and $v \leq z^*(v, b) \leq b$ such that

$$\int_a^v \lambda(t_1, z) \cdots \lambda(t_{n-1}, z) \cdot \pi(z|t)\mathrm{d}z = \lambda(t_1, z^*(a, v)) \cdots \lambda(t_{n-1}, z^*(a, v)) \int_a^v \pi(z|t)\mathrm{d}z$$

and

$$\int_v^b \lambda(t_1, z)) \cdots \lambda(t_{n-1}, z) \cdot \pi(z|t)\mathrm{d}z = \lambda(t_1, z^*(v, b)) \cdots \lambda(t_{n-1}, z^*(v, b)) \int_v^b \pi(z|t)\mathrm{d}z.$$

Thus,

$$\Pi^n(v|t_1, \ldots, t_{n-1}) = \frac{\lambda(t_1, z^*(a, v)) \cdots \lambda(t_{n-1}, z^*(a, v)) \cdot \int_a^v \pi(z|t)\mathrm{d}z}{\lambda(t_1, z^*(a, v)) \cdots \lambda(t_{n-1}, z^*(a, v)) \cdot \int_a^v \pi(z|t)\mathrm{d}z + \lambda(t_1, z^*(v, b)) \cdots \lambda(t_{n-1}, z^*(v, b)) \cdot \int_v^b \pi(z|t)\mathrm{d}z}$$
$$\leq \int_a^v \pi(z|t)\mathrm{d}z = \Pi(v|t).$$

Since $\lambda(t_1, z) \cdots \lambda(t_{n-1}, z)$ is an increasing function of z,

$$\lambda(t_1, z^*(a, v)) \cdots \lambda(t_{n-1}, z^*(a, v)) \leq \lambda(t_1, z^*(v, b)) \cdots \lambda(t_{n-1}, z^*(v, b)),$$

and, therefore, Inequality (5.72) is justified.

□

Example 5.3 Suppose that $F(t, z)$ is an exponential distribution with parameter $\lambda(t, z) = z\lambda$ and let $\pi(z)$ be an exponential pdf in $[0, \infty)$ with parameter θ. Then direct integration in (5.11) gives: $\lambda_m(t) = \lambda/(\lambda t + \theta)$. Observe that

$$\pi^n(z|t_1, \ldots, t_{n-1}) \equiv \frac{\lambda(t_1, z) \cdots \lambda(t_{n-1}, z) \exp\{-\int_0^t \lambda(s, z)\mathrm{d}s\} \cdot \pi(z)}{\int_a^b \lambda(t_1, z) \cdots \lambda(t_{n-1}, z) \exp\{-\int_0^t \lambda(s, z)\mathrm{d}s\} \cdot \pi(z)dz}$$
$$= \frac{(z\lambda)^{n-1} \exp\{-z\lambda t\} \cdot \theta \exp\{-\theta z\}}{\int_0^\infty (z\lambda)^{n-1} \exp\{-z\lambda t\} \cdot \theta \exp\{-\theta z\}\mathrm{d}z},$$

and, from (5.69) and (5.70),

$$\lambda_m^n(t) = \frac{\int_0^\infty (z\lambda)^n \exp\{-(\lambda t + \theta)z\}\mathrm{d}z}{\int_0^\infty (z\lambda)^{n-1} \exp\{-(\lambda t + \theta)z\}\mathrm{d}z} = n\frac{\lambda}{\lambda t + \theta}.$$

Finally,

$$\lambda_1 = \sum_{n=1}^{\infty} n\frac{\lambda}{\lambda t + \theta} I(T_{n-1} \le t < T_n), \quad T_0 \equiv 0.$$

Thus, $\lambda_m(t) \le \lambda_t$, $t \ge 0$, holds.

Denote by $H_m(t)$ and $H_\lambda(t)$ the mean numbers of repairs (failures) in $[0, t)$ that correspond to the minimal repair processes of type 1 and type 2, respectively. The following result obviously follows from Theorem 5.9: $H_m(t) \le H_\lambda(t)$.

5.9 Preventive Maintenance in Heterogeneous Populations

The previous section dealt with the minimal repair as a specific type of corrective maintenance (CM). Now we will consider the preventive maintenance in heterogeneous populations. Our presentation mostly follows Cha and Finkelstein [21], whereas the developed approach is related to that of Sect. 5.8.

Preventive maintenance (PM) for non-repairable systems is a schedule of planned maintenance actions aimed at the prevention of breakdowns and failures of deteriorating systems. By "non-repairable" in this context we mean that the failure of a system is considered as an 'end event' and, therefore, the CM is not performed. We shall use this term in the defined sense throughout this section. Detailed surveys on the PM models for deteriorating systems can be found in, e.g., Valdez-Flores and Feldman [60] and Wang [65]. However, almost all models, procedures, and approaches described in the literature and those applied in reliability practice deal *only* with the case when the items come from homogeneous populations. Therefore, as in the case of the minimal repair in the previous section, it is quite a challenge to generalize PM to the case of heterogeneous populations of items.

As previously, we deal with the population described by the continuous mixtures setting (5.10)–(5.12). If the items are not maintained during operation, then their susceptibility to failures can be described by the 'ordinary' failure rate (2.4) (homogeneous case) or (5.12) (heterogeneous case). However, when maintenance actions that can affect reliability of items are performed, the corresponding effects should be taken into account. In the following, we will assume that the times of maintenance are negligible.

Consider first, reliability of a non-repairable item from a homogeneous population under PM (without CM). As PM affects its lifetime, we need to define new reliability measures in this case. Let T_P be the time to failure of item 'under preventive maintenance' and H_t be the maintenance history in $[0, t)$, i.e., the times of maintenance actions and the stochastic effects of the corresponding maintenances. Then, in order to describe the susceptibility to failure at time t, it is natural to define the following conditional failure rate:

$$\lambda_c(t) \equiv \lim_{\Delta t \to 0} \frac{\Pr[t < T_P \le t + \Delta t | H_t, T_P > t]}{\Delta t}, \quad t \ge 0 \tag{5.73}$$

Note that when maintenance is deterministic (times and effect), $\lambda_c(t)$ is also deterministic. However, if, e.g., times of maintenances are random, then $\lambda_c(t)$ is the stochastic process. The following example for the 'homogeneous items' is crucial for our further discussion:

Example 5.4 A non-repairable item with a lifetime described by the *increasing* failure rate $\lambda(t)$ starts its operation at $t = 0$. If it operable, it is preventively maintained at times kt_{PM}, $k = 1, 2, \ldots$. Assume that each preventive maintenance does not change the 'shape' of the function $\lambda(t)$, but the age of the item is reduced in accordance with the factor $0 < \alpha < 1$ (the reduced age is called the 'virtual age'). Therefore, PM has the effect of decreasing the failure rate as compared to an item that is not preventively maintained [28, 42]). Under these assumptions, the 'virtual age' of the item just after the first PM is αt_{PM}, just after the second PM is $\alpha(\alpha t_{PM} + t_{PM}) = \alpha t_{PM} + \alpha^2 t_{PM}, \ldots$, and the virtual age just after the $(n-1)$th PM, is

$$\begin{aligned} t_{n-1} &= \alpha t_{PM} + \alpha^2 t_{PM} + \ldots + \alpha^{n-1} t_{PM} \\ &= [\alpha(1 - \alpha^{n-1})/(1 - \alpha)] t_{PM}, \quad n = 2, 3, \ldots. \end{aligned} \tag{5.74}$$

Suppose that the item under this PM schedule has not failed until time t, $t \in [(n-1)t_{PM}, nt_{PM})$ meaning that it has been preventively maintained for $(n-1)$times at kt_{PM}, $k = 1, 2, \ldots, (n-1)$, whereas the last PM was performed at $(n-1)t_{PM}$. Thus, the virtual age of this item at time t is given by $t_{n-1} + (t - (n-1)t_{PM})$. Due to the PM assumptions, the statistical state of the maintained item at time t is the same as that of an identical (without maintenance) item with age $t_{n-1} + (t - (n-1)t_{PM})$. Accordingly, the conditional failure rate (5.73) that takes into account the described specific history H_t is given by

$$\lambda_c(t) = \lambda(t_{n-1} + (t - (n-1)t_{PM})), \quad t \in [(n-1)t_{PM}, nt_{PM}).$$

or, equivalently, letting $t_0 \equiv 0$:

$$\lambda_c(t) = \sum_{n=1}^{[t/t_{PM}]+1} \lambda(t_{n-1} + (t - (n-1)t_{PM})) I((n-1)t_{PM} \le t < nt_{PM}). \tag{5.75}$$

where $I(\cdot)$ is the corresponding indicator and $[t/t_{PM}]$ denotes the integer part of t/t_{PM}. Therefore, if the original failure rate $\lambda(t)$ is increasing, then $\lambda_c(t) \le \lambda(t)$, for all t and accordingly, PMs increase reliability of our item, i.e.,

$$\exp\left\{-\int_0^t \lambda_c(u)\mathrm{d}u\right\} \le \exp\left\{-\int_0^t \lambda(u)\mathrm{d}u\right\}.$$

We will now study the PM considered in Example 5.4, but for items from a heterogeneous population described by (5.10)–(5.12). Suppose that an item is randomly selected from this population and is preventively maintained at times kt_{PM}, $k = 1, 2, \ldots$. Preventive maintenance does not change the shape of the failure rate of an item but reduces its age in the same way as described by (5.74). Then, following the similar reasoning as in Example 5.4, one may construct the conditional failure rate by simply replacing $\lambda(t)$ in (5.75) with $\lambda_m(t)$:

$$\lambda_c(t) = \sum_{n=1}^{[t/t_{\mathrm{PM}}]+1} \lambda_m(t_{n-1} + (t - (n-1)t_{\mathrm{PM}}))I((n-1)t_{\mathrm{PM}} \le t < nt_{\mathrm{PM}}). \quad (5.76)$$

However, distinct from the homogeneous case, it is now not clear at all how this age reducing operation can be performed. In what follows, we will investigate the appropriateness of $\lambda_c(t)$ in (5.76) in defining the actual susceptibility of the survived item to failure at time t. For this purpose, we will suggest the operational profile that results in (5.76) and explain why it is unrealistic in practice. Then, we will suggest alternative profile with a different form of the conditional failure rate, which can be already justified in practice. Finally, the corresponding comparison of two profiles will be performed.

Operation profile 1 *An item is chosen at random from our population and starts operation at* $t = 0$. *Furthermore, a statistically identical "NEW" population is 'switched on' at time* $t_{\mathrm{PM}} - \alpha t_{\mathrm{PM}}$ *(the delayed start). At time* $t = t_{\mathrm{PM}}$, *if the selected item has not failed yet, it is replaced by an item randomly selected from the "delayed" population with age* αt_{PM}. *Then the replaced one starts its operation. At time* $t = 2t_{\mathrm{PM}}$, *if the replaced item has not failed yet, it is replaced by an item randomly selected from another 'delayed' population that started its operation at* $2t_{\mathrm{PM}} - (\alpha^2 t_{\mathrm{PM}} + \alpha t_{\mathrm{PM}})$ *and, therefore, its age is now* $\alpha^2 t_{\mathrm{PM}} + \alpha t_{\mathrm{PM}}$. *Then the replaced item starts its operation, and so on.*

We will construct the corresponding conditional failure rate for the described Operation profile 1 and will show that it is eventually given by Eq. (5.76). First, it is necessary to have in mind that the conditional failure rate defined in (5.73) can be expressed for the heterogeneous case as

$$\begin{aligned}\lambda_c(t) &= \lim_{\Delta t \to 0} \frac{E[\Pr[t < T_P \le t + \Delta t | H_t, T_P > t, Z]]}{\Delta t} \\ &= E\left[\lim_{\Delta t \to 0} \frac{\Pr[\ t < T_P \le t + \Delta t | H_t, T_P > t, Z]}{\Delta t}\right] \\ &= E[\lambda_{t,Z}],\end{aligned} \quad (5.77)$$

where the expectation is with respect to the conditional distribution $Z|(H_t, T_P > t)$ and

$$\lambda_{t,Z} \equiv \lim_{\Delta t \to 0} \frac{\Pr[t < T_P \le t + \Delta t | H_t, T_P > t, Z]}{\Delta t}. \quad (5.78)$$

Then $\lambda_{t,z}(Z = z)$ in (5.78) can be interpreted as the conditional (with respect to Z in addition to H_t) failure rate of the item, indexed by the frailty Z.

Denote by $\lambda_m^1(t)$ the failure rate $\lambda_c(t)$ in the interval $[0, t_{\mathrm{PM}})$ (defined by (5.77) for the Operation profile 1). It obviously equals the mixture failure rate in this interval, i.e.,

$$\lambda_c(t) \equiv \lambda_m^1(t) = \lambda_m(t), \ \text{ in } [0, t_{\mathrm{PM}}),$$

as information at hand is just the initial distribution $\pi(z)$ (and the fact that the item has survived in $[0, t)$).

As the survived item is replaced by an item *randomly selected* from the statistically identical population (but with the initial age αt_{PM}) at $t = t_{\mathrm{PM}}$, the conditional failure rate $\lambda_{t,Z}$ in $[t_{\mathrm{PM}}, 2t_{\mathrm{PM}})$ is

$$\lambda_{t,Z} = \lambda(\alpha t_{\mathrm{PM}} + (t - t_{\mathrm{PM}}), Z), \tag{5.79}$$

where Z is the frailty randomly selected at the previous PM. Consider now the conditional distribution $Z|(H_t, T_P > t)$. Note that at $t = t_{\mathrm{PM}}$, the initial distribution of Z is

$$\pi_{02}(z) = \frac{\bar{F}(\alpha t_{\mathrm{PM}}, z)\pi(z)}{\int_0^\infty \bar{F}(\alpha t_{\mathrm{PM}}, z)\pi(z)\mathrm{d}z} \tag{5.80}$$

and we know that the item has additionally survived in $(t_{\mathrm{PM}}, t]$. Therefore, the corresponding survival function (for $Z = z$) is

$$\frac{\bar{F}(\alpha t_{\mathrm{PM}} + (t - t_{\mathrm{PM}}), z)}{\bar{F}(\alpha t_{\mathrm{PM}}, z)}.$$

After updating, the conditional distribution $Z|(H_t, T_P > t)$ becomes

$$\frac{\bar{F}(\alpha t_{\mathrm{PM}} + (t - t_{\mathrm{PM}}), z)\pi(z)}{\int_0^\infty \bar{F}(\alpha t_{\mathrm{PM}} + (t - t_{\mathrm{PM}}), z)\pi(z)\mathrm{d}z}. \tag{5.81}$$

Therefore, in accordance with (5.78), the failure rate $\lambda_m^2(t)$ in $[t_{\mathrm{PM}}, 2t_{\mathrm{PM}})$ for the described operation is

$$\begin{aligned}\lambda_c(t) &\equiv \lambda_m^2(t) \\ &= \int_0^\infty \lambda(\alpha t_{\mathrm{PM}} + (t - t_{\mathrm{PM}}), z)\frac{\bar{F}(\alpha t_{\mathrm{PM}} + (t - t_{\mathrm{PM}}), z)\pi(z)}{\int_0^\infty \bar{F}(\alpha t_{\mathrm{PM}} + (t - t_{\mathrm{PM}}), z)\pi(z)\mathrm{d}z}\mathrm{d}z, \ \text{ in } [t_{\mathrm{PM}}, 2t_{\mathrm{PM}}).\end{aligned}$$

Similar to (5.81), the conditional distribution $Z|(H_t, T_P > t)$ for the interval $t \in [(n-1)t_{\mathrm{PM}}, nt_{\mathrm{PM}})$ is

$$\frac{\bar{F}(t_{n-1} + (t - (n-1)t_{\mathrm{PM}}), z)\pi(z)}{\int_0^\infty \bar{F}(t_{n-1} + (t - (n-1)t_{\mathrm{PM}}), z)\pi(z)\mathrm{d}z}$$

and we eventually arrive at

$$\lambda_c(t) \equiv \lambda_m^n(t)$$
$$= \int_0^\infty \lambda(t_{n-1} + (t-(n-1)t_{\mathrm{PM}}), z) \frac{\bar{F}(t_{n-1} + (t-(n-1)t_{\mathrm{PM}}), z)\pi(z)}{\int_0^\infty \bar{F}(t_{n-1} + (t-(n-1)t_{\mathrm{PM}}), z)\pi(z)\mathrm{d}z} \mathrm{d}z, \text{ in } [(n-1)t_{\mathrm{PM}}, nt_{\mathrm{PM}}),$$

$n = 1, 2, 3, \ldots$, where $t_0 \equiv 0$ and t_{n-1} are defined in (5.74).

Taking into account Eq. (5.12),

$$\lambda_m^n(t) = \lambda_m(t_{n-1} + (t-(n-1)t_{\mathrm{PM}})), \quad n = 1, 2, 3, \ldots \tag{5.82}$$

and thus, $\lambda_c(t)$ for the Operation profile 1 is given by (5.76). However, this strategy can hardly be realized in the PM practice for many reasons. For instance, even if the item selected at time $t = 0$ has been described by the frailty $Z = z_1$, its value can be changed to $Z = z_2$, $z_1 \neq z_2$ just after the first PM at t_{PM}, which is unrealistic.

Then, what is the proper conditional failure rate for our PM policy? It is more realistic to assume that the original frailty variable $Z = z$ selected at time $t = 0$ is *preserved* throughout the whole operation of an item:

Operation profile 2 *An item is chosen at random from our population and starts operation at $t = 0$. The original frailty that is 'acquired' at $t = 0$ is preserved during the PM actions that follow the pattern of the 'virtual age structure' defined in (5.74).*

As the PMs are applied to the same item, this operation profile is definitely more adequate than the first one. However, the construction of the corresponding failure rate is completely different in this case.

In $[0, t_{\mathrm{PM}})$, the failure rate is still the same:

$$\lambda_c(t) \equiv \lambda_m^1(t) = \lambda_m(t), \text{ in } [0, t_{\mathrm{PM}}),$$

as the information at hand is the same as before.

Consider now the second cycle $[t_{\mathrm{PM}}, 2t_{\mathrm{PM}})$. As the survived item was randomly selected at time $t = 0$ from the heterogeneous population, the conditional failure rate $\lambda_{t,Z}$ in $[t_{\mathrm{PM}}, 2t_{\mathrm{PM}})$ is given by (5.79), where Z is the frailty 'randomly selected' at $t = 0$. At $t = t_{\mathrm{PM}}$, the survived item has the frailty $Z = z$ with the pdf that in accordance with (5.12) is

$$\pi_{02}(z) \equiv \frac{\bar{F}(t_{\mathrm{PM}}, z)\pi(z)}{\int_0^\infty \bar{F}(t_{\mathrm{PM}}, z)\pi(z)\mathrm{d}z}.$$

We also have the information that the item with the decreased age αt_{PM} after the PM has additionally survived in $(t_{\mathrm{PM}}, t]$. Therefore, the corresponding survival function is

$$\frac{\bar{F}(t_1 + (t - t_{\mathrm{PM}})), z)}{\bar{F}(t_1, z)}.$$

In accordance with (5.12), the conditional distribution $Z|(H_t, T_P > t)$ is given now by

$$\frac{[\overline{F}(t_1+(t-t_{\mathrm{PM}})),z)/\overline{F}(t_1,z)]\cdot\pi_{02}(z)}{\int_0^\infty[\overline{F}(t_1+(t-t_{\mathrm{PM}})),z)/\overline{F}(t_1,z)]\cdot\pi_{02}(z)\mathrm{d}z}$$

and the failure rate $\lambda_m^2(t)$ in $[t_{\mathrm{PM}}, 2t_{\mathrm{PM}})$, in accordance with (5.77), is

$$\begin{aligned}\lambda_m^2(t) &= \int_0^\infty \lambda(t_1+(t-t_{\mathrm{PM}})),z)\cdot\frac{[\overline{F}(t_1+(t-t_{\mathrm{PM}})),z)/\overline{F}(t_1,z)]\cdot\pi_{02}(z)}{\int_0^\infty[\overline{F}(t_1+(t-t_{\mathrm{PM}})),z)/\overline{F}(t_1,z)]\cdot\pi_{02}(z)\mathrm{d}z}\mathrm{d}z\\ &= \int_0^\infty \lambda(t_1+(t-t_{\mathrm{PM}})),z)\cdot\frac{[\overline{F}(t_1+(t-t_{\mathrm{PM}})),z)/\overline{F}(t_1,z)]\cdot\overline{F}(t_{\mathrm{PM}},z)\pi(z)}{\int_0^\infty[\overline{F}(t_1+(t-t_{\mathrm{PM}})),z)/\overline{F}(t_1,z)]\cdot\overline{F}(t_{\mathrm{PM}},z)\pi(z)\mathrm{d}z}\mathrm{d}z.\end{aligned}$$

In a similar way, for $t\in[(n-1)t_{\mathrm{PM}}, nt_{\mathrm{PM}})$,

$$\begin{aligned}\lambda_m^n(t) = \int_0^\infty \lambda(t_{n-1}+(t-(n-1)t_{\mathrm{PM}})),z)\cdot\frac{[\overline{F}(t_{n-1}+(t-(n-1)t_{\mathrm{PM}})),z)/\overline{F}(t_{n-1},z)]}{\int_0^\infty[\overline{F}(t_{n-1}+(t-(n-1)t_{\mathrm{PM}})),z)/\overline{F}(t_{n-1},z)]}\\ \times\frac{\overline{F}(t_{\mathrm{PM}},z)\cdot\frac{\overline{F}(t_1+t_{\mathrm{PM}},z)}{\overline{F}(t_1,z)}\cdots\frac{\overline{F}(t_{n-2}+t_{\mathrm{PM}},z)}{\overline{F}(t_{n-2},z)}\pi(z)}{\overline{F}(t_{\mathrm{PM}},z)\cdot\frac{\overline{F}(t_1+t_{\mathrm{PM}},z)}{\overline{F}(t_1,z)}\cdots\frac{\overline{F}(t_{n-2}+t_{\mathrm{PM}},z)}{\overline{F}(t_{n-2},z)}\pi(z)\mathrm{d}z},\end{aligned} \tag{5.83}$$

where t_{n-1}, $n=1,2,3,\ldots$, $(t_0\equiv 0)$ are defined in (5.74).

Observe that conditional failure rates for both operation profiles can now be uniformly written as

$$\lambda_c^J(t) = \sum_{n=1}^{[t/t_{\mathrm{PM}}]+1}\lambda_{mJ}^n(t)I((n-1)t_{\mathrm{PM}}\le t< nt_{\mathrm{PM}}),\ J=I,II,$$

where $I(\cdot)$ is the corresponding indicator and $J=I,II$ refers to the number of the profile. Thus, $\lambda_{mI}^n(t)$ corresponds to $\lambda_m^n(t)$ in (5.82) and $\lambda_{mII}^n(t)$ to $\lambda_m^n(t)$ in (5.83).

Therefore, in practice, $\lambda_c^{II}(t)$ (not $\lambda_c^I(t)$) should be applied for the described type of PM. However, assume that the user, who is performing the PM (via reducing the age of items by the method described previously), does not know (or does not take into account) the heterogeneity structure of the population and considers it as homogeneous with the corresponding time to failure distribution $F_m(t)$ and the failure rate $\lambda_m(t)$. Then he is using the failure rate $\lambda_c^I(t)$ to assess reliability of items in operation. What is the consequence of this error? The following theorem answers to this question.

Theorem 5.10 *Let the values of $\lambda(t,z)$. be ordered with respect to z: for all* $z_1, z_2 \in [0,\infty]$, $t \geq 0$.

$$\lambda(t,z_1) < \lambda(t,z_2) \text{ if } z_1 < z_2.$$

Then

$$\lambda_c^{II}(t) \leq \lambda_c^{I}(t), \text{ for all } t \geq 0.$$

The *proof* of this theorem is rather straightforward (although technical) and can be found in Cha and Finkelstein [21].

□

It follows from this theorem that using $\lambda_c^I(t)$ instead of the 'proper' $\lambda_c^{II}(t)$ eventually results in the overestimation of the failure rate of items under operation. Practically, this may cause unnecessary frequent PMs and therefore, additional redundant costs.

Example 5.5 Suppose that $\lambda(t,z)$ is strictly increasing in t for each z (e.g., $\lambda(t,z) = z\lambda t, \lambda > 0$). An item is randomly selected from the heterogeneous population and it is preventively maintained at times kt_{PM}, $k = 1,2,\ldots$. Let τ be the mission time of the item in field operation. If the mission is successful, a gain $K > 0$ is obtained, whereas if the mission is not completed (a failure in $[0,\tau)$), a cost $c_f > 0$ is incurred ($K > c_f$). Furthermore, the cost for each PM is $c_p > 0$. Then, the following cost function, which is the function of t_{PM}, can be constructed.

$$\begin{aligned} c(t_{\text{PM}}) &= \left\langle \frac{\tau}{t_{\text{PM}}} \right\rangle c_p + c_f \cdot P(T_p \leq \tau) - K \cdot P(T_p > \tau) \\ &= \left\langle \frac{\tau}{t_{\text{PM}}} \right\rangle c_p - (K + c_f) \cdot \exp\left\{ -\int_0^\tau \lambda_c^{II}(u)\text{d}u \right\} + c_f, \end{aligned} \tag{5.84}$$

where $<\tau/t_{\text{PM}}>$ is the largest integer which is *strictly* less than τ/t_{PM}. The problem is to find the optimal t^*_{PM} which satisfies

$$c(t^*_{\text{PM}}) = \min_{t_{\text{PM}} \in (0,\infty)} c(t_{\text{PM}})$$

It is reasonable to consider only $t_{\text{PM}} \in (0,\tau]$ as $c(t_{\text{PM}}) = c(\tau)$, for all $t_{\text{PM}} \in (\tau,\infty)$. When $t_{\text{PM}} \to 0$ ($[\tau/t_{\text{PM}}] \to \infty$), obviously, $\exp\{-\int_0^\tau \lambda_c^{II}(u)\text{d}u\} \to \exp\{-\lambda_m(0)\tau\}$, which implies that $\lim_{t_{\text{PM}}\to 0} c(t_{\text{PM}}) = \infty$. On the other hand, $c(\tau) = c_f - (K + c_f)\exp\{-\int_0^\tau \lambda_m(u)\text{d}u\}$. Therefore, there should be an optimal $t^*_{\text{PM}} \in (0,\tau)$ depending on the parameters involved, e.g., when τ is large enough and K is relatively large compared with c_f and c_p.

5.10 Population Mortality at Advanced Ages (Demographic Application)

In Sects. 5.4 and 5.5, we have briefly discussed asymptotic behavior of mixture failure rates as $t \to \infty$. In the current section, we will deal with this problem from a different view point and in more detail [31].

The shape of the failure rate (force of mortality) at advanced ages especially for human populations has attracted a considerable interest in the last decades when more and more centenarians and super centenarians have been recorded. The International Database on Longevity (http://www.supercentenarians.org/) offers the detailed information on thoroughly validated cases of super centenarians. Gampe [35] has used these data to estimate the human force of mortality after the age of 110. Her analysis revealed that human mortality between ages 110 and 114 levels off regardless of gender. The widely used explanation of this fact is by employing the corresponding fixed frailty models that account for heterogeneity of populations. Beard [7, 8] (see also Vaupel et al. [63]) has considered the Gompertz (baseline)-gamma-frailty model, which results in the asymptotically flat hazard rate. Note that, the exponentially increasing hazard rate of the Gompertz distribution is the only baseline function that can 'produce' this shape in the framework of the multiplicative frailty model (see Sect. 5.3.1), which can be considered as another justification of the uniqueness and importance of this distribution for human mortality modeling. As follows from the results of Sect. 5.4, the gamma distribution of frailty is not so unique in this respect and all probability density functions $f(z)$ that behave as $z^{\alpha}, \alpha > 1$ when $z \to 0$ are equivalent in this sense.

The intuitive meaning of the deceleration of mortality at advanced ages in this context is simple and meaningful at the same time: the oldest-old mortality in *heterogeneous* populations with properly ordered subpopulations is defined by the small values of frailty, as the subpopulations with larger values of frailty (and, therefore, larger values of the failure rate) are dying out first.

The first question to be answered is what *common* statistical distributions are characterized by the asymptotically flat failure rate? The exponential distribution that is often used for statistical analysis of non-degrading objects is obviously not relevant for our topic. The most popular distribution of the desired type is the inverse Gaussian distribution. It is well known that it describes the distribution of the first passage time for the Wiener process with drift. Although its sample paths are nonmonotone and even can be nonpositive, the inverse Gaussian distribution was widely used, e.g., in reliability analysis of stochastic deterioration (aging) in engineering objects. It was also applied in vitality models for modeling the lifespan of organisms [3, 45], where the initial vitality (resource) of organisms is 'consumed' in the course of life in accordance with the Wiener process with drift. This model was also studied in the path-breaking papers by Aalen and Gjessing [1] and Steinsaltz and Evans [55] as an example highlighting the meaning and properties of the corresponding quasistationary distributions for this particular case. Our goal in this section is more modest: to exploit further some relevant distributional properties in the context of stochastic ordering of lifetimes of subpopulations in heterogeneous populations. However, the combination of these two approaches can hopefully be considered as the basis for the future research on statistical inference in heterogeneous populations with underlying stochastic processes (e.g., the Wiener process).

The other example of a distribution with asymptotically flat hazard rate is the Birnbaum-Saunders distribution [12] that was also derived as a distribution of the first passage time for the corresponding deterioration process and, therefore, is a good candidate for vitality models as well. We also consider the gamma process as a possible model of deterioration (with monotone sample paths), although the failure rate in this case is decreasing to 0 as $t \to \infty$. It should be noted, however, that the initial increase in the failure rates for all these models is not exponential, as in the case of the Gompertz distribution and, therefore, the possibilities of the corresponding mortality modeling for *human populations* for intermediate ages (30–90 years) are obviously limited.

5.10.1 Fixed and Evolving (Changing) Heterogeneity

Let $F(t)$, $f(t)$, and $\lambda(t)$ be the Cdf, the pdf, and the failure rate (force of mortality) for some infinite *homogeneous* population that characterize the corresponding random lifetime $T \geq 0$. As previously, by heterogeneity of a population we mean that it consists of a finite or non-finite number of homogeneous subpopulations that differ in some respect to be discussed. For instance, in the multiplicative frailty model of the form $\lambda(t, Z) = Z\lambda(t)$, the difference between subpopulations is modeled directly by the differences in failure rates: for two realizations $z_2 > z_1$, this difference is $(z_2 - z_1)\lambda(t)$. Thus, the multiplicative frailty model describes the ordering of subpopulations in the sense of the *hazard rate ordering* (2.70). More generally, the smaller is the value of z, the larger is the lifetime of the subpopulation T_z in the appropriate stochastic sense (e.g., (2.69), (2.70) or (2.71)):

$$T_{z_1} \geq T_{z_2}; \quad z_1 \leq z_2. \tag{5.85}$$

As previously in this chapter, we will understand the fixed heterogeneity (frailty) of a population as:

Heterogeneity in lifetimes of the corresponding homogeneous subpopulations that is defined by the appropriate stochastic ordering.

This also means that, if randomization of a parameter (parameters) of a lifetime distribution leads to the corresponding stochastic ordering, which formally is not always the case, then this operation can be also interpreted in terms of the fixed frailty modeling. For example, the Gompertz Cdf $F(t, a, b))$ is a function of two parameters, and the corresponding failure rate is:

$$\lambda(t, a, b) = ae^{bt}. \tag{5.86}$$

If we randomize a, whereas b is fixed, then (taking care, of course, of the corresponding baseline constant), we obviously arrive at the multiplicative frailty model (and to the asymptotically flat rate when the distribution of frailty is, e.g., gamma), which illustrates ordering (2.70). We just want to emphasize the fact that in this specific model, frailty acts multiplicatively and directly on the failure rate, which is not the case in general even when the hazard rate ordering (2.70) holds. Some relevant aspects of frailty modeling for the bivariate case will be considered later.

In accordance with our definition, the fixed heterogeneity (frailty) is described only by ordered subpopulation lifetimes. What can happen, if apart from the information on failure times (the black box point of view), we possess some information or adopt a model on a failure process or mechanism (the process point of view)? In this case, another type of heterogeneity, which is usually referred to as evolving (or changing) (see, e.g., Li and Anderson [45]) comes into play. This type of heterogeneity usually does not lead to ordering of lifetimes in the described here sense. However, it characterizes an important feature of a model, which can be useful for further analysis.

In order to illustrate our point, consider the model for vitality loss (fixed initial value) that will be treated in detail further in this section. The loss of vitality of an organism (deterioration) is modeled by the Wiener process with negative drift, in which the time to death is determined by the first passage time to the zero boundary. It is well known that the variance of the Wiener process is increasing linearly in time and if the drift is positive, the mean is also linearly increasing. However, due to the boundary, the most vulnerable organisms (or items in reliability engineering applications) are dying out first and linear functions that correspond to the non-boundary case 'decelerate'. Actual shapes depend on parameters of the model (see the graphs in Li and Anderson [45] for the corresponding shapes for the specific values of parameters). Thus we do not see here any frailty parameters or ordered (in the defined in this section sense) lifetimes, but we observe the changing in time mean and variability in the survived population. And this is how the evolving heterogeneity should be understood:

Variability in sample paths of the underlying process of deterioration.

In this section, however, we are mostly interested in the fixed heterogeneity of lifetimes and the evolving heterogeneity of processes will be 'hidden' in lifetime distributions. We feel that this 'distributional approach' in the context of randomization of parameters and of the corresponding ordering of lifetimes was not sufficiently elaborated in the literature so far. For instance, for the first passage time models to be considered further, randomization of the initial vitality of an organism and of the corresponding drift parameter of the Brownian motion definitely illustrates this ordering, as the larger is the vitality and (or) the smaller is the drift parameter, the larger is the lifetime in some suitable stochastic sense to be discussed. Note that, there can be other situations when randomization is relevant but does not lead to the ordered subpopulations.

5.10.2 Fixed Heterogeneity

Equations (5.10)–(5.12) describe the standard statistical mixture (or the fixed frailty) model for an item and for the collection of items (population) as well. As was discussed in the previous subsection, we understand heterogeneity as the property of a population that consists of ordered homogeneous subpopulations (ordered lifetimes T_z, defined by Inequality (5.85)). But what type of ordering is sufficient for our reasoning? As we are looking at the failure rates, the first guess would be that this should be (2.70). How can we interpret in mathematical terms the well-known and intuitively clear property: "the weakest populations are dying out first" and the resulting mortality deceleration with time? To answer these questions, denote, as previously, by $\Pi(z)$ the Cdf of Z and by $\Pi(z|t)$ the Cdf that corresponds to the density $\pi(z|t)$. Therefore, the deceleration can be a consequence of the increasing in t distribution function $\Pi(z|t)$ [28]. This would mean that $\Pi(z|t)$ tends to be more concentrated around small values of $Z \geq 0$ as time increases, which corresponds to stronger populations. The following theorem proves this result.

Theorem 5.11 *Let stochastic ordering* (5.85) *in the sense of the failure rates hold. Then* $\Pi(z|t)$ *is a non-decreasing function of t for each fixed z.*

Proof. It follows from (5.12) that

$$\Pi(z|t) = \frac{\int_0^z \bar{F}(t,u)\pi(u)\mathrm{d}u}{\int_0^z \bar{F}(t,u)\pi(u)\mathrm{d}u}.$$

It is easy to see that the derivative of this function is nonpositive if

$$\frac{\int_0^z \bar{F}_t'(t,u)\pi(u)\mathrm{d}u}{\int_0^z \bar{F}(t,u)\pi(u)\mathrm{d}u} \geq \frac{\int_0^\infty \bar{F}_t'(t,u)\pi(u)\mathrm{d}u}{\int_0^\infty \bar{F}(t,u)\pi(u)\mathrm{d}u}.$$

Therefore, it is *sufficient* to show that the function:

$$A(t,z) = \frac{\int_0^z \bar{F}_t'(t,u)\pi(u)\mathrm{d}u}{\int_0^z \bar{F}(t,u)\pi(u)\mathrm{d}u}$$

is nonincreasing in z. As $\bar{F}'_t(t,z) = -\mu(t,z)\bar{F}(t,z)$, inequality $A_z'(t,z) \leq 0$ is equivalent to the following one:

$$\mu(t,z)\int_0^z \bar{F}(t,u)\pi(u)\mathrm{d}u \geq \int_0^z \mu(t,u)\bar{F}(t,u)\pi(u)\mathrm{d}u,$$

which obviously follows from Ordering (5.85) which should be understood in the sense of the hazard rate ordering.

□

Consider now the bivariate frailty model. We will need the following considerations for analyzing asymptotic failure rates for vitality models of the next subsection. Let Z_1 and Z_2 be interpreted as non-negative random variables with supports in $[0, \infty)$. Similar to the univariate case,

$$P(T \le t|Z_1 = z, Z_2 = z_2) \equiv P(T \le t|z_1, z_2) = F(t, z_1, z_2)$$

and

$$\lambda(t, z_1, z_2) = \frac{f(t, z_1, z_2)}{\bar{F}(t, z_1, z_2)}$$

Assume that Z_1 and Z_2 have the joint pdf $\pi(z_1, z_2)$. The mixture failure rate is defined in this case as [28]:

$$\begin{aligned} \lambda(t) = \frac{f(t)}{F(t)} &= \frac{\int_0^\infty \int_0^\infty f(t, z_1, z_2)\pi(z_1, z_2)\mathrm{d}z_1 dz_2}{\int_0^\infty \int_0^\infty \bar{F}(t, z_1, z_2)\pi(z_1, z_2\mathrm{d}z_1\mathrm{d}z_2} \\ &= \int_0^\infty \int_0^\infty \lambda(t, z_1, z_2)\pi(z_1, z_2|t)\mathrm{d}z_1\mathrm{d}z_2, \end{aligned} \tag{5.87}$$

where the corresponding conditional pdf (on condition $T > t$) is

$$\pi(z_1, z_2|t) = \pi(z_1, z_2)\frac{\bar{F}(t, z_1, z_2)}{\int_0^\infty \int_0^\infty \bar{F}(t, z_1, z_2)\pi(z_1, z_2)\mathrm{d}z_1\mathrm{d}z_2}. \tag{5.88}$$

Equation (5.87) is a general result and can be analyzed for some specific cases. For instance, it can be easily shown that when we assume the independence of frailties:

$$\pi(z_1, z_2) = \pi_1(z_1)\pi_2(z_2)$$

and the competing risks for the failure model:

$$F(t, z_1, z_2) = 1 - \bar{F}_1(t, z_1)\bar{F}_2(t, z_2),$$

the population failure rate is just the sum $\lambda(t) = \lambda_1(t) + \lambda_2(t)$ of the corresponding 'univariate failure rates'.

Although it is difficult to analyze $\lambda(t)$ in (5.87) in full generality, certain qualitative considerations that will be very helpful in the next subsection can be stated. Indeed, let us first fix the second frailty $Z_2 = z_2$. Then the corresponding failure rate is defined by the univariate frailty model

$$\lambda(t, z_2) = \int_0^\infty \mu(t, z_1, z_2)\pi(z_1, z_2|t)\mathrm{d}z_1. \tag{5.89}$$

Thus, at the *first stage,* we have selected from our overall heterogeneous population the *heterogeneous subpopulation* that corresponds to $Z_2 = z_2$

$(z_2 < Z_2 \leq z_2 + dz_2)$ and have defined its failure rate. As our goal is to analyze the failure rate, at the *second stage,* we consider our overall population as a 'continuous collection' of *homogeneous subpopulations* with failure rates given by (5.89). Then we can analyze $\lambda(t)$ again in the univariate manner. For instance, assume that the family $\lambda(t, z_2)$ is ordered in z_2 (the smaller values of z_2 correspond to the smaller values of $\lambda(t, z_2)$). Therefore, the deceleration in mortality due to 'the weakest populations are dying out first' takes place. Specifically, let $\lambda(t, z_2)$ for each z_2 decreases (nonincreases) at least, asymptotically when $t \to \infty$. It is well known that the corresponding population (mixture) failure rate is strictly decreasing in this case (see, e.g., Ross [54]). Thus, we have described the following result [31]:

Theorem 5.12 *Let frailty $Z_1 = z_1$ $(Z_2 = z_2)$ in the bivariate frailty model be first fixed. Assume that the corresponding univariate frailty model (with respect to Z_2 (Z_1) results in the decreasing ordered failure rates for all subpopulations.*

Then 'allowing' random Z_1 (Z_2), results in the strictly decreasing population failure rate.

The formal *proof* of the validity of the two-stage procedure is straightforward and is based on the representation of the bivariate density $\pi(z_1, z_2)$ as a product $\pi_1(z_1|Z_2 = z_2)\pi_2(z_2)$ and on the similar representation for the conditional density:

$$\pi(z_1, z_2|t) = \pi_1(z_1|Z_2 = z_2, T > t)\pi_2(z_2|t).$$

The latter seems intuitively evident, and can be immediately obtained formally from Eqs. (5.87), (5.88). Theorem 2 then follows, as the (univariate) mixture of distributions with decreasing (nonincreasing) failure rates is characterized by the strictly decreasing failure rate.

□

Example 5.6 An important application that illustrates Theorem 2 deals with the Gompertz law of mortality (5.86). It is well known that randomization of a (e.g., via the gamma distribution of the frailty) results in the mortality plateau as $t \to \infty$. Thus, randomization of b (second stage) results in the decreasing force of mortality as $t \to \infty$. Therefore, if we observe the mortality plateau for some population that follows the Gompertz-gamma model, then there should not be noticeable heterogeneity in this population due to parameter b.

The described multistage approach can be applied in a similar way to the case when there are more than 2 frailties or parameters of distributions that can be randomized. It is possible that all failure rates from the ordered family converge asymptotically (as $t \to \infty$) to one curve (specifically, to a constant). Therefore, the population failure rate also tends to this curve which will also be illustrated in the next subsection.

The foregoing discussion will help us to analyze the shape of the failure rate for some examples of vitality models. We will focus mostly on the vitality model described by the Wiener process with drift [3, 45, 64]. Parameters of lifetime distributions after randomization will act as fixed frailties that define the corresponding ordered subpopulations. This interpretation adds some simple and useful additional reasoning from the *distributional point of view* to the *process point of view* approach developed by Aalen and Gjessing [1] and Steinsaltz and Evans [55].

5.10.3 Vitality Models and Lifetime Distributions

Linear process of degradation. We start with the simplest vitality model that will be used as an explanatory example for highlighting certain properties and approaches.

Let $v_0 > 0$ be the deterministic initial (at $t = 0$) vitality of an organism, which is monotonically decreasing with t in accordance with the simplest stochastic process:

$$V_t = v_0 - Rt, \tag{5.90}$$

where R is a positive random variable with the Cdf $S(t)$. For each realization $R = r$, (5.90) can model the linear decline in physiological functions of organisms noted by Strehler and Mildvan [57] and in numerous subsequent publications. However, exponential and logarithmic models for this decline can be also considered.

Death occurs when V_t reaches 0. Denote the corresponding lifetime by T_R. Therefore, the Cdf that describes this lifetime is

$$F_R(t) = \Pr[T_R \le t] = \Pr[R \ge v_0/t] = 1 - S(v_0/t).$$

Assume that R is gamma-distributed with the pdf $a^{\eta}x^{\eta-1}e^{-ax}/\Gamma(\eta)$ with the scale parameter $a > 0$ and the shape parameter $\eta > 0$. Then the pdf $f_R(t) = F'_R(t)$ has the form of the inverse gamma distribution:

$$f_R(t) = \frac{(v_0 a)}{\Gamma(\eta)} t^{-\eta-1} e^{-v_0 a/t}. \tag{5.91}$$

We will analyze the shape of the corresponding hazard rate using the 'classic' Glazer's theorem [37], formulated in a slightly more general form by Marshall and Olkin [48] as can be seen from Theorem 2.1 in Chap. 2. We will intensively use this result and other relevant considerations in what follows.

The essential fact to be exploited is that the behavior of the failure rate $\lambda(t)$ is related to the behavior of the derivative of the logarithm of the density of a lifetime distribution $F(t)$, namely,

$$g(t) = -\frac{d\log f(t)}{dt} = -\frac{f'(t)}{f(t)}.$$

The failure rate $\lambda_R(t)$ that corresponds to (5.91) can be easily analyzed with the help of Theorem 2.1. Indeed, as $\lim_{t\to\infty} f_R(t) = 0$, it follows that $\lim_{t\to\infty} \lambda_R(t) = 0$, whereas

$$\lim_{t\to\infty} \lambda_R(t) = \lim_{t\to\infty} f_R(t)/\bar{F}_R(t) = \lim_{t\to\infty} -\frac{\mathrm{d}\log f_R(t)}{\mathrm{d}t} = 0$$

and $\lambda_R(t)$ is bell-shaped with a maximum at $t = 2v_0 a/(\eta + 1)$.

This simple example, however, can be helpful for discussing the notion of heterogeneity that we adopt. If we consider the model as a black box with the lifetime described by the Cdf $F_R(t)$, then by definition, the corresponding population is homogeneous. However, in view of the model (5.90), we can identify the corresponding subpopulations for each value of $R = r$ that will be definitely ordered (in this case the lifetimes that correspond to each realization $R = r$ are deterministic, and therefore, can be ordered accordingly). Thus, our infinite population can be considered as heterogeneous in the described sense.

The considered vitality model results in the vanishing at the infinity failure rate. If we are interested in explaining mortality plateaus that has been observed in human and other populations, then we must look at other, more realistic vitality models. The first candidate for that is when the simplest stochastic process Rt is substituted by the more advanced stochastic model given by the Wiener process with drift.

Wiener process with drift. We modify the degradation model (5.90) with the fixed initial vitality v_0 to

$$\begin{aligned} V_t &= v_0 - R_t, \\ R_t &= rt + W_t, \end{aligned} \tag{5.92}$$

where $R_t, t \geq 0$ is the Wiener process with drift, r is a drift parameter and $W_t, t \geq 0$ is the standard Wiener process with normally distributed values (for each fixed t) with mean 0 and variance $\sigma^2 t$.

It is well known (see, e.g., [24]) that the probability distribution for the first passage time (when R_t reaches the boundary v_0 for the first time) is defined by the inverse Gaussian distribution with the pdf:

$$f_R(t) \equiv f_R(t; v_0, r, \sigma) = \frac{v_0}{\sigma\sqrt{2\pi}} t^{-3/2} \exp\left\{ -\frac{(v_0 - rt)^2}{2\sigma^2 t} \right\}. \tag{5.93}$$

The exact expression for the corresponding failure rate, $\lambda_R(t) \equiv \mu_R(t; v_0, r, \sigma)$, is complicated and, therefore, as our goal is just to analyze its *shape*, we will use Theorem 2.1. It is easy to derive from (5.93) that

$$g_R(t) = -\frac{\mathrm{d}\log f_R(t)}{\mathrm{d}t} = \frac{3}{2t} + \frac{r^2}{2\sigma^2} - \frac{v_0^2}{2\sigma^2 t^2}. \tag{5.94}$$

Note that, (5.93) is written in parameterization v_0, r, σ. However, reparameterization: $\lambda = r^2/\sigma^2$, $\omega = rv_0/\sigma^2$ leads to the standard two-parameter form of the inverse Gaussian distribution (which we need for stating some useful properties):

$$f_R(t;\lambda,\omega) = \frac{\lambda\omega}{\sigma\sqrt{2\pi}}(\lambda t)^{-3/2}\exp\left\{-\frac{(\omega-\lambda t)^2}{2\lambda t}\right\}. \tag{5.95}$$

It immediately follows from (5.94) that the failure rate tends to a constant when $t \to \infty$(mortality plateau):

$$\lim_{t\to\infty}\lambda_R(t) = \lim_{t\to\infty} -\frac{\mathrm{d}\log f_R(t)}{\mathrm{d}t} = \frac{\lambda}{2} = \frac{r^2}{2\sigma^2}. \tag{5.96}$$

It is also obvious that $\lim_{t\to 0}\lambda_R(t) = 0$. The 'rest of the shape' of $\lambda_R(t)$ is defined by Theorem 2.1: $\lambda_R(t)$is increasing for $t \in [0 \le t_2]$, where $t_2 \le t_1 = 2v_0^2/3\sigma^2$ and is asymptotically decreasing to the plateau for $t \ge t_2$. This form of the hazard rate for the inverse Gaussian distribution was first described by Chhikara and Folks [22] using straightforward calculus and asymptotic bounds. We, however, rely on a general Theorem 2.1 that can be used for analysis of other distributions as well.

Although the 'underlying physics' of the inverse Gaussian distribution is given by the Wiener process with drift, we cannot identify now the corresponding subpopulations in the sense that we have defined earlier. Therefore, the corresponding population in this 'black-box' analysis should be considered as *homogeneous* and there is no (fixed) heterogeneity in the defined sense so far.

From (5.95) it follows that λ is the scale parameter. Therefore, obviously, the corresponding lifetimes are decreasing in λ in the sense of the usual stochastic ordering (2.69), i.e., for the fixed ω:

$$F_R(\lambda_1 t;\omega) \le F(\lambda_2 t;\omega); \quad \lambda_1 \le \lambda_2,\ t \in [0,\infty). \tag{5.97}$$

This is a simple general fact. However, for the specific case of inverse Gaussian distribution, it can be shown that the stronger hazard rate ordering (2.70) also takes place [48], which means:

$$\begin{aligned}&\mu_R(t;\lambda_1,\omega) = \lambda_1\mu_R(\lambda_1 t;\omega_1) \le \lambda_2\mu_R(\lambda_2 t;\omega_2) = \lambda_R(t;\lambda_2,\omega);\\ &\quad \lambda_1 \le \lambda_2,\ t \in [0,\infty).\end{aligned}$$

As $\lambda = r^2/\sigma^2$, the distribution of the first passage time $f_R(t;\lambda,\omega)$ does not change when we change r and σ proportionally. Thus the mechanism of the failure process driven by the Wiener process with drift is such that, e.g., the increase in the drift parameter is compensated by the proportional increase in the standard deviation σ. This is a rather unexpected observation, however, as stated, it is a consequence of the considered specific setting. Strictly speaking, as parameters λ and ω are 'dependent' the foregoing orders hold only asymptotically for large t and this is how we will understand it in what follows.

After discussing the issue of stochastic ordering, we can now *qualitatively* analyze the shape of $\lambda_R(t;\lambda,\omega)$ for large t with respect to the randomized parameters r and σ (v_0 is fixed so far) to be denoted by R and Σ, respectively. Note that, Aalen and Gjessing [1], have performed the necessary derivations assuming that R is normally distributed and σ is fixed. However, as the drift $(-r)$ can be positive in this case, the resulting survival distribution is defective. These distributions are often used for describing the corresponding 'cure models'.

Assume that R and Σ are non-negative random variables with supports in $[0,\infty)$. Thus, the bivariate frailty model discussed in Sect. 3 can be applied. We proceed as described there: fixing $\Sigma=\sigma$ and considering subpopulations with one frailty parameter R. At the *first stage,* we select from the overall heterogeneous population the heterogeneous (with respect to different values of r) *subpopulation* that corresponds to $\Sigma=\sigma$ and define its failure rate. As the corresponding homogeneous 'sub-subpopulations' (for different fixed values of r) are ordered in the sense of the hazard rate ordering and 'have' the shapes of the failure rates described above (increasing and then decreasing to a plateau), this heterogeneous subpopulation has asymptotically decreasing to 0 failure rate [54]. Now, at the second stage, as these failure rates are ordered with respect to the values of the second frailty $\Sigma=\sigma$, we can use Theorem 5.12, which means that the population failure rate is also decreasing as $t\to\infty$ (and in our specific case, it is decreasing to 0).

Thus, *mortality plateaus cannot occur in the described frailty model.* However, this can still happen, if the supports of frailties R and Σ are modified to $[a,\infty]$ and $[0,b]$, respectively. Then the population failure rate tends to the failure rate of the strongest subpopulation which is, in accordance with (5.96) [31],

$$\lim_{t\to\infty}\lambda_R(t)=\frac{a^2}{2b^2}. \tag{5.98}$$

We are ready now to add variability to the initial vitality. Denote the corresponding random variable by $V_0\geq 0$ (fixed frailty). It immediately follows from (5.96) that, in contrast to the other considered fixed frailties, the effect of the initial vitality vanishes as $t\to\infty$. Therefore, it has no effect asymptotically on the shape of the failure rate. This was analytically shown and discussed using the concept of quasisationary distributions in Aaalen and Gjessing [1], Steinsaltz and Evans [55], and Li and Anderson [45].

Gamma process and the Birnbaum-Saunders distribution. The Wiener process is often criticized as a model for degradation and aging as its sample paths are not necessarily positive and strictly increasing. On the other hand, the gamma process always possesses these properties. Therefore, let $R_t, t\geq 0$ be now the stationary gamma process with the following density for each t:

$$f_{R_t}(x)=Ga(x|r^2t/\sigma^2,r/\sigma^2),\ \mu,\sigma>0, \tag{5.99}$$

$$E[R_t]=rt,\quad \mathrm{Var}(R_t)=\sigma^2t,$$

where $Ga(x|\alpha,\beta)$ denotes the gamma distribution with shape parameter α and scale parameter β. We see that the mean and the variance of this process have the same functional form as for the corresponding Brownian motion with drift. The first passage time distribution function for the vitality model with initial value v_0 is

$$F_{R_t}(t) = \Pr[T_R \le t] = \Pr[R_t \ge v_0] = \int_{v_0}^{\infty} f_{R_t}(x)dx = \frac{\Gamma(r^2t/\sigma^2, v_0r/\sigma^2)}{\Gamma(r^2t/\sigma^2)}, \quad (5.100)$$

where $\Gamma(a,x) = \int_x^\infty z^{a-1}e^{-z}dz$ is the incomplete gamma function for $x \ge 0$ and $a > 0$. This function can be calculated numerically [61]. It is shown by Liao et al. [46] that the corresponding failure rate is increasing, whereas Abdel-Hameed [2] proves that it tends to infinity as $t \to \infty$, which means that the *mortality plateau cannot occur* in accordance with this model.

Park and Padgett [53] have derived a very complex exact expression for the pdf $f_R(t)$. Therefore, a simpler meaningful approximation for (5.100) was suggested by these authors in the form of the Birnbaum-Saunders distribution that can be already effectively analyzed. In a general form, this distribution is given by

$$F_{BS}(t;\lambda,\alpha) = \Phi(\alpha^{-1}h(\lambda t)), \quad t > 0, \quad (5.101)$$

where $\lambda, \alpha > 0$; $\Phi(\cdot)$ is a standard normal distribution function and $h(t) = t^{1/2} - t^{-1/2}$. For our specific case, the corresponding approximation reads [61]:

$$F_{R_t}(t) \approx \Phi\left(\sqrt{\frac{v_0 r}{\sigma^2}}\left[\sqrt{\frac{rt}{v_0}} - \sqrt{\frac{v_0}{rt}}\right]\right). \quad (5.102)$$

It was obtained by Park and Padgett [53] via discretization of the first passage time and then using the central limit theorem. The error of the approximation was not assessed, however, it was stated that it can be used at least for the case when $r >> \sigma$. On the other hand, it should be noted that approximation of distribution functions does not necessarily mean that the tails of the failure rate functions are also approximated. Therefore, given our interest in asymptotic behavior of failure rates, why not to start directly from distribution (5.102) that, similar to the inverse Gaussian distribution, also has a meaningful *process point of view* interpretation. To see this, consider the following damage accumulation model. Let R_t in (5.92) be modeled by the following shock process: suppose that shocks occur at regular intervals at times Δ, 2Δ, 3Δ, ... Let each shock causes a random damage $Y_i > 0$: i.i.d with $E[Y_i] = \Delta\mu$, $\mathrm{Var}(Y_i) = \Delta\sigma^2$. Damages accumulate additively and the k-th shock is survived if the accumulated damage is less than the initial vitality v_0, i.e., $\sum_1^k Y_i \le v_0$. Then, letting $\Delta \to 0$ and using the central limit theorem, after straightforward derivations [48] one can obtain the lifetime distribution (5.100), where

$$\alpha = \sigma/\sqrt{\mu v_0}, \quad \lambda = \mu/v_0. \tag{5.103}$$

Differentiation of (5.101) results in the following density

$$f_{BS}(t;\lambda,\alpha) = \frac{\lambda}{2\alpha\sqrt{2\pi}}\left[\frac{1}{\sqrt{\lambda t}}\left(1+\frac{1}{\lambda t}\right)\right]\exp\left\{-\frac{1}{2\alpha^2}\left(\lambda t - 2 + \frac{1}{\lambda t}\right)\right\}. \tag{5.104}$$

Obviously, $\lim_{t\to 0}\lambda_{BS}(t;\lambda,\alpha) = 0$. Using Theorem 2.1, it can be shown now that the failure rate is bell-shaped [9] and is decreasing to a constant as $t \to \infty$(mortality plateau):

$$\begin{aligned}\lim_{t\to\infty}\lambda_{BS}(t;\lambda,\alpha) &= \lim_{t\to\infty} - \frac{\mathrm{d}\log f_{BS}(t;\lambda,\alpha)}{\mathrm{d}t} \\ &= \frac{\lambda}{2\alpha^2} = \frac{\mu^2}{2\sigma^2}.\end{aligned} \tag{5.105}$$

It follows from (5.105) that, as previously, the effect of initial vitality v_0 is vanishing as $t \to \infty$. Similar to the case of the inverse Gaussian distribution, it can be seen from (5.104) that $\lambda = \mu/v_o$ is a *scale parameter* and, therefore, the usual stochastic ordering (and the hazard rate ordering) holds, i.e., if $v_o(\mu)$ is fixed, then the larger values of $\mu(v_o)$ will result in the larger (smaller) values of the failure rate in $[0,\infty)$.

The possibility of ordering with respect to the values of σ for a general case is not clear (it is an open question in the theory of this distribution). On the other hand, as follows from (5.105), this ordering exists *asymptotically.* Assume now that μ is a realization of a random variable M, whereas σ is a realization of a random variable Σ with support to $[0,\infty]$. Then, similar to the case of the inverse Gaussian distribution, the randomization results in the asymptotically decreasing to 0 population failure rate. Mortality plateaus are theoretically possible in this model only when the supports of the frailties M and Σ are $[a,\infty]$ and $[0,b]$, respectively.

5.11 On the Rate of Aging in Heterogeneous Populations

In this section, we will consider another application of heterogeneity modeling to demography [30]. It should be noted that because of the existing heterogeneity, e.g., in populations for different countries, statistical models describing this property are crucial for this discipline.

Non-parametric classes of lifetime distributions were extensively studied in numerous publications of the last decades (see e.g., the excellent encyclopedic monograph by Lai and Xie [43] and the references therein). One of the main properties of a lifetime random variable that defines the corresponding non-parametric class is a property of *stochastic aging*. This notion can be understood in many ways. The most intuitively evident and the first to be considered in the literature was the class of aging distributions with increasing (nondecreasing) failure rate (IFR) (see, e.g., Barlow and Proschan [6] for this and other basic classes).

Let $T \geq 0$ be a lifetime with an absolutely continuous Cdf $F(t)$, pdf $f(t)$ and the failure rate $\lambda(t) = f(t)/(1 - F(t))$. As in the previous section, we will use the terms *failure rate* and *mortality rate* interchangeably employing the first one mostly for a more general reasoning and the second one in a demographic context. Assume that the derivative $\lambda'(t)$ exists. Then, obviously, $F(t) \in IFR$, if $\lambda'(t) \geq 0, t \geq 0$. We can compare the 'extent of aging' described by different IFR distributions by the value of this derivative at each instant of time. However, this is not always the right thing to do, as intuitively, it is clear that at many instances in order to compare aging for different lifetimes some 'relative reasoning' should be also employed.

In life sciences (e.g., in demography), the *rate of aging* $R(t)$ is usually defined as

$$R(t) \equiv \frac{\mathrm{d}\ln\lambda(t)}{\mathrm{d}t} = \frac{\lambda'(t)}{\lambda(t)}. \tag{5.106}$$

This characteristic already describes the *relative* change in the failure (mortality) rate in an infinitesimally small unit interval of time. It takes into account the value of $\lambda(t)$, as intuition prompts that this measure should often depend not only on the derivative but on the value of the failure rate itself. Indeed, consider, for instance, two failure rates $\lambda(t)$ and $\lambda(t) + c$, where c is a constant. It is clear that the relative change for the second failure rate decreases as c increases and when c is large, the change in the failure rate can be negligible compared with the failure rate itself.

Thus, not only the change in the derivative is important, but also the level of the failure rate as well. Formal definition (5.106) is the simplest way to implement this relative concept. As most of simple definitions that are trying to describe complex properties, it has its *pros* and *contras* (e.g., De Gray [25] mostly focuses on the *contras*). However, this approach to defining the rate of aging is well justified in demography, as for the Gompertz law of mortality (5.86) that describes mortality rate at adult ages, it is a constant, i.e., $R(t) = b$. Thus, in practical demography, b is usually estimated as the slope of the Gompertz regression, i.e., the slope of $\ln\lambda(t)$. It should be understood, however, that $R(t)$ is just a useful (at least, for the Gompertz law) statistical measure, which describes in some 'integrated way' the real aging processes that are manifested by the changes in probabilities of failure (death) over time.

The foregoing considerations refer to the homogeneous populations, where obviously, b can be also regarded as the *individual rate of aging*. However, human populations are heterogeneous, and it is interesting to consider the rate of aging for this case. The general mixture model is described in Sect. 5.1 given by Eqs. (5.10)–(5.12). In what follows, we will focus on the specific multiplicative model (5.17). We will also need the following example:

Example 5.6 Let the frailty Z be a gamma-distributed random variable with shape parameter α and scale parameter β, whereas the baseline distribution be an arbitrary distribution with the failure rate $\lambda(t)$. It is well known [28] that (5.21) is generalized in this case to

$$\lambda_m(t) = \frac{\alpha\lambda(t)}{\beta + \Lambda(t)}, \tag{5.107}$$

where $\Lambda(t)$ is the cumulative failure rate $\Lambda(t) = \int_0^t \lambda(u)du$. Therefore,

$$E[Z|t] = \frac{\alpha}{\beta + \Lambda(t)}.$$

As $E[Z] = \alpha/\beta$ and $\mathrm{Var}(Z) = \alpha/\beta^2$, Eq. (5.107) can now be written in terms of $E[Z]$ and $\mathrm{Var}(Z) \equiv \sigma^2$ in the following way:

$$\lambda_m(t) = \lambda(t)\frac{E^2[Z]}{E[Z] + \sigma^2\Lambda(t)}, \tag{5.108}$$

which, for the specific case $E[Z] = 1$, gives the result of Vaupel et al. [63] that is widely used in demography:

$$\lambda_m(t) = \frac{\lambda(t)}{1 + \sigma^2\Lambda(t)}. \tag{5.109}$$

We will use Eq. (5.109) for analyzing the rate of aging as a function of parameters of the baseline and frailty distributions.

We start analyzing the rate of aging in heterogeneous populations with the specific gamma-Gompertz multiplicative model with the failure rate given by Eq. (5.21). Therefore,

$$\ln \lambda_m(t) = \ln a + bt - \ln\left[1 + (a\sigma^2/b)(\exp\{bt\} - 1)\right] \tag{5.110}$$

and the corresponding rate of aging is

$$R_m(t) = (\ln \lambda_m(t))' = b - \frac{a\sigma^2 \exp\{bt\}}{1 + (a\sigma^2/b)(\exp\{bt\} - 1)}. \tag{5.111}$$

Equation (5.111) states a simple and expected fact that the observed (population) rate of aging $R_m(t)$ is smaller than the individual rate of aging b. The latter, as was staed, corresponds to the homogeneous case. It can be also clearly seen that when σ^2 increases, $R_m(t)$ decreases. Therefore, the following hypothesis makes sense: *the increase in the rate of aging observed in the previous century in the developed countries could be attributed to the decreasing heterogeneity in mortality of populations in these countries.*

Another important feature that follows from (5.111) is that the increase in parameter a also results in the decrease in $R_m(t)$, which can be interpreted as some kind of negative correlation between a of the Gompertz mortality law and the rate of aging.

In the case of arbitrary lifetimes, (5.109) results in

$$\begin{aligned} R_m(t) &= (\ln \lambda_m(t))' \\ &= \frac{\lambda'(t)}{\lambda(t)} - \sigma^2 \frac{\lambda(t)}{1+\sigma^2\Lambda(t)} = R(t) - \sigma^2\lambda_m(t) \end{aligned} \tag{5.112}$$

and, obviously, the rate of aging is also decreasing as a function of variance of the gamma-distributed frailty (for the fixed expectation $E[Z] = 1$). The similar conclusion was made in Yashin et al. [67].

Consider now a general case of the multiplicative model (5.17) not restricting ourselves to the gamma-distributed frailty. It can be shown [30] that

$$\begin{aligned} R_m(t) = (\ln \lambda_m(t))' &= \frac{\lambda'(t)E[Z|T>t] + \lambda(t)E'[Z|T>t]}{\lambda(t)E[Z|T>t]} \\ &= \frac{\lambda'(t)}{\lambda(t)} + \frac{E'[Z|T>t]}{E[Z|T>t]} \\ &= R(t) - \lambda(t)\frac{Var(Z|T>t)}{E[Z|T>t]}. \end{aligned} \tag{5.113}$$

Thus, as previously, the observed (mixture) rate of aging $R_m(t)$ is smaller than the individual rate of aging $R(t)$ defined for the baseline distribution with the failure rate $\lambda(t)$. A similar result using a different approach for derivations was independently recently obtained by Vaupel and Zhang [62]. As we are focusing on the specific multiplicative model (5.17), Eq. (5.113) is very helpful in analyzing a 'proportional effect of environment' on mortality rates.

Suppose now we have two heterogeneous populations with the *same baseline* $\lambda(t)$ and different frailties Z_1, Z_2. In other words, compositions of populations are different. Let

$$\frac{\mathrm{Var}(Z_2|T>t)}{E[Z_2|T>t]} \le \frac{\mathrm{Var}(Z_1|T>t)}{E[Z_1|T>t]}, \ t>0. \tag{5.114}$$

Then it is easy to see that the corresponding rates of aging are ordered as $R_{2m}(t) \ge R_{1m}(t)$. Thus, the rate of aging decreases as the *relative variance* increases, i.e.,

$$R_{2m}(t) - R_{1m}(t) = \lambda(t)\left[\frac{\mathrm{Var}(Z_1|T>t)}{E[Z_1|T>t]} - \frac{\mathrm{Var}(Z_2|T>t)}{E[Z_2|T>t]}\right] \ge 0, \quad \forall t \ge 0.$$

Inequality (5.114) defines a new class of stochastic ordering of random variables that can be called ordering in the sense of the relative variance [30]. The corresponding measure depends not only on the variance (variability), but on the mean as well.

References

1. Aalen OO, Gjessing HK (2001) Understanding the shape of the hazard rate: a process point of view. Stat Sci 16:11–22
2. Abdel-Hameed M (1975) A gamma wear process. IEEE Trans Reliab 24:152–153
3. Anderson JJ (2000) A vitality-based model relating stressors and environmental properties to Arjas E, Norros I (1989). Change of life distribution via a hazard transformation: an inequality with application to minimal repair. Math Oper Res 14:355–361
4. Aven T, Jensen U (1999) Stochastic models in reliability. Springer, New York
5. Aven T, Jensen U (2000) A general minimal repair model. J Appl Probab 37:187–197
6. Barlow R, Proschan F (1975). Statistical theory of reliability and life testing. Holt, Renerhart & Winston, New York
7. Beard RE (1959) Note on some mathematical mortality models. In: Woolstenholme GEW, O'Connor M (eds) The lifespan of animals. Little, Brown and Company, Boston, pp 302–311
8. Beard RE (1971) Some aspects of theories of mortality, cause of death analysis, forecasting and stochastic processes. In: Brass W (ed) Biological aspects of demography. Taylor & Francis, London, pp 57–68
9. Bebbington M, Lai CD, Zitikis R (2008) A proof of the shape of the Birnbaum-Saunders hazard rate function. Math Sci 33:49–56
10. Bergman B (1985) Reliability theory and its applications. Scand J Stat 12:1–41
11. Bingham NH, Goldie CM, Teugels JL (1987) Regular Variation. Cambridge University Press, Cambridge
12. Birnbaum ZW, Saunders SC (1969) A new family of life distributions. J Appl Probab 6:319–327
13. Block HW, Joe H (1997) Tail behaviour of the failure rate functions of mixtures. Lifetime Data Anal 3:269–288
14. Block HW, Li Y, Savits TH (2003) Initial and final behavior of failure rate functions for mixtures and systems. J Appl Probab 40:721–740
15. Block HW, Li Y, Savits TH (2003) Preservation of properties under mixture. Probab Eng Inf Sci 17:205–212
16. Block HW, Li Y, Savits TH, Wang J (2008) Continuous mixtures with bathtub-shaped failure rates. J Appl Probab 45:260–270
17. Block HW, Mi J, Savits TH (1993) Burn-in and mixed populations. J Appl Probab 30:692–702
18. Block HW, Savits TH, Wondmagegnehu ET (2003) Mixtures of distributions with increasing linear failure rates. J Appl Probab 40:485–504
19. Boland PJ, El-Neweihi E (1998) Statistical and information based minimal repair for k out of n systems. J Appl Probab 35:731–740
20. Cha JH, Finkelstein M (2011) Stochastic intensity for minimal repairs in heterogeneous populations. J Appl Probab 48:868–876
21. Cha JH, Finkelstein M (2012) Stochastic analysis of preventive maintenance in heterogeneous populations. Oper Res Lett 40:416–421
22. Chhikara RS, Folks JL (1977) The inverse Gaussian distribution as a lifetime model. Technometrics 19:461–468
23. Clarotti CA, Spizzichino F (1990) Bayes burn-in and decision procedures. Probab Eng Inf Sci 4:437–445
24. Cox DR, Miller HD (1965) Theory of stochastic processes. Methuen & Co., London
25. De Gray DNJ (2005) "The rate of aging": a counterproductively undefinable term. Rejuvenation Res 8(2):77–78
26. Finkelstein M (1992) Some notes on two types of minimal repair. Adv Appl Probab 24:226–228
27. Finkelstein M (2004) Minimal repair in heterogeneous populations. J Appl Probab 41:281–286

28. Finkelstein M (2008) Failure rate modelling for reliability and risk. Springer, London
29. Finkelstein M (2009) Understanding the shape of the mixture failure rate (with engineering and demographic applications). Appl Stoch Models Bus Ind 25:643–663
30. Finkelstein M (2011) On the 'rate of aging' in heterogeneous populations. Math Biosci 22:20–23
31. Finkelstein M (2012) On ordered subpopulations and population mortality at advanced ages. Theor Popul Biol 81:292–299
32. Finkelstein M, Esaulova V (2001) Modelling a failure rate for the mixture of distribution functions. Probab Eng Inf Sci 15:383–400
33. Finkelstein M, Esaulova V (2006) Asymptotic behavior of a general class of mixture failure rates. Adv Appl Probab 38:244–262
34. Finkelstein M, Vaupel JW (2006) The relative tail of longevity and the mean remaining lifetime. Demogr Res 14(6):111–138
35. Gampe J (2010) Human mortality beyond age 110. In: Maier H, Gampe J, Jeune B, Robine J-M, Vaupel JW (eds) Supercentenarians, demographic research monographs. Springer, Berlin, pp 219–230
36. Gavrilov NA, Gavrilova NS (2001) The reliability theory of ageing and longevity. J Theor Biol 213:527–545
37. Glaser RE (1980) Bathtub and related failure rate characterizations. J Am Stat Assoc 75:667–672
38. Gupta RC, Gupta PL (1996) Ageing characteristics of the Weibull mixtures. Probab Eng Inf Sci 10:591–600
39. Gupta RC, Warren R (2001) Determination of change points of nonmonotonic failure rates. Commun Stat Theory Methods 30:1903–1920
40. Gurland J, Sethuraman J (1995) How pooling failure data may reverse increasing failure rate. J Am Stat Assoc 90:1416–1423
41. Hougaard P (2000) Analysis of multivariate survival data. Springer-Verlag, New York
42. Kijima M (1989) Some results for repairable systems with general repair. J Appl Probab 26:89–102
43. Lai CD, Xie M (2006) Stochastic ageing and dependence for reliability. Springer, New York
44. Li Y (2005) Asymptotic baseline of the hazard rate function of mixtures. J Appl Probab 42:892–901
45. Li T, Anderson JJ (2009) The vitality model: a way to understand population survival and demographic heterogeneity. Theor Popul Biol 76:118–131
46. Liao H, Elsayed EA, Chan LY (2006) Maintenance of continuously monitored degrading systems. Eur J Oper Res 175:821–835
47. Lynn NJ, Singpurwalla ND (1997) Comment: "burn-in" makes us feel good. Stat Sci 12:13–19
48. Marshall AW, Olkin I (2007) Life distributions. Springer, New York
49. Missov TI, Finkelstein M (2011) Admissible mixing distributions for a general class of mixture survival models with known asymptotics. Theor Popul Biol 80:64–70
50. Natvig B (1990) On information-based minimal repair and reduction in remaining system lifetime due to a failure of a specific module. J Appl Probab 27:365–375
51. Navarro J, Hernandez PJ (2004) How to obtain bathtub-shaped failure rate models from normal mixtures. Probab Eng Inf Sci 18:511–531
52. Navarro J, Hernandez PJ (2008) Mean residual life functions of finite mixtures, order statistics and coherent systems. Metrica 67:277–298
53. Park C, Padgett WJ (2005) Accelerated degradation models for failure based on geometric Brownian motion and gamma process. Lifetime Data Anal 11:511–527
54. Ross SM (1996) Stochastic processes, 2nd edn. Wiley, New York
55. Steinsaltz D, Evans S (2004) Markov mortality models: implications of quasistationarity and varying initial distributions. Theor Popul Biol 65:319–337
56. Steinsaltz D, Wachter KW (2006) Understanding mortality rate deceleration and heterogeneity. Math Popul Stud 13:19–37

57. Strehler L, Mildvan AS (1960) General theory of mortality and aging. Science 132:14–21
58. Thatcher RE (1999) The long-term pattern of adult mortality and the highest attained age. J Roy Stat Soc 162:5–43
59. Tsiatis A (1975) A nonidentifiability aspect of the problem of competing risks. Proc Natl Acad Sci U S A 72:20–22
60. Valdez-Flores C, Feldman RM (1989) A survey of preventive maintenance models for stochastically deteriorating single-unit systems. Naval Res Logist 36:419–446
61. Van Nortwijk JM (2009) A survey of the application of gamma processes in maintenance. Reliab Eng Syst Saf 94:2–21
62. Vaupel JW, Zhang Z (2010) Attrition in heterogeneous cohorts. Demogr Res 23:737–749
63. Vaupel JW, Manton KG, Stallard E (1979) The impact of heterogeneity in individual frailty on the dynamics of mortality. Demography 16:439–454
64. Weitz JS, Fraser HB (2001) Explaining mortality rates plateaus. PNAS 98:15383–15386
65. Wang H (2002) A survey of maintenance policies of deteriorating systems. Eur J Oper Res 139:469–489
66. Yashin AI, Manton KG (1997) Effects of unobserved and partially observed covariate processes on system failure: a review of models and estimation strategies. Stat Sci 12:20–34
67. Yashin AI, Ukraintseva SV, Boiko SI, Arbeev KG (2002) Individual aging and mortality rate: how are they related. Soc Biol 49:206–217

Chapter 6
The Basics of Burn-in

In this chapter, we introduce the concept of burn-in and review initial research in this area. Burn-in is a method of 'elimination' of initial failures (infant mortality) of components before they are shipped to customers or put into field operation. Usually, to burn-in a component or a system means to subject it to a fixed time period of simulated use prior to the actual operation. That is, before delivery to the customers, the components are exposed to electrical or thermal conditions that approximate the working conditions in field operation. Those components which fail during the burn-in procedure will be scrapped or repaired and only those, which have survived the burn-in procedure will be considered to be of the satisfactory quality. An introduction to this important area of reliability engineering can be found in Jensen and Petersen [24] and Kuo and Kuo [28]. Surveys of research on different aspects of burn-in can be found in Leemis and Beneke [29], Block and Savits [9], Liu and Mazzuchi [30], and Cha [15].

Burn-in has been widely accepted as an effective method of screening out these initial failures due to the large failure rate at early stages of component's life. The failure rate is often initially large, but decreases more or less steeply as the component goes into its useful life period, where it is usually relatively small and nearly constant. This is illustrated by the first part of the traditional bathtub-shaped curve (see Fig. 6.1).

An important question arises: why does the failure rate initially decrease? It is observed that a population of the manufactured items is often composed of two subpopulations—the subpopulation with normal lifetimes (Main Distribution) and the subpopulation with relatively shorter lifetimes ('Freak' Distribution). In practice, items belonging to the 'freak distribution' can be produced along with the items of the main distribution due to, for example, defective resources and components, human errors, unstable production environment caused by uncontrolled significant quality factors, etc. (see, [24, 26]). In this case, the freak distribution generally exhibits the larger failure rate than the main distribution, which results in a mixture of stochastically ordered subpopulations (see Chap. 5). As stated in the previous chapter, the mixture of ordered failure rates is the main cause of the decreasing population failure rate (see also [1, 20]). Therefore, as will be discussed

M. Finkelstein and J. H. Cha, *Stochastic Modeling for Reliability*,
Springer Series in Reliability Engineering, DOI: 10.1007/978-1-4471-5028-2_6,

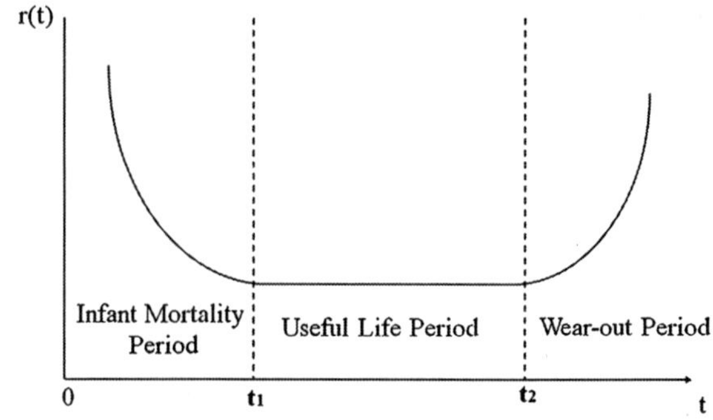

Fig. 6.1 Bathtub-shaped failure rate function

later in this book, the burn-in procedure needs to be studied under the mixture setting. However, the initial research in this area was mostly done based only on the given (merged) population failure rate without considering, e.g., the cause of its initial decrease. Therefore, the objective of this chapter is to introduce basic concepts of this 'classical' burn-in based on the given population failure rate.

As most electronic or mechanical devices often exhibit initially decreasing failure rate, the goal of the burn-in procedure for these items is to shift the failure rate function to the left and to avoid in this way its initially large values. It can be achieved by the fixed time period of simulated use prior to the actual operation. This is the basic logic of the burn-in procedure.

If burn-in is too short, then the items with shorter lifetimes will still remain in the population. On the other hand, if the procedure is too long, then it shortens the lifetime of the items with normal lifetimes. Therefore, to determine the length of the burn-in period (to be called the 'burn-in time') is the most important issue for the corresponding modeling. The best time to stop the burn-in procedure for a given criterion to be optimized is called the *optimal burn-in time*. As burn-in is generally a costly procedure, certain cost structures have been proposed and the corresponding problem of finding the optimal burn-in time has been intensively studied in the literature.

In this chapter, we will provide a detailed background on burn-in. By investigating the relationship between the population failure rate and the performance quality measures, we illustrate how the burn-in procedure can be justified for items with initially decreasing failure rates. We will review some methods for optimizing the performance criteria and that for minimizing various cost functions. It should be noted that latent failures or weak components of highly reliable products require usually a long time to detect or identify. Thus, as stated in Block and Savits [9], for decreasing the length of this procedure, burn-in is often performed in an accelerated environment. In the last part of this chapter, the stochastic models for accelerated burn-in procedures will be introduced.

6.1 Population Distribution for Burn-in

As discussed in the previous section, it is widely believed that many products can be characterized by the bathtub-shaped failure rate functions. This belief is supported by the extensive data from industry.

Definition 6.1 The failure rate function $r(t)$ is said to have a bathtub shape if there exists $0 \leq t_1 \leq t_2 \leq \infty$ such that

(i) $r(t)$ strictly decreases when $0 \leq t \leq t_1$
(ii) $r(t)$ is a constant when $t_1 \leq t \leq t_2$
(iii) $r(t)$ strictly increases when $t_2 \leq t$.

The time instants t_1 and t_2 are called the first and the second change points, respectively. The time interval $[0,\ t_1]$ is called the infant mortality period; the interval $[t_1,\ t_2]$, where $r(t)$ is flat is called the normal operating life period (useful life period); the interval $[t_2,\ \infty)$ is called the wear-out period. In practice, the failure rate during the second period is often only approximately constant. Observe that the above defined bathtub-shaped failure rate function has IFR $(t_1 = 0)$, CFR $(t_1 = 0,\ t_2 = \infty)$ and DFR $(t_1 = \infty)$ as special cases. The typical shape of the bathtub-shaped failure rate function is shown in Fig. 6.1.

Although lifetime distribution functions with the bathtub-shaped failure rates are of importance for burn-in, most popular lifetime distributions do not exhibit this property. However, they can result from the operation of mixing. As discussed in detail in Chap. 5, mixtures can result in different shapes of failure rates [43]. For example, in Glaser [21] it is shown that under appropriate conditions the mixture of two gamma distribution function exhibits a bathtub-shaped failure rate function. Rajarshi and Rajarshi [42] review bathtub distributions and give many references on this topic (see also Sects. 5.1–5.3).

The following is a simple example of a mixture which yields a bathtub-shaped failure rate function.

Example 6.1 Let the population be composed of two subpopulations with subpopulation failure rates $r_1(t) = 0.01t + 0.01$ and $r_2(t) = r_1(t) + 1$. The corresponding mixture failure rate is given in Fig. 6.2.

In this case, the failure rate is bathtub-shaped with $t_1 = t_2$.The mixture pdf is given in Fig. 6.3.

From Fig. 6.3, it follows that the two subpopulation distributions (the 'Freak' and 'Normal' distributions) are well separated in this case. The mean residual lifetime function for this case is given in Fig. 6.4.

From Fig. 6.4, the relationship between the bathtub-shaped failure rate and the corresponding mean residual lifetime function can be observed, which corresponds to our general Theorem 2.4 in Sect. 2.3.

In addition to the traditional bathtub-shaped failure rate, there is also so-called the modified bathtub-shaped failure rate.

Fig. 6.2 Mixture failure rate

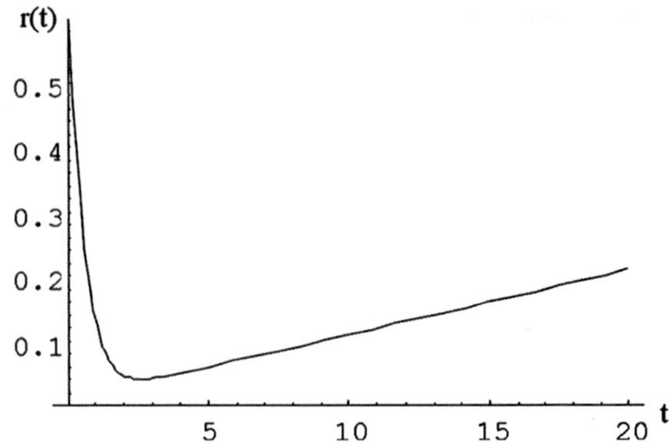

Fig. 6.3 Mixture pdf

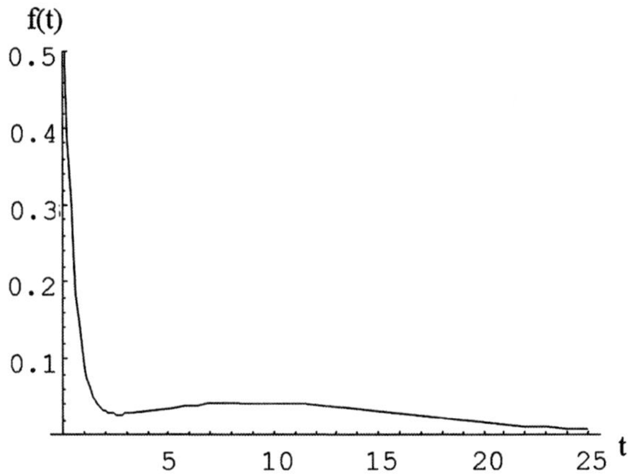

Fig. 6.4 Mean residual lifetime (MRL)

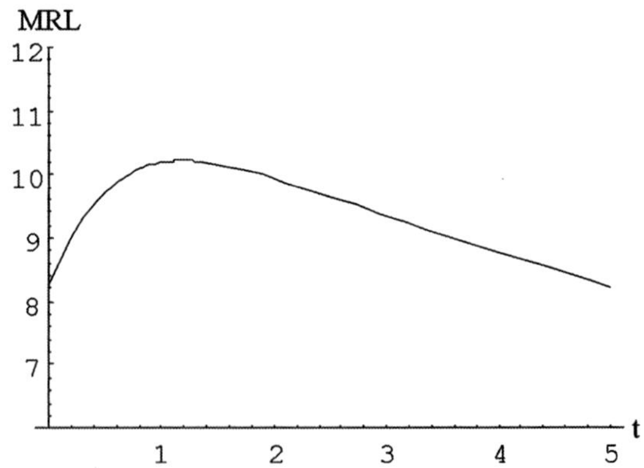

Definition 6.2 The failure rate function $r(t)$ is said to have the modified bathtub shape if there exist $0 \leq t_0 \leq t_1 \leq t_2 \leq \infty$ such that $r(t)$ is strictly increasing in $t \in [0,\ t_0]$, and has a bathtub shape with change points t_1 and t_2 on the interval $t \in [t_0,\ \infty)$.

The modified bathtub-shaped failure rate can be also obtained from the mixture of a distribution for strong components (the 'Main' distribution) and that of weak components (the 'Freak' distribution) [23]. The typical shape of the modified bathtub-shaped failure rate is given in Fig. 6.5.

There has been much research on the shape of failure rates of mixed distributions. For instance, in Block et al. [7, 8] and Klutke et al. [27], the shape of failure rates of mixture distributions, which is neither of the traditional bathtub shape nor of the modified bathtub shape are investigated. Klutke et al. [27] pointed out that the assumption of the traditional bathtub-shaped failure rate could be rather restrictive for burn-in procedures. Kececioglu and Sun [25] state that the bathtub-shaped failure rate is relevant only for 10–15 % of practical applications.

Thus, it is natural to consider a more general form of the failure rate that can describe a wider class of failure rates [18].

Definition 6.3 The failure rate $r(t)$ is eventually increasing if there exists $0 \leq x_0 < \infty$ such that $r(t)$ strictly increases in $t > x_0$. For the eventually increasing failure rate $r(t)$, the first and the second wear-out points t^* and t^{**} are defined by

$$t^* = \inf\{t \geq 0 | r(x) \text{ is non-decreasing in } x \geq t\},$$

$$t^{**} = \inf\{t \geq 0 | r(x) \text{ strictly increases in } x \geq t\}.$$

Obviously, $0 \leq t^* \leq t^{**} \leq x_0 < \infty$ if $r(t)$ is eventually increasing. Observe that if $r(t)$ has a bathtub shape with change points $t_1 \leq t_2 < \infty$, or $r(t)$ has a modified bathtub shape with $0 \leq t_0 \leq t_1 \leq t_2 < \infty$, then it is eventually increasing with $t^* = t_1$ and $t^{**} = t_2$. Therefore, the eventually increasing failure rate includes both the traditional bathtub-shaped and the modified bathtub-shaped failure rates as special cases.

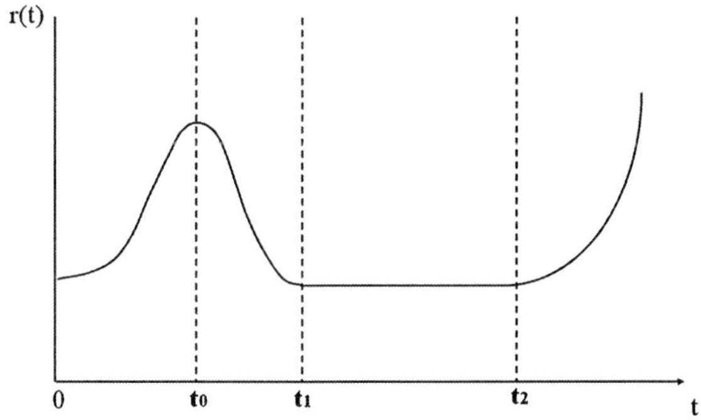

Fig. 6.5 Modified bathtub-shaped failure rate

Definition 6.4 The failure rate $r(t)$ is initially decreasing if there exists $0 < x_0 \leq \infty$ such that $r(t)$ strictly decreases in $t \in [0,\ x_0]$. For an initially decreasing failure rate $r(t)$ the first and second infancy points t_* and t_{**} are defined by

$$t_* = \sup\{t \geq 0 | r(x) \text{ strictly decreases in } x \leq t\},$$

$$t_{**} = \sup\{t \geq 0 | r(x) \text{ is non-decreasing in } x \leq t\}.$$

Obviously, $0 < x_0 \leq t_* \leq t_{**} \leq \infty$ if $r(t)$ is initially decreasing. Mi [38] and Cha [13, 14] studied the problem of determining the optimal burn-in time assuming eventually increasing failure rate function.

6.2 Optimal Burn-in for Performance Criteria

In this section, we will consider burn-in procedures for optimizing system performance measures. There can be different performance measures to be optimized.

a. **Mean Remaining Lifetime**

We first consider burn-in for maximizing the mean remaining lifetime of a system. That is, the MRL after burn-in should be maximized. Watson and Wells [46] initially considered this problem aiming at obtaining the MRL larger than the initial mean life. Essentially, they considered IFR and DFR distributions and showed that if the lifetime distribution is IFR (DFR), then the MRL is always shorter (longer) than the initial mean life. Park [41] examined the effect of burn-in on the MRL of an item with a bathtub-shaped failure rate. It is shown that the first change point does not maximize the mean residual life although the failure rate achieves its minimum value at the point.

Let $\overline{F}(t)$ and $r(t)$ be the survival function and the failure rate function of the lifetime of the system and $M(b)$ be the MRL after the burn-in procedure with burn-in time b. Then, in accordance with (2.6),

$$M(b) = \int_0^\infty \frac{\overline{F}(b+t)}{\overline{F}(b)} \mathrm{d}t = \exp\{\Lambda(b)\} \int_b^\infty \exp\{-\Lambda(t)\}\, \mathrm{d}t,$$

where $\Lambda(t) \equiv \int_0^t r(u)\mathrm{d}u$. We will now find the optimal burn-in time b^* which satisfies

$$M(b^*) = \max_{b \geq 0} M(b).$$

When the population failure rate has the bathtub shape with $0 \leq t_1 \leq t_2 \leq \infty$, we have the following result.

Theorem 6.1 *Suppose that the failure rate* $r(t)$ *is bathtub-shaped with* $0 \leq t_1 \leq t_2 \leq \infty$.

(i) *If* $t_1 = 0$ *and* $t_2 < \infty$, *then* $b^* = 0$.
(ii) *If* $t_1 > 0$ *and* $t_2 = \infty$, *then* $b^* = t_1$.
(iii) *If* $0 < t_1 \leq t_2 < \infty$, *then* $b^* \leq t_1$.

Proof Observe that

$$\begin{aligned} M'(b) &= r(b)\exp\{\Lambda(b)\}\int_b^{\infty}\exp\{-\Lambda(t)\}\mathrm{d}t - 1 \\ &= \exp\{\Lambda(b)\}\int_b^{\infty}(r(b) - r(t))\exp\{-\Lambda(t)\}\mathrm{d}t. \end{aligned}$$

(i) If $t_1 = 0$ and $t_2 < \infty$, then $r(b) - r(t) \leq 0$ for all $t > b$ and for each $b \geq 0$ there exists $t_0 \geq t_2$ such that $r(b) - r(t) < 0$ for all $t \geq t_0$. Therefore, we can conclude that $M'(b) < 0$ and $M(b)$ is strictly decreasing.
(ii) Suppose that $t_1 > 0$ and $t_2 = \infty$. For $0 \leq b < t_1, r(b) - r(t) > 0$ for all $t > b$, and for $b \geq t_1, \lambda(b) - \lambda(t) = 0$ for all $t > b$. Thus, $M(b)$ is strictly increasing in $b \in [0,\ t_1]$ and is constant in $[t_1,\ \infty]$. Therefore, $b^* = t_1$.
(iii) Suppose that $0 < t_1 \leq t_2 < \infty$. In this case, similar to the case (i), it can be shown that $M'(b) < 0$ for all $b \geq t_1$ and therefore we have $b^* \leq t_1$.

□

From the above results, we can conclude that it is not necessary to burn-in products longer than the first change point t_1. More detailed discussions on the relationship between the shape of failure rate and that of the mean residual lifetime function can be found in Mi [36] and Finkelstein [20].

b. **Probability of a Mission Success**

In practice, the systems often perform a given mission in field operation. Let T be the lifetime of the system and τ be the given mission time. The system is required to complete the mission without failure. Then the corresponding success probability for the original system is given by $\overline{F}(\tau)$. If the system is burned-in for a time b, the corresponding success probability is

$$\frac{\overline{F}(b+\tau)}{\overline{F}(b)} = \exp\left\{-\int_b^{b+\tau} r(u)\mathrm{d}u\right\}. \tag{6.1}$$

Then it is desirable to maximize the success probability in (6.1). The set of all optimal burn-in times is defined by

$$B^* = \left\{b^* \geq 0 \left| \frac{\overline{F}(b^*+\tau)}{\overline{F}(b^*)} = \max_{b\,\geq\,0}\frac{\overline{F}(b+\tau)}{\overline{F}(b)}\right.\right\}.$$

It is clear that B^* can equivalently be expressed as

$$B^* = \left\{ b^* \geq 0 \left| \int\limits_{b^*}^{b^*+\tau} r(u)\,\mathrm{d}u = \min_{b \geq 0} \int\limits_{b}^{b+\tau} r(u)\,\mathrm{d}u \right. \right\}.$$

The following theorem characterizes the set B^* when the failure rate has a bathtub shape.

Theorem 6.2 *Let the continuous failure rate function $r(t)$ have a bathtub shape with change points t_1 and t_2, and $\tau > 0$ be a given mission time.*

(i) *If $\tau \leq t_2 - t_1$, then $B^* = [t_1,\ t_2 - \tau]$.*
(ii) *If $\tau > t_2 - t_1$, then $B^* = \{b^*\}$, where $b^* \in [0,\ t_1]$.*

Proof This theorem is proved based on the following property, which can be intuitively well understood. Let $r(t)$ be a continuous bathtub-shaped failure rate function. If there exist $0 \leq b_1 < b_2$ such that $r(b_1) = r(b_2)$ and $b_2 - b_1 = \tau$, then

$$\int\limits_{b}^{b+\tau} r(t)\,\mathrm{d}t \geq \int\limits_{b_1}^{b_2} r(t)\,\mathrm{d}t,\ \forall\, b \geq 0.$$

Then the theorem can be rather straightforwardly proved. The details can be found in Mi [35].

□

Remark 6.1 In practice, the burn-in cost is proportional to the total burn-in time. Therefore, it is reasonable to define the optimal burn-in time as $b^* = \inf B^*$. Then the above theorem states that it is always true that $b^* \leq t_1$.

In certain cases, the mission time may be random. In this case, the optimal burn-in time is studied by the following theorem.

Theorem 6.3 *Let T be a lifetime with bathtub-shaped failure rate function $r(t)$. Let M_1 and M_2 be two random mission times having distribution functions G_1 and G_2, respectively. Let $T_b =_d (T - b|T > b)$ and M_1, M_2 are independent of T. Then $P(T_b > M_i)$ attains its maximum at some finite $b_i^* \in [0,\ t_1]$, $i = 1, 2$. If in addition $M_1 \leq_{st} M_2$, then*

$$\max_{b \geq 0} P(T_b > M_1) \geq \max_{b \geq 0} P(T_b > M_2).$$

Proof The function $\int_b^{b+\tau} r(t)\mathrm{d}t$ is increasing in $b > t_1$ for any τ since $r(t)$ is increasing in $t \geq t_1$. From this, it can be shown that

$$\int\limits_0^\infty \frac{\overline{F}(t_1 + \tau)}{\overline{F}(t_1)} \mathrm{d}G_i(\tau) \geq \int\limits_0^\infty \frac{\overline{F}(b + \tau)}{\overline{F}(b)} \mathrm{d}G_i(\tau),\ \forall b > t_1.$$

Therefore, the continuous function

$$\int_0^\infty \frac{\overline{F}(b+\tau)}{\overline{F}(b)} \mathrm{d}G_i(\tau)$$

must attain its maximum value at some $b_i \in [0,\ t_1]$. The second part of the theorem can also be easily shown. For more details, see Mi [34, 35].

□

c. Mean Number of Failures

Suppose now that in field operation, we replace a failed component by an identical component. Then the number of failures in the time interval $[0,\ t]$ follows a renewal process $\{N_b(t),\ \ t \geq 0\}$, where the subscript b is used to denote that these i.i.d. components have a common survival function $\overline{F}_b(t) = \overline{F}(b+t)/\overline{F}(b)$, i.e., they have survived the same burn-in time b. The problem is to minimize the mean number of failures during a given interval $[0,\ \tau]$, which is given by

$$m_b(\tau) \equiv E[N_b(\tau)] = \sum_{k=1}^{\infty} F_b^{(k)}(\tau),$$

where $F_b^{(k)}$ denotes the k-fold convolution of F_b with itself. Then, we have the following theorem about the optimal burn-in time.

Theorem 6.4 *Let*

$$B^* = \left\{ b^* \geq 0 |\ m_{b^*}(\tau) = \max_{b\,\geq\,0} m_b(\tau) \right\}.$$

Then $B^* \cap [t_2,\ \infty] = \emptyset$. *In particular,*

(i) *if* $\tau > t_2 - t_1$, *then optimal burn-in occurs no later than* t_1, i.e., $B^* \subseteq [0, t_1]$,
(ii) *if* $\tau \leq t_2 - t_1$, *then optimal burn-in occurs at each point of* $[t_1, t_2 - \tau]$, i.e., $B^* = [t_1, t_2 - \tau]$.

Proof The proof of the theorem uses the following intuitively clear property. Suppose that $\overline{F}_1(t) < \overline{F}_2(t)$, for all $t \geq 0$, i.e., $F_1 <_{st} F_2$. Then $F_1^{(k)}(t) > F_2^{(k)}(t)$, for all $t \geq 0$, where $F_i^{(k)}(t)$ is the k-fold convolution of $F_i(t)$, $i = 1,\ 2$. Based on this basic property, the theorem can be proved. The details of the proof are given in Mi [35].

□

d. System Availability

An important measure of performance for a repairable system which can be in one of two states, namely, "up (on)" and "down (off)", is *availability*. Here, by "up" we obviously mean the system is functioning and by "down" we mean that the system is not functioning. Let the state of the system be given by a binary variable

$$X(t) = \begin{cases} 1, & \text{if the systemis up at time } t \\ 0, & \text{otherwise} \end{cases}$$

Then the instant availability at time t (or point availability) is defined by

$$A_t = P(X(t) = 1).$$

Reliability can be obviously considered as a measure of system's effectiveness. However, it is well-known that availability is a more appropriate measure of effectiveness for repairable systems, as it takes into account its maintainability.

As it is very difficult to obtain explicit expressions for $A(t)$ except for a few simple cases, other measures of availability have been proposed. Engineers are often interested in the limiting behavior of this quantity, i.e., the extent to which the system will be available after it has been run for a long time. One of these measures is the steady-state availability (or limiting availability) of a system, which is defined by

$$A = \lim_{t \to \infty} A_t,$$

if the limit exists. Some other types of availability measures that are useful in practical applications can be found in Birolini [5, 6] and Høyland and Rausand [22]. For example, average availability in the interval $(0, t]$ is defined as

$$A_{av}(t) = \frac{1}{t} \int_0^t A_u \, du,$$

which can also be interpreted as the mean fraction of the time interval where the system is functioning during $(0, t]$ (Barlow and Proschan [2]). Note that its limit, $\lim_{t \to \infty} A_{av}(t)$, exists and equals A if $\lim_{t \to \infty} A_t$ exists.

Denote the Cdf and the survival function of the system by $F(t)$ and $\overline{F}(t) = 1 - F(t)$, respectively. The system is replaced by a new identical system on failure. Assume that the repair time distribution is $G(t)$ with mean η. In this case, it is well-known that

$$A = \lim_{t \to \infty} A_t = \frac{\mu}{\mu + \eta},$$

where $\mu = \int_0^\infty \overline{F}(u) du$ is the mean value of the system lifetime. If the system is burned-in for time b, then its limiting availability as the function of burn-in time b is

$$A(b) = \lim_{t \to \infty} A_t = \frac{M(b)}{M(b) + \eta},$$

where

$$M(b) = \int_0^\infty \frac{\overline{F}(b+t)}{\overline{F}(b)} \mathrm{d}t.$$

The objective in this case is to find the optimal burn-in time which maximizes the limiting availability. However, as

$$A(b) = \frac{1}{1 + \eta / M(b)},$$

this problem is equivalent to finding the optimal burn-in time which maximizes the mean remaining lifetime $M(b)$. This problem was already considered previously. Therefore, we will consider now a more general model for systems with two types of failures: the Type I failure (the minor failure that can be 'removed' by a minimal repair), which occurs with probability $1 - p(t)$, where t is the age of the system at failure and the Type II failure (the catastrophic failure that can be 'removed' only by a complete repair or a replacement), which occurs with probability $p(t)$. This model is usually called the general failure model [3, 4, 20].

A new system is burned-in for time b, and it will be put in field use if it survives burn-in. In the field use, the system is replaced by another system, which has survived the same burn-in time b, at the "field use age" T or at the time of the first Type II failure, whichever occurs first. For each Type I failure occurring during field use, only minimal repair will be performed. Denote the lifetime of a system, its distribution function, density function, and the failure rate by X, $F(t)$, $f(t)$ and $r(t)$, respectively.

Assume first that the repair times are negligible. Let the random variable Y_b be the time from 0 to the first Type II failure of a burned-in system, and denote the distribution and the survival functions of Y_b as $G_b(t)$, and $\overline{G}_b(t)$, respectively. Then [3],

$$\begin{aligned}\overline{G}_b(t) =& P(Y_b > t) = \exp\left\{ - \int_0^t p(b+u)\, r(b+u) \mathrm{d}u \right\} \\ =& \exp\left\{ - \left[\Lambda_p(b+t) - \Lambda_p(b) \right] \right\},\ \forall t \geq 0,\end{aligned}$$

where $\Lambda_p(t) \equiv \int_0^t p(u)\, r(u) \mathrm{d}u$. Define $Z_b = \min\{Y_b,\ T\}$. It is easy to see that

$$E(Z_b) = \int_0^T \bar{G}_b(t) \mathrm{d}t.$$

Let $N(b;\ T)$ be the total number of minimal repairs of a burned-in system which occur in the interval $[0,\ Z_b]$, then [3] the expectation of $N(b;\ T)$ is

$$E[N(b;\ T)] = \int_0^T r(b+t)\,\bar{G}_b(t)\,\mathrm{d}t - G_b(T).$$

To consider the system availability, we now assume that the repair times are not negligible. Let η_1, η_2, and η_3 be the means of the minimal repair time, the time for an unplanned replacement caused by the Type II failure, and the mean time for a replacement at field use age T (preventive maintenance), respectively. We further assume that $\int_0^\infty p(u)\,r(u)\,\mathrm{d}u = \infty$. Then, by similar arguments to those described in Cha and Kim [17], it can be shown that the steady-state availability of the system under the policy $(b,\ T)$ is given by

$$\begin{aligned} A(b,\ T) &= \frac{E(\text{total up time in a renewal cycle})}{E(\text{the length of a renewal cycle})} \\ &= \frac{\int_0^T \bar{G}_b(t)\,\mathrm{d}t}{\int_0^T \bar{G}_b(t)\mathrm{d}t + \left[\int_0^T r(b+t)\,\bar{G}_b(t)\,\mathrm{d}t - G_b(T)\right]\eta_1 + G_b(T)\,\eta_2 + \bar{G}_b(T)\,\eta_3} \end{aligned}$$

We first consider the simpler case when only the burn-in procedure is applied but no preventive maintenance is performed, i.e., when $T = \infty$. In this case, the steady-state availability is given by

$$A(b) = \frac{\int_0^\infty \bar{G}_b(t)\,\mathrm{d}t}{\int_0^\infty \bar{G}_b(t)\,\mathrm{d}t + \left[\int_0^\infty r(b+t)\,\bar{G}_b(t)\,\mathrm{d}t\right]\eta_1 + (\eta_2 - \eta_1)}. \tag{6.2}$$

The objective is to find the optimal burn-in time b^* such that

$$A(b^*) = \max_{b\,\geq\,0} A(b).$$

We make the following assumptions:

Assumption 1 The lifetime distribution function $F(t)$ has a bathtub-shaped failure rate function $r(t)$ which has change points $0 \leq s_1 \leq s_2 \leq \infty$.

Assumption 2

$$\eta_2 > \eta_1.$$

Theorem 6.5 *Suppose that the lifetime distribution function $F(t)$ has a bathtub-shaped failure rate function $r(t)$ which has change points $0 \leq s_1 \leq s_2 < \infty$. Let the set V be*

$$V \equiv \{t : p(u)\,r(u) \text{ is nondecreasing for all } u \geq t\}$$

and define $v_1 \equiv \inf\ V$, $w_1 \equiv \max\{s_1,\ v_1\}$, where $v_1 \equiv \infty$ if $V = \phi$. Then the optimal burn-in time, b^* agrees with the following inequality: $b^* \leq w_1$.

If, in addition,

$$\eta_1 r(0) + (\eta_2 - \eta_1)p(0)\, r(0) > \frac{(\eta_2 - \eta_1) + \eta_1 \int_0^\infty r(t) \exp\{-\Lambda_p(t)\}\, \mathrm{d}t}{\int_0^\infty \exp\{-\Lambda_p(t)\}\, \mathrm{d}t}, \quad (6.3)$$

then $b^* > 0$.

Proof Observe that maximization of $A(b)$ is equivalent to minimization of

$$\begin{aligned}
\varphi(b) \equiv \frac{1}{A(b)} - 1 &= \frac{\left[\int_0^\infty r(b+t)\, \bar{G}_b(t)\, \mathrm{d}t\right] \eta_1 + (\eta_2 - \eta_1)}{\int_0^\infty \bar{G}_b(t)\, \mathrm{d}t} \\
&= \frac{\left[\int_0^\infty r(b+t) \exp\left\{-\left[\Lambda_p(b+t) - \Lambda_p(b)\right]\right\} \mathrm{d}t\right] \eta_1 + (\eta_2 - \eta_1)}{\int_0^\infty \exp\left\{-\left[\Lambda_p(b+t) - \Lambda_p(b)\right]\right\} \mathrm{d}t} \\
&= \frac{\left[\int_b^\infty r(t) \exp\left\{-\left[\Lambda_p(t) - \Lambda_p(b)\right]\right\} \mathrm{d}t\right] \eta_1 + (\eta_2 - \eta_1)}{\int_b^\infty \exp\left\{-\left[\Lambda_p(t) - \Lambda_p(b)\right]\right\} \mathrm{d}t}.
\end{aligned} \quad (6.4)$$

Differentiating $\varphi(b)$, we obtain

$$\begin{aligned}
\varphi'(b) = \frac{1}{\left(\int_b^\infty \exp\{-[\Lambda_p(t) - \Lambda_p(b)]\}\mathrm{d}t\right)^2} \Bigg[&\eta_1 \int_b^\infty r(t) \exp\{-[\Lambda_p(t) - \Lambda_p(b)]\}\mathrm{d}t \\
&- \eta_1 \int_b^\infty r(b) \exp\{-[\Lambda_p(t) - \Lambda_p(b)]\}\mathrm{d}t + (\eta_2 - \eta_1) \\
&- (\eta_2 - \eta_1)p(b)r(b) \int_b^\infty \exp\{-[\Lambda_p(t) - \Lambda_p(b)]\}\mathrm{d}t \Bigg].
\end{aligned}$$

Note that, inequality

$$\int_b^\infty r(t) \exp\{-[\Lambda_p(t) - \Lambda_p(b)]\}\mathrm{d}t - \int_b^\infty r(b) \exp\{-[\Lambda_p(t) - \Lambda_p(b)]\}\mathrm{d}t > 0 \quad (6.5)$$

holds for all $b \geq s_1$, and

$$p(b)r(b) \int_b^\infty \exp\{-[\Lambda_p(t) - \Lambda_p(b)]\}\mathrm{d}t \leq \int_b^\infty p(t)r(t) \exp\{-[\Lambda_p(t) - \Lambda_p(b)]\}\mathrm{d}t = 1, \quad (6.6)$$

holds for all $b \geq v_1$ since $\int_0^\infty p(u)\, r(u)\, \mathrm{d}u = \infty$. This implies that $\varphi'(b) > 0$, $\forall\, b \geq w_1$. Therefore, we can conclude that $b^* \leq w_1$.

We can also see that

$$\varphi'(0) = \frac{1}{\left(\int_0^\infty \exp\{-\Lambda_p(t)\}\mathrm{d}t\right)^2}\left[\eta_1 \int\limits_0^\infty r(t)\, \exp\{-\Lambda_p(t)\}\mathrm{d}t \right.$$

$$-\eta_1 r(0) \int\limits_0^\infty \exp\{-\Lambda_p(t)\}\mathrm{d}t + (\eta_2 - \eta_1)$$

$$\left. -(\eta_2 - \eta_1)\, p(0)\, r(0) \int\limits_0^\infty \exp\{-\Lambda_p(t)\}\mathrm{d}t \right].$$

Thus, we obtain $\varphi'(0) < 0$ by (6.3). This means that $\varphi(b)$ is strictly decreasing in the right neighborhood of $b = 0$. Therefore, $b^* > 0$. □

Remark 6.2 Theorem 6.5 indicates that the large value of the initial failure rate, $r(0)$ 'justifies' the positive burn-in time (i.e., $b^* > 0$).

Remark 6.3 If $p(t)$ is eventually increasing, then the set V in Theorem 6.5 is not empty and, therefore, b^* has a nontrivial upper bound.

If the Type II failure probability function $p(t)$ satisfies certain special conditions, the upper bound for the optimal burn-in time can be found more easily than in the case of Theorem 6.5. The following results of Corollaries 6.1 and 6.2 discuss this problem.

Corollary 6.1 *Suppose that the lifetime distribution function $F(t)$ has a bathtub-shaped failure rate function $r(t)$ which has change points $0 \leq s_1 \leq s_2 < \infty$ and $p(t)$ is the bathtub-shaped function with change points u_1 and u_2, $0 \leq u_1 \leq u_2 \leq \infty$. Let $t_1 = \max\{s_1,\ u_1\}, t_2 = \min\{s_2,\ u_2\}$, and assume that $t_1 \leq t_2$ holds. Then $b^* \leq t_1$. Specifically, if $p(t)$ is nondecreasing in $t \geq 0$, then $b^* \leq s_1 < \infty$.*

Proof Under the assumptions, the function $p(u)\, r(u)$ has a bathtub shape with change points $t_1,\ t_2$. Thus, $V = [t_1,\ \infty)$ and $v_1 = t_1$. From Theorem 6.5 we have: $b^* \leq \max\{s_1,\ v_1\} = v_1 = t_1$. If $p(t)$ is nondecreasing, then, obviously, $v_1 \leq s_1$. Thus, $b^* \leq s_1$. □

Definition 6.5 A function $h(t)$ is *eventually nonconstant* if, for any $t' \geq 0$, there exists $t'' \geq t'$ such that $h(t') \neq h(t'')$.

Corollary 6.2 *Suppose that the lifetime distribution function $F(t)$ has a bathtub-shaped failure rate $r(t)$ with change points $0 \leq s_1 \leq s_2 \leq \infty$ and the Type II failure probability $p(t)$ is eventually nonconstant. Let*

$$V^* \equiv \{t : p(u) \text{ is nondecreasing for all } u \geq t\},$$

and $v_1^* \equiv \inf V^*$. Then $b^* \leq \max\{s_1,\ v_1^*\}$.

Proof First note that inequality (6.6) holds strictly for $b \geq \max\{s_1, v_1^*\}$ since $p(u)$ is eventually nonconstant. This implies that the result of Theorem 6.5 still holds, even though now s_2 can be ∞. It is easy to see that $p(u)r(u)$ is nondecreasing in $u \geq \max\{s_1, v_1^*\}$ and thus, $v_1 \leq \max\{s_1, v_1^*\}$ and consequently, $b^* \leq \max\{s_1, v_1\} \leq \max\{s_1, v_1^*\}$. □

We now consider some particular cases of the considered model. First, let $F(t)$ be exponential; that is, its failure rate function is given by $r(t) = r_0$, $\forall t \geq 0$. In this case, from (6.4) we see that

$$\varphi(b) = r_0\eta_1 + \frac{\eta_2 - \eta_1}{\int_0^\infty \exp\{-r_0 \int_b^{b+t} p(u)\mathrm{d}u\}\mathrm{d}t}. \tag{6.7}$$

Theorem 6.6 *Suppose that the two change points of $r(t)$ satisfy: $s_1 = 0$ and $s_2 = \infty$; that is, $F(t)$ is an exponential distribution with $r(t) = r_0$, $\forall t \geq 0$.* (i) *If $p(t)$ is a nonincreasing and eventually nonconstant function of t, then $b^* = \infty$; and* (ii) *if $p(t)$ is a nondecreasing and nonconstant function of t, then $b^* = 0$.*

Proof We prove (i). The result of (ii) can be shown in the similar way. If $p(t)$ is a nonincreasing and eventually nonconstant, then $\int_b^{b+t} p(u)\mathrm{d}u$ is nonincreasing and eventually strictly decreasing in b for each fixed $t > 0$. This implies that the same properties hold for $\varphi(b)$ and, therefore, $b^* = \infty$. □

Generally it is widely believed that if the lifetime of the system follows an exponential distribution, then the burn-in procedure is not necessary ($b^* = 0$). However, the following theorem shows that if there are two types of failure, the burn-in procedure may have to be applied (i.e., $b^* > 0$), even though the distribution of the system is exponential.

Theorem 6.7 *Suppose that the two change points of $r(t)$ are $s_1 = 0$ and $s_2 = \infty$ and $p(t)$ is a bathtub-shaped function with change points $u_1 \leq u_2$.* (i) *If $0 < u_1 \leq u_2 < \infty$, then $0 \leq b^* \leq u_1$;* (ii) *if $0 < u_1 \leq u_2 = \infty$, then b^* can be any value from $[u_1, \infty)$;* (iii) *if $p(t)$ is not a constant function and $p(\infty) \leq p(0)$, then $u_0 \leq b^* \leq u_1$, where $u_0 < u_1$ is uniquely determined by $p(u_0) = p(\infty)$.*

Proof To prove (i), note that (6.6) holds strictly since $u_2 < \infty$. Hence it is still true that $b^* \leq \max\{s_1, v_1\} = v_1$. However, $u_1 \in V$ so $v_1 \leq u_1$ and consequently $b^* \leq u_1$.

In the case of (ii) we see that the left side of (6.5) equals 0 for any $b \geq 0$, and in (6.6) the equality holds for all $b \geq u_1$, but the strict inequality holds for all $b \in [0, u_1)$. This implies that $A(b)$ strictly increases in $b \subset [0, u_1]$ and is a constant in $[u_1, \infty)$ and, therefore, (ii) is true.

Now we consider (iii). From the assumptions, we must have $0 < u_1 \leq u_2 < \infty$. The result of (i) shows that $b^* \leq u_1$. By (6.7), $b^* \geq u_0$. Therefore, $u_0 \leq b^* \leq u_1$. □

In the next theorems we consider the special case when $p(t) = p$.

Theorem 6.8 *Suppose that the lifetime distribution function $F(t)$ has a bathtub-shaped failure rate function $r(t)$ which has change points $0 < s_1 \le s_2 < \infty$ and $p(t) = p$, $0 < p < 1$, that is, the Type II failure probability function is a constant function of t. Then,* (i) *the optimal burn-in time satisfies $0 \le b^* \le s_1$;* (ii) *if we further assume $r(\infty) \le r(0)$, then the optimal burn-in time b^* satisfies $s_0 \le b^* \le s_1$, where s_0 is uniquely determined by $r(s_0) = r(\infty)$.*

Proof When $p(t) = p, 0 < p < 1$, by Corollary 6.1, we have $b^* \le s_1$. Moreover, if $r(\infty) \le r(0)$, we can also show that

$$\varphi'(b) < 0, \quad \forall 0 \le b \le S_0$$

and so $s_0 \le b^* \le s_1$. □

Theorem 6.9 *Suppose that the lifetime distribution function $F(t)$ has a bathtub-shaped failure rate function $r(t)$ which has change points $0 \le s_1 \le s_2 \le \infty$ and $p(t) = p$, $0 < p < 1$. Then the following hold:* (i) *When $s_1 = \infty$,* i.e., *$r(t)$ is strictly DFR, the optimal burn-in time $b^* = \infty$.* (ii) *When $s_1 = 0$, the optimal burn-in time $b^* = 0$.* (iii) *When $s_1 > 0$ and $s_2 = \infty$ the optimal burn-in time b^* could be any value in $[s_1, \infty)$.*

Proof For the case (i), from (6.5) and (6.6), it can be easily shown that $\varphi'(b) < 0$ for all $b \ge 0$. Hence the desired result follows. Similarly, the cases (ii) and (iii) can be proved. □

Remark 6.4 In Theorem 6.8, we assume that $0 < p < 1$. Two special cases are worthy of note.

(i) When $p(t) = 0$, $\forall t \ge 0$, the steady-state availability $A(b)$ does not exist for all $b \ge 0$.
(ii) When $p(t) = 1$, $\forall t \ge 0$, the steady-state availability of the model is given by

$$A(b) = \frac{\mu(b)}{\mu(b) + \eta_2},$$

where $\mu(b) = \int_0^\infty \bar{F}(b+t)/\bar{F}(b)dt$. In this case, the problem of maximizing the steady-state availability is equivalent to maximizing the MRL $\mu(b)$. The latter was discussed in Park [41] and Mi [36].

Remark 6.5 (Optimal Burn-in Time and Preventive Maintenance Policy). When both the burn-in procedure and the replacement policy are applied, the problem of finding the optimal burn-in time b^* and optimal replacement policy T^* such that

$$A(b^*, T^*) = \max_{b \ge 0, T \ge 0} A(b, T)$$

is equivalent to minimizing

$$\varphi(b, T) \equiv \frac{1}{A(b, T)} - 1.$$

Note that, $\varphi(b,\ T)$ is given by

$$\frac{\left[\int_0^T r(b+t)\bar{G}_b(t)\mathrm{d}t - G_b(T)\right]\eta_1 + G_b(T)\eta_2 + \bar{G}_b(T)\eta_3}{\int_0^T \bar{G}_b(t)\mathrm{d}t},$$

which has the same form as the cost function $c(b,\ T)$ in Cha [12]. Therefore, when both the failure rate function $r(t)$ and $p(t)$ are bathtub-shaped, the properties of optimal burn-in time and replacement policy can be easily obtained from the results of Cha [12]. In some other more special cases, i.e., when $p(t) = p,\ 0 \le p \le 1$, the explicit results could be similarly derived from Mi [34] and Cha [10, 11]. We will discuss these problems in Chap. 7.

6.3 Optimal Burn-in for Minimizing Costs

As burn-in is usually a costly procedure, several cost structures have been considered in the literature to determine the optimal burn-in time. Many different cost functions for burn-in are discussed in the review papers of Kuo and Kuo [28] and Leemis and Beneke [29]. Nguyen and Murthy [40] examine the optimal burn-in time to achieve a trade-off between the reduction in the warranty cost and the increase in the manufacturing cost. In this section, we discuss relatively recent research on optimal burn-in which deals with cost functions.

a. **Loss Function Approach**

Let T denote the random failure time of the component of interest and $F(t)$ its distribution function. Clarotti and Spizzichino [19] considered the following choice of decisions in the burn-in problem:

- a_0 The component is immediately put into operation
- a_b The component is tested for the time b (burn-in time) and it is put into operation if it survives the test
- a_∞ As above with $b = \infty$; infinite duration of the test means that the component is judged not suitable for its mission

Let τ be the component mission time. Then the loss function $l(a_b; T)$ is assumed to have the following form:

$$l(a_b;\ T) = \begin{cases} c_1, & \text{if } T < b \\ C, & \text{if } b < T < b+\tau \\ -K, & \text{if } T > b+\tau \end{cases}$$

$0 < c_1 < C,\ K > 0$. Then the expected loss is

$$\begin{aligned}\phi(b) =&E[l(a_b;T)] = c_1F(b) + C[F(b+\tau) - F(b)] - K[1 - F(b+\tau)]\\ =&c_1 + (C - c_1)\overline{F}(b) - (K + C)\overline{F}(b+\tau).\end{aligned}$$

The properties of the optimal decision rule are given in the following theorem.

Theorem 6.10 *Suppose that $g(b) \equiv f(b+\tau)/f(b)$ is strictly increasing in b. Then*

(i) a_∞ *is optimal if and only if* $\lim_{b\to\infty} g(b) < (C - c_1)/(C + K)$.
(ii) a_0 *is optimal if and only if* $g(0) \geq (C - c_1)/(C + K)$.
(iii) $a_{b^*}\,(0 < b^* < \infty)$ *is optimal if and only if* $g(b^*) = (C - c_1)/(C + K)$.

Proof It is easy to see that

$$\phi'(b) \leq 0 \text{ if and only if } f(b+\tau)/f(b) \leq (C - c_1)/(C + K).$$

The condition $\lim_{b\to\infty} g(b) < (C - c_1)/(C + K)$ implies that $\phi'(b) < 0$ for all $b > 0$, and thus, $\phi(b)$ is strictly decreasing in $b > 0$ in this case. On the other hand, the condition $g(0) \geq (C - c_1)/(C + K)$ implies that $\phi'(b) > 0$ for all $b > 0$, and thus, $\phi(b)$ is strictly increasing in $b > 0$ in this case. Finally, if there exists $0 < b^* < \infty$ such that $g(b^*) = (C - c_1)/(C + K)$ then this means that $\phi'(b) < 0$, for $b < b^*$, and $\phi'(b) > 0$, for $b > b^*$. Therefore, this b^* is the unique optimal burn-in time. □

Example 6.2 Let T be distributed according to the Weibull distribution with the scale parameter $\beta, 0 < \beta < 1$:

$$f(t) = \lambda\beta t^{\beta-1}\exp\{-\lambda t^\beta\},\ t > 0,$$

$$g(b) = \left(1 + \frac{\tau}{b}\right)^{\beta-1}\exp\{-\lambda[(b+\tau)^\beta - b^\beta\}.$$

It is easy to see that $g(b)$ is strictly increasing. In this case, a_{b^*} $(0 < b^* < \infty)$ is optimal if there exists $0 < b^* < \infty$ such that

$$\left(1 + \frac{\tau}{b^*}\right)^{\beta-1}\exp\{-\lambda[(b^*+\tau)^\beta - b^{*\beta}\} = (C - c_1)/(C + K).$$

b. **Average Cost for Non-repairable Systems**

In Mi [33, 36], the problems of minimizing the cost functions that are defined via the cost of the burn-in procedure as such and of the gain obtained from field operation were considered.

Consider the fixed burn-in time b and begin to burn-in a new device. If the device fails before the time b, then it is repaired with the shop repair cost c_s, and the repaired device is burned-in again and so on. It is assumed that the repair is complete, i.e., the repaired device is as good as new. If the device survives the burn-in time b, then it is put into field operation. The cost for burn-in is assumed to be proportional to the total burn-in time with proportionality constant c_0. We will derive now the average cost incurred for obtaining the first component surviving the burn-in. Let $h(b)$ be the total cost incurred until the first component surviving burn-in is obtained. Let X_1 be the time to failure of a new component, which is first subject to the burn-in procedure, and $F(t)$ be its distribution function. Then, by conditioning, $E\,[h(b)]$ can be derived as follows. Given the event $\{X_1 > b\}$ (the new component survives burn-in time b at the first trial), the conditional expectation $E\,[h\,(b)|X_1 > b]$ is

$$E\,[h\,(b)|X_1 > b] = c_0 b = c_0 E[\min\{b, X_1\}|X_1 > b]. \tag{6.8}$$

On the other hand, given the event $\{X_1 \le b\}$ (the new component does not survive burn-in time b at the first trial), the conditional expectation $E\,[h\,(b)|X_1 \le b]$ is

$$\begin{aligned} E[h(b)|X_1 \le b] &= c_0 E[X_1|X_1 \le b] + c_s + E[h(b)] \\ &= c_0 E[\min\{b,\ X_1\}|X_1 \le b] + c_s + E[h(b)]. \end{aligned} \tag{6.9}$$

From (6.8) and (6.9), the following equation holds:

$$E\,[h\,(b)] = c_0 E\,[\min\{b, X_1\}] + c_s F(b) + E\,[h\,(b)]F(b). \tag{6.10}$$

Then from (6.10), $E\,[h\,(b)]$ is given by

$$E\,[h\,(b)] = \frac{1}{\overline{F}(b)}\left[c_0 \int_0^b \overline{F}(t)\mathrm{d}t + c_s F(b)\right],$$

where $\overline{F}(t) = 1 - F(t)$.

Let τ be the mission time. In field operation, the cost C is incurred if the burned-in component fails before τ. On the other hand, if the burned-in component survives the mission time τ, then the gain K is obtained. Then the complete cost function is given by

$$c(b) = -c_s + \frac{c_0 \int_0^b \overline{F}(t)\mathrm{d}t + c_s}{\overline{F}(b)} + C\frac{\overline{F}(b) - \overline{F}(b+\tau)}{\overline{F}(b)} - K\frac{\overline{F}(b+\tau)}{\overline{F}(b)}. \tag{6.11}$$

For this cost function, the following intuitively obvious result had been shown in Mi [33]: the optimal burn-in time b^* which minimizes $c(b)$ in (6.11) never exceeds the first change point t_1 if $F(t)$ is described by the bathtub-shaped failure rate.

We consider now a different cost structure for field operation. The average cost of the burn-in procedure is the same as before. The second part, which is the gain part, is proportional to the mean life of the component used in field operation. That is, if we denote the proportionality constant by $K > 0$, then the gain is given by

$$K \frac{\int_b^\infty \overline{F}(t)\mathrm{d}t}{\overline{F}(b)}.$$

Thus, the cost function in this case has the following form:

$$c(b) = -c_s + \frac{c_0 \int_0^b \overline{F}(t)\mathrm{d}t + c_s}{\overline{F}(b)} - K \frac{\int_b^\infty \overline{F}(t)\mathrm{d}t}{\overline{F}(b)}. \tag{6.12}$$

Clearly, the term

$$\frac{c_0 \int_0^b \overline{F}(t)\mathrm{d}t + c_s}{\overline{F}(b)}$$

in (6.12) is strictly increasing in $b \geq 0$. According to the result in Park [41] (see also Theorem 6.1 in Sect. 6.2) the mean remaining life

$$\mu(b) = \frac{\int_b^\infty \overline{F}(t)\mathrm{d}t}{\overline{F}(b)}$$

is decreasing in $b \geq t_1$ if the failure rate function $r(t)$ has the bathtub shape with the first change point t_1. Hence, the minimum value for $c(b)$ in (6.12) cannot be obtained in the interval (t_1, ∞), i.e., $b^* \leq t_1$. Therefore, we have obtained the following result.

Theorem 6.11 *Suppose that the failure rate function $r(t)$ has a bathtub shape. Then for the cost function given in* (6.12), *the optimal burn-in time b^* is unique and* $0 \leq b^* \leq t_1$. *If, in addition, $\mu(0)K - c_s > 0$ and $r(0) > (c_0 + K)/(\mu(0)K - c_s)$, then $b^* > 0$.*

c. **Average Cost for Systems with Replacement**

Consider the case when the system can be replaced at each failure during field operation. The corresponding cost structure is described as follows. As before, we burn-in components until we obtain the one that survives burn-in. This component is then put into field operation. If it fails during field operation, it is *replaced* by another burned-in component at a cost c_f. We assume that $c_f > c_s$, where c_s is the cost of each shop repair. In this case, it is clear that, by the theory of renewal reward process (see, e.g, Ross [43]), the long-run average cost rate is given by

$$
\begin{aligned}
c(b) &= \left(-c_s + \frac{c_0 \int_0^b \overline{F}(t)\mathrm{d}t + c_s}{\overline{F}(b)} + c_f\right) \times \left(\frac{\int_b^\infty \overline{F}(t)\mathrm{d}t}{\overline{F}(b)}\right)^{-1} \\
&= \frac{c_f - (c_f - c_s)F(b) + c_0 \int_0^b \overline{F}(t)\mathrm{d}t}{\int_b^\infty \overline{F}(t)\mathrm{d}t}.
\end{aligned}
\tag{6.13}
$$

In this case, we have the following result for the optimal burn-in time. The proof of the following theorem can be found in Mi [36].

Theorem 6.12 *Let the failure rate function $r(t)$ be continuous and have a bathtub shape with change points t_1 and t_2. If $c_f > c_s$, then the optimal burn-in time b^* minimizing the cost function $c(b)$ in (6.13) is unique and satisfies $0 \le b^* \le t_1$. Furthermore, $b^* > 0$ if and only if*

$$
r(0) > \frac{c_0\mu(0) + c_f}{(c_f - c_s)\mu(0)}.
$$

Remark 6.6 For the cost structures considered in Theorems 6.11 and 6.12, we can see that the large initial failure rate $r(0)$ 'justifies' burn-in, i.e., $b^* > 0$.

6.4 Models for Accelerated Burn-in Procedures

Burn-in is generally considered to be expensive and, therefore, the duration of burn-in is typically limited. Furthermore, for today's highly reliable products, many latent failures or weak components require a long time to detect or identify. Thus, as stated in Block and Savits [9], for decreasing the length of this procedure, burn-in is often performed in an accelerated environment. Some real examples of accelerated burn-in procedures in electronic industry can be found in Kuo and Kuo [28] and Usami and Yoshioka [44] (see also [23–26, 47]). However, most of stochastic modeling in the literature has been performed only for the normal stress levels during burn-in. Recently, Cha [13] proposed a new stochastic model for the accelerated burn-in procedure based on the concept of virtual age. In Cha and Finkelstein [16], the model of Cha [13] has been extended to the case of proportional (additive) hazards during burn-in. In this section, we will consider approaches developed in these two papers.

6.4.1 Failure Rate Model for Accelerated Burn-in Procedure

This subsection is devoted to constructing the probabilistic frame for accelerated burn-in procedure, which employs the basic statistical property commonly used in accelerated life tests (ALT). Accelerated life tests are frequently used in practice to obtain timely information on the life distribution or performance over time of highly reliable products in an affordable amount of testing time. Test units are used more frequently than usual or are subjected to larger than usual levels of stress or stresses like temperature and voltage. Then the information obtained from the test performed under a larger stress is used to predict actual product performance under a normal (usual) stress. Nelson [39] provides an extensive and comprehensive source for practical methodology, basic theory, and examples for accelerated testing. Meeker and Escobar [31] present a good review paper on these issues.

Let X denote the lifetime of a component (with the Cdf $F(t)$) used under the usual stress. We assume that X is an absolutely continuous nonnegative random variable and thus the pdf $f(t)$ and the failure rate $r(t)$ exist. Also denote by X_A the lifetime of a component that operates in the accelerated environment and is characterized by $F_A(t), f_A(t), r_A(t)$ accordingly. The Accelerated Failure Time (AFT) regression model is the most widely used parametric failure time regression model in ALT. Under this model, the larger stress has the effect of 'shrinking' time through a scale factor. This can be modeled as

$$F_A(t) = F(\rho \cdot t), \ \forall t \geq 0, \tag{6.14}$$

where ρ is a constant that depends on the accelerated stresses. As given in Sect. 3 of Meeker and Escobar [31], a more general model can be defined as

$$F_A(t) = F(\rho(t)), \ \forall t \geq 0, \tag{6.15}$$

where $\rho(t)$ depends on the accelerated environment. Since the accelerated environment gives rise to larger stresses than the usual environment, $\rho \geq 1$ for model (6.14) and $\rho(t) \geq t$ for all $t > 0$ and $\rho(0) = 0$, for model (6.15). Furthermore, we assume that $\rho(t)$ in (6.15) is strictly increasing, continuous, and differentiable. Then, (6.14) and (6.15) imply that $X_A \leq_{st} X$. Here, the notation $\leq_{st}$ denotes the usual stochastic order, that is, we say that Z_1 is said to be *smaller than* Z_2 *in the usual stochastic order* denoted $Z_1 \leq_{st} Z_2$ if $F_2(t) \leq F_1(t)$, for all $t \geq 0$, where $F_1(t)$ and $F_2(t)$ are the distribution functions of Z_1 and Z_2, respectively (see Sect. 2.8). From (6.15), the failure rate function in the accelerated environment is given by

$$r_A(t) = \frac{\rho'(t)f(\rho(t))}{1 - F(\rho(t))} = \rho'(t)r(\rho(t)).$$

On the other hand, right after a new component has been burned-in during a fixed burn-in time b under the accelerated environment, its 'virtual age' [20], transformed to the usual level of stress should be larger than b. Assume that the survival function in the normal environment of the burned-in (accelerated burn-in during time b) component is given by

$$\bar{F}_b(t) \equiv \exp\left(-\int_0^t r(a(b) + u)\mathrm{d}u\right) = \frac{\bar{F}(a(b) + t)}{\bar{F}(a(b))}, \tag{6.16}$$

where the function $a(b)$ satisfies $a(b) \geq b$ for all $b \geq 0$, $a(0) = 0$, and is strictly increasing and differentiable. Equation (6.16) implies that the performance of a component with accelerated burn-in time b is the same as that of a component that has been operated under the usual stress during the time $a(b)$. From (6.16), it is easy to see that the burned-in component with the accelerated burn-in time b and the 'field use age' u has the failure rate

$$r(a(b) + u), \forall u \geq 0.$$

Combining the accelerated burn-in phase and the field use phase, the failure rate function of a component under accelerated burn-in time b, which is denoted by $\lambda_b(t)$, can be defined as

$$\lambda_b(t) = \begin{cases} \rho'(t)\, r(\rho(t)), & \text{if } 0 \leq t \leq b \text{ (burn-in phase)} \\ r(a\,(b) + (t - b)), & \text{if t} \geq \text{b (field use phase).} \end{cases} \tag{6.17}$$

Generally, the shapes of $\rho(t)$ and $a\,(b)$ depend on the level(s) of stress(es) during the accelerated burn-in process. Larger levels of stresses would yield rapidly increasing functions $\rho(t)$and $a(b)$, whereas smaller levels of stresses would result in slowly increasing $\rho(t)$and $a(b)$.

Similar to the cumulative exposure model described in Nelson [39], assume now that the virtual age $a(t)$ in the normal environment 'produces' the same population cumulative fraction of units failing as the age t does in the accelerated environment. Formally, it means that

$$F(a(t)) = F_{\mathrm{A}}(t). \tag{6.18}$$

Applying the inverse operator F^{-1} to both sides of (6.18):

$$a(t) = F^{-1}(F_{\mathrm{A}}(t)) = F^{-1}(F(\rho(t))) = \rho(t), \forall t \geq 0.$$

Therefore, $a(t) = \rho(t), \forall t \geq 0$. (See Finkelstein [20] for a similar reasoning). In what follows, unless otherwise specified, we will implicitly assume this relationship.

Since the conditions on the functions $\rho(t)$ and $a(b)$ are not too restrictive, the failure rate model in (6.17) can be considered as a general one. It can be used in a wide range of applications. Also note that, if the burn-in procedure is performed under normal stresses, then obviously, $\lambda_b(t) = r(t)$ for all $t \geq 0$. Therefore, the accelerated burn-in model under consideration is a generalization of the burn-in model without acceleration.

6.4.2 *Optimal Burn-in Time*

In this section we consider the following burn-in procedure under an accelerated environment.

- Burn-in procedure: Fix a burn-in time b and begin to burn-in a new component under the accelerated environment. If the component fails before the burn-in time b, then repair it completely with shop repair cost c_s and then burn-in the repaired component again, and so on.

During the complete (perfect) repair, the failed component is repaired to the "as good as new" state. This means that the lifetime of the repaired component is independent of the lifetime of the original component and has the same distribution function. Note that the burn-in procedure stops when there is no failure during the fixed burn-in time for the first time. We assume that the cost for accelerated burn-in is proportional to the total burn-in time with proportionality constant c_1.

Let $h(b)$ be the total cost incurred until the first component surviving burn-in is obtained. Then, following the procedures similar to those described in the previous section, $E[h(b)]$ can be obtained by

$$E[h(b)] = \frac{1}{\overline{F}_A(b)}\left[c_1\int_0^b \overline{F}_A(t)\mathrm{d}t + c_s F_A(b)\right],$$

where

$$\overline{F}_A(t) \equiv 1 - F_A(t) = \exp\left\{-\int_0^t \rho'(u) r(\rho(u))\mathrm{d}u\right\}.$$

In the following we discuss three burn-in models, which can be considered as generalizations of those studied by Mi [37].

a. **Model 1: Gain Due to No Failure within Mission Time**

Many practical problems require a component to accomplish a task in field operation with a given mission time $\tau > 0$. This means that the given mission is accomplished when the component operates continuously without any failure for the time τ. The corresponding cost function consists in this case of the following three parts:

(i) the mean cost $E[h(b)]$ for obtaining a component that survives the accelerated burn-in time b;
(ii) the cost C incurred by the event $\{X_b \le \tau\}$;
(iii) the gain K that results from the event $\{X_b > \tau\}$,

where X_b is the time of failure of the component that has survived the accelerated burn-in time b and thus is described the distribution function $F_b(t) \equiv 1 - \overline{F}_b(t)$. Then the cost function $c(b)$ is given by

$$
\begin{aligned}
c(b) &= E[h(b)] + C F_b(\tau) - K\overline{F}_b(\tau) \\
&= \frac{1}{\overline{F}_A(b)}\left[c_1\int_0^b \overline{F}_A(t)\mathrm{d}t + c_s F_A(b)\right] - (C+K)\frac{\overline{F}(a(b)+\tau)}{\overline{F}(a(b))} + C.
\end{aligned} \tag{6.19}
$$

Let b^* be optimal burn-in time that minimizes $c(b)$ in Equation (6.19). Then the following result gives an upper bound for the optimal burn-in time b^*.

Theorem 6.13 *Suppose that the failure rate function $r(t)$ is eventually increasing with the first wear-out point t^*. Then $a^{-1}(t^*)$ is an upper bound for the optimal burn-in time b^* satisfying $c(b^*) = \min_{b\geq 0} c(b)$, that is, $b^* \leq a^{-1}(t^*) < \infty$, where $a^{-1}(t^*)$ is the unique solution of the equation $a(t) = t^*$. In addition, if*

$$(C+K)a'(0)\exp\{-\Lambda(\tau)\} - c_s\rho'(0) > 0$$

and

$$r(0) > \frac{c_1 + (C+K)a'(0)r(\tau)\exp\{-\Lambda(\tau)\}}{(C+K)a'(0)\exp\{-\Lambda(\tau)\} - c_s\rho'(0)},$$

then, $b^* > 0$, where $\Lambda(t) \equiv \int_0^t r(u)\mathrm{d}u$.

Proof Observe that the cost function $c(b)$ in (6.19) can be rewritten as

$$
\begin{aligned}
c(b) = c_1 &\int_0^b \exp\{-\Lambda(\rho(t))\}\mathrm{d}t \cdot \exp\{\Lambda(\rho(b))\} + c_s\exp\{\Lambda(\rho(b))\} \\
&- (C+K)\times\exp\{-[\Lambda(a(b)+\tau) - \Lambda(a(b))]\} + (C - c_s),
\end{aligned}
$$

where $\Lambda(t) \equiv \int_0^t r(u)\mathrm{d}u$. To prove $0 \leq b^* \leq a^{-1}(t^*)$, it suffices to show that $c(b)$ strictly increases for all $b \in \{b : a(b) > t^*\}$. Then

$$
\begin{aligned}
c'(b) = c_1 &\left[\rho'(b)\, r(\rho(b))\exp\{\Lambda(\rho(b))\} \times \int_0^b \exp\{-\Lambda(\rho(t))\}\mathrm{d}t + 1\right] \\
&+ c_s\rho'(b)\, r(\rho(b)) \times \exp\{\Lambda(\rho(b))\} \\
&+ (C+K)\, a'(b)\, [r(a(b)+\tau) - r(a(b))] \times \exp\{-[\Lambda(a(b)+\tau) - \Lambda(a(b))]\}.
\end{aligned}
$$

Since the functions $\rho(t)$ and $a(t)$ are strictly increasing and $[r(a(b)+\tau) - r(a(b))] \geq 0$ for all b such that $a(b) > t^*$ by the eventually increasing failure rate assumption, it holds that $c'(b) > 0$ for all b such that $a(b) > t^*$. This means that $c(b)$ is strictly increasing for all $b \in \{b : a(b) > t^*\}$.

For the second part of the theorem, consider the derivative of $c(b)$ evaluated at $b = 0$. It is easy to check that

$$c'(0) = c_1 + c_s\rho'(0)r(0) + (C + K)a'(0)[r(\tau) - r(0)]\exp\{-\Lambda(\tau)\}\,.$$

Assume that

$$(C + K)a'(0)\exp\{-\Lambda(\tau)\} - c_s\rho'(0) > 0$$

and

$$r(0) > \frac{c_1 + (C + K)a'(0)r(\tau)\exp\{-\Lambda(\tau)\}}{(C + K)a'(0)\exp\{-\Lambda(\tau)\} - c_s\rho'(0)},$$

then $c'(b) < 0$ holds. This means that $c(b)$ is strictly decreasing in a right-hand neighborhood of $b = 0$. Therefore $b^* > 0$.

□

b. Model 2: Gain Proportional to the Mean Time to the Failure

In the second model, the cost structure that contains the following two parts will be considered:

(i) the average cost $E[h(b)]$ incurred during the burn-in;
(ii) the gain that is proportional to the mean time to failure in field operation with proportionality constant K.

Thus, the objective cost function $c(b)$ has the following form:

$$c(b) = E[h(b)] - K\int_0^\infty \overline{F}_b(t)\mathrm{d}t = \frac{1}{\overline{F}_A(b)}\left[c_1\int_0^b \overline{F}_A(t)\mathrm{d}t + c_sF_A(b)\right] - K\frac{\int_{a(b)}^\infty \overline{F}(t)\mathrm{d}t}{\overline{F}(a(b))}. \qquad (6.20)$$

Let b^* be the optimal burn-in time that minimizes $c(b)$ in Eq. (6.20). Then we have the following result.

Theorem 6.14 *Suppose that the failure rate function $r(t)$ is eventually increasing with the first wear-out point t^*. Then $a^{-1}(t^*)$ is an upper bound for optimal burn-in time b^*, which satisfies $c(b^*) = \min_{b\geq 0} c(b)$, that is, $b^* \leq a^{-1}(t^*) < \infty$, where $a^{-1}(t^*)$ is the unique solution of the equation $a(t) = t^*$. In addition, if*

$$Ka'(0)\int_0^\infty \exp\{-\Lambda(t)\}\mathrm{d}t - c_s\rho'(0) > 0$$

and

$$r(0) > \frac{c_1 + Ka'(0)}{Ka'(0)\int_0^\infty \exp\{-\Lambda(t)\}\mathrm{d}t \; - c_s\rho'(0)},$$

then $b^* > 0$.

Proof Observe that the cost function $c(b)$ in (6.20) can be rewritten as

$$\begin{aligned} c(b) = {} & c_1 \int_0^b \exp\{-\Lambda(\rho(t))\}\mathrm{d}t \cdot \exp\{\Lambda(\rho(b))\} + c_s \exp\{\Lambda(\rho(b))\} \\ & - K \int_{a(b)}^\infty \exp\{-\Lambda(t)\}\mathrm{d}t \cdot \exp\{\Lambda(a(b))\} - c_s. \end{aligned}$$

By differentiating $c(b)$,

$$\begin{aligned} c'(b) = {} & c_1 \left[\rho'(b)\, r(\rho(b)) \exp\{\Lambda(\rho(b))\} \times \int_0^b \exp\{-\Lambda(\rho(t))\}\mathrm{d}t + 1\right] \\ & + c_s\rho'(b)\, r(\rho(b)) \exp\{\Lambda(\rho(b))\} + Ka'(b) \\ & \times \left[1 - r(a(b)) \int_{a(b)}^\infty \exp\{-\Lambda(t)\}\mathrm{d}t \exp\{\Lambda(a(b))\}\right], \end{aligned}$$

where, by the eventually increasing failure rate assumption, for all b such that $a(b) > t^*$,

$$\begin{aligned} & Ka'(b)\left(1 - r(a(b)) \int_{a(b)}^\infty \exp\{-\Lambda(t)\}\mathrm{d}t \cdot \exp\{\Lambda(a(b))\}\right) \\ & \geq Ka'(b)\left(1 + \int_{a(b)}^\infty -r(t) \exp\{-\Lambda(t)\}\mathrm{d}t \cdot \exp\{\Lambda(a(b))\}\right) \\ & = Ka'(b)(1 + [\exp\{-[\Lambda(t) - \Lambda(a(b))]\}]_{a(b)}^\infty) = 0 \end{aligned}$$

hold. Therefore, $c'(b) > 0$ for all b such that $a(b) > t^*$. This means that $c(b)$is strictly increasing for all $b \in \{b : a(b) > t^*\}$. Thus, we can conclude that $b^* \leq a^{-1}(t^*)$. For the second part of the theorem, consider the derivative of $c(b)$ evaluated at $b = 0$. It is easy to check that

$$c'(0) = c_1 + c_s\rho'(0)\, r(0) + Ka'(0)\left[1 - r(0)\int_0^\infty \exp\{-\Lambda(t)\}\mathrm{d}t\right].$$

If

$$Ka'(0)\int_0^\infty \exp\{-\Lambda(t)\}\mathrm{d}t - c_s\rho'(0) > 0$$

and

$$r(0) > \frac{c_1 + Ka'(0)}{Ka'(0)\int_0^\infty \exp\{-\Lambda(t)\}\mathrm{d}t - c_s\rho'(0)},$$

then $c'(0) < 0$. This means that $c(b)$ is strictly decreasing in the right-hand neighborhood of $b = 0$. Therefore, $b^* > 0$.

□

c. Model 3: Replacement at Failure During Field Operation

The cost structure that is considered in this model is described as follows:

(i) as in the preceding model, burn-in the components until the one that survives the burn-in is obtained. Then put this component into field operation;
(ii) if the component fails during the field operation, it is replaced by another burned-in component at the cost c_f.

We assume that $c_f > c_s$, where c_s is the cost of each shop repair. During field operation, a failure of a system causes its unavailability and thus it generally incurs additional high penalty cost. Therefore, the constraint of $c_f > c_s$ can be considered as a reasonable assumption. Let $R(t)$ be the total operational cost in the field operational interval $[0,\ t]$. Then, by the theory of renewal reward process (see, e.g., [43]), the long-run average cost rate as a function of the burn-in time b is given by

$$c(b) = \lim_{t\to\infty} \frac{E[R(t)]}{t} = \frac{E[\text{Total cost in a renewal cycle}]}{E[\text{The length of a renewal cycle}]} = \frac{E[h(b)] + c_f}{\int_0^\infty \overline{F}_b(t)\mathrm{d}t}. \quad (6.21)$$

Let b^* be optimal burn-in time that satisfies $c(b^*) = \min_{b\geq 0} c(b)$. Then the following result gives an upper bound for optimal burn-in.

Theorem 6.15 *Suppose that the failure rate function $r(t)$ is eventually increasing with the first wear-out point t^*. Then the optimal burn-in time b^* satisfies $0 \leq b^* \leq a^{-1}(t^*)$, where $a^{-1}(t^*) < \infty$ is the unique solution of the equation $a(t) = t^*$. In addition, if*

$$a'(0)\mu(0)c_f - \rho'(0)\mu(0)c_s > 0$$

and

$$r(0) > \frac{c_1\mu(0) + a'(0)c_f}{a'(0)\mu(0)c_f - \rho'(0)\mu(0)c_s},$$

then $b^* > 0$, *where* $\mu(0) \equiv \int_0^\infty \exp\{-\Lambda(t)\}\mathrm{d}t$.

Proof Observe that the cost rate $c(b)$ in (6.21) can be rewritten as

$$c(b) = \frac{\eta(b)}{\mu(b)},$$

where $\mu(b)$ is given by

$$\mu(b) \equiv \exp\{\Lambda(a(b))\}\int_{a(b)}^\infty \exp\{-\Lambda(t)\}\mathrm{d}t$$

and $\eta(b)$ is defined by

$$\eta(b) \equiv c_1 \exp\{\Lambda(\rho(b))\}\int_0^b \exp\{-\Lambda(\rho(t))\}\mathrm{d}t + c_s \exp\{\Lambda(\rho(b))\} + (c_f - c_s).$$

It is clear that $\eta(b)$ is strictly increasing in $b > 0$. On the other hand, by the eventually increasing failure rate function assumption, it can be shown that $\mu'(b) < 0$ for all b such that $a(t) > t^*$. These consequently imply that $c(b)$ strictly increases for all $b \in \{b : a(b) > t^*\}$. Therefore, we can conclude that $b^* \leq a^{-1}(t^*)$.

For the second part of the theorem, consider the derivative of $c(b)$ evaluated at $b = 0$. It is easy to see that

$$c'(0) = \frac{1}{[\mu(0)]^2} \times \Phi(0),$$

where $\Phi(0)$ is given by

$$\Phi(0) = [c_1 + c_s\rho'(0)r(0)]\mu(0) - [a'(0)r(0)\mu(0) - a'(0)]c_f.$$

If

$$a'(0)\mu(0)c_f - \rho'(0)\mu(0)c_s > 0$$

and

$$r(0) > \frac{c_1\mu(0) + a'(0)c_f}{a'(0)\mu(0)c_f - \rho'(0)\mu(0)c_s},$$

then $c'(0) < 0$ holds. This means that $c(b)$ is strictly decreasing in the right-hand neighborhood of $b = 0$. Therefore, $b^* > 0$. □

Remark 6.7 In each of the Theorems 6.13–6.15, the sufficient conditions for a positive burn-in (i.e., $b^* > 0$) have been obtained in the form of two inequalities. From these conditions, we can see that (i) the large field cost $(C,\ c_f)$, the large field reward (K),and the small shop repair cost (c_s) and (ii) the large initial failure rate $r(0)$ justify the positive burn-in. In particular, if $r(t)$ has the bathtub shape with two change points t_1 and t_2, then the upper bounds for the optimal burn-in time b^* in the considered models are given by $a^{-1}(t_1)$. Mi [37] considered the optimal burn-in time for various additive cost models under the usual level of stress. We can see that the considered burn-in models can be reduced to those studied by Mi [37] if we set $\rho(t) = t$ for all $t \geq 0$ and $a(b) = b$ for all $b \geq 0$.

Remark 6.8 When $\rho(t) = \rho t$, the following simple relationship holds:

$$E[X_A] = \int_0^\infty \overline{F}_{\mathrm{A}}(t)\mathrm{d}t = \int_0^\infty \overline{F}(\rho t)\mathrm{d}t = \frac{1}{\rho}E[X],$$

where parameter ρ has a clear 'physical' meaning. Furthermore, it follows from Theorems 6.13–6.15 that the upper bound for the optimal burn-in time in this case is simply given by $a^{-1}(t^*) = (1/\rho)t^*$.

6.4.3 Proportional Hazards and Additive Hazards Models

In this subsection, the extended model in Cha and Finkelstein [16] will be introduced. Observe that ALM (ALT) (6.15) is not the only way of modeling the impact of a severer (accelerated) environment. Consider the *proportional hazards* (PH) model for describing the failure rates in both environments. This model is used in numerous applications:

$$r_{\mathrm{A}}(t) = \gamma r(t), \forall t \geq 0, \tag{6.22}$$

where $\gamma \geq 1$. Then obviously,

$$r(t) \leq r_A(t), \forall t \geq 0, \text{i.e.}, X_A \leq_{fr} X$$

and the failure rate ordering of the corresponding lifetimes (see, e.g., [43] and Sect. 2.8) holds. In accordance with (6.18) and (6.22), we can formally define the corresponding virtual age from the following equation:

$$
\begin{aligned}
F(a(t)) &= 1-\exp\{-\int_0^{a(t)} r(u)\mathrm{d}u\} = 1-\exp\{-\Lambda(a(t))\}, \\
&= 1-\exp\{-\gamma\Lambda(t)\} = 1-\exp\{-\gamma\int_0^{t} r(u)\mathrm{d}u\} = F_{\mathrm{A}}(t),
\end{aligned} \tag{6.23}
$$

where $\Lambda(t) \equiv \int_0^t r(u)\mathrm{d}u$. Then

$$a(t) = \Lambda^{-1}(\gamma\Lambda(t)), \forall t \geq 0,$$

and the combined failure rate function, similar to (6.17), is defined as

$$\lambda_b(t) = \begin{cases} \gamma r(t), & 0 \leq t < b, \\ (\Lambda^{-1}(\gamma\Lambda(b)) + (t-b)), & t \geq b. \end{cases}$$

Similar to Cha [13], consider the following setting. Let the cost for the accelerated burn-in is proportional to the total burn-in time with proportionality constant c_1. Then, the expected cost during burn-in is given by

$$\frac{1}{\overline{F}_A(b)}\left[c_1\int_0^b \overline{F}_A(t)\mathrm{d}t + c_s F_A(b)\right],$$

where c_s, as previously, is the cost of a complete repair (shop repair price). Given the mission time τ, the cost function consists of the following three parts :

(i) The mean 'aggregated' cost for 'obtaining' a component that survives the accelerated burn-in time b;
(ii) The cost C incurred by the event $\{X_b \leq \tau\}$ (Failure of the Mission);
(iii) The gain K that results from the event $\{X_b > \tau\}$ (Success of the Mission),

where X_b is the time to failure of the component which survived the accelerated burn-in procedure during time b. Then the corresponding total expected cost function $c(b)$ for the proportional hazards model is

$$
\begin{aligned}
c(b) = c_1 &\int_0^b \exp\{-\gamma\Lambda(t)\}\mathrm{d}t \cdot \exp\{\gamma\Lambda(b)\} + c_s\exp\{\gamma\Lambda(b)\} \\
&- (C+K)\exp\{-[\Lambda(a(b)+\tau) - \Lambda(a(b))]\} + (C-c_s),
\end{aligned} \tag{6.24}
$$

where $a(b) = \Lambda^{-1}(\gamma\Lambda(b))$.

Theorem 6.16 *Suppose that the failure rate function $r(t)$ is eventually increasing with the first wear-out point t^*. Then $a^{-1}(t^*) = \Lambda^{-1}\left(\frac{1}{\gamma}\Lambda(t^*)\right)$ is an upper bound* *for optimal burn-in time* b^*, *which minimizes* (6.24), *that is,* $b^* \leq \Lambda^{-1}(\frac{1}{\gamma}\Lambda(t^*)) < \infty$. *In addition, if*

$$(C+K)a'(0)\exp\{-\Lambda(\tau)\} - c_s\gamma > 0$$

and

$$r(0) > \frac{c_1 + (C+K)a'(0)r(\tau)\exp\{-\Lambda(\tau)\}}{(C+K)a'(0)\exp\{-\Lambda(\tau)\} - c_s\gamma},$$

then $b^* > 0$.

Proof The cost function in (6.24) is composed of two parts: the average cost during burn-in and that during field operation. Note that the average cost during burn-in is obviously strictly increasing for all $b > 0$. Similar to the proof of Theorem 1 in Cha [13], it can be shown that, by the eventually increasing failure rate assumption, the average cost during field operation is strictly increasing for all b such that $a(b) > t^*$. This means that $c'(b) > 0$ for all b such that $a(b) > t^*$, which implies the first result.

On the other hand,

$$c'(0) = c_1 + c_s\gamma r(0) + (C+K)a'(0)[r(\tau) - r(0)]\exp\{-\Lambda(\tau)\}$$

and it is easy to see that if the two given conditions are satisfied, then $c'(0) < 0$, which finally implies that the optimal burn-in time b^* is positive. □

Example 6.3 Suppose that the failure rate $r(t)$ is given by

$$r(t) = \begin{cases} -2t+2, & 0 \leq t \leq 1, \\ 1, & 1 \leq t \leq 10, \\ t-9, & 10 \leq t. \end{cases}$$

Let $\gamma \equiv 2.0$. Clearly it is a traditional bathtub-shaped failure rate and therefore it is eventually increasing with the first wear-out point $t^* = 1.0$. Then $\Lambda(t^*) = 3/2$ and the upper bound is given by

$$a^{-1}(t^*) = \Lambda^{-1}(\frac{1}{\gamma}\Lambda(t^*)) = 1/2.$$

Another specific case that can be used for ordering lifetimes in normal and accelerated environments is the additive hazards (AH) model, which is also widely used in survival analysis:

$$r_A(t) = r(t) + q(t), \forall t \geq 0, \tag{6.25}$$

where $q(t) \geq 0, \forall t \geq 0$. From (6.25),

$$\overline{F}_A(t) = \exp\{-\int_0^t r(u) + q(u)du\} = \exp\{-\Lambda(t) - Q(t)\},$$

where $\Lambda(t) \equiv \int_0^t r(u)du$ and $Q(t) = \int_0^t q(u)du$. Similar to (6.23),

$$\begin{aligned} F(a(t)) &= 1 - \exp\{-\int_0^{a(t)} r(u)du\} = 1 - \exp\{-\Lambda(a(t))\} \\ &= 1 - \exp\{-\Lambda(t) - Q(t)\} = 1 - \exp\left\{-\int_0^t r(u) + q(u)du\right\} = F_A(t), \end{aligned}$$

and the virtual age is given by

$$a(t) = \Lambda^{-1}(\Lambda(t) + Q(t)), \forall t \geq 0,$$

whereas the combined failure rate for this case is defined as

$$\lambda_b(t) = \begin{cases} r(t) + q(t), & 0 \leq t < b, \\ r(\Lambda^{-1}(\Lambda(b) + Q(b)) + (t-b)), & t \geq b. \end{cases}$$

The corresponding cost function can be expressed as

$$\begin{aligned} c(b) =& c_1 \int_0^b \exp\{-\Lambda(t) - Q(t)\}dt \cdot \exp\{\Lambda(b) + Q(b)\} + c_s \exp\{\Lambda(b) + Q(b)\} \\ &- (C+K)\exp\{-[\Lambda(a(b)+\tau) - \Lambda(a(b))]\} + (C - c_s), \end{aligned} \tag{6.26}$$

where $a(b) = \Lambda^{-1}(\Lambda(b) + Q(b))$. Then, similar to Theorem 6.16, the following theorem can be proved:

Theorem 6.17 *Suppose that the failure rate function $r(t)$ is eventually increasing. Then $a^{-1}(t^*) = (\Lambda + Q)^{-1}(\Lambda(t^*))$, where $(\Lambda + Q)^{-1}(t)$ is the inverse function of $\Lambda(t) + Q(t)$, is an upper bound for optimal burn-in time b^* which minimizes* (6.26), *that is, $b^* \leq (\Lambda + Q)^{-1}(\Lambda(t^*)) < \infty$. In addition, if*

$$(C+K)a'(0)\exp\{-\Lambda(\tau)\} - c_s > 0$$

and

$$r(0) > \frac{c_1 + c_s q(0) + (C+K)a'(0)r(\tau)\exp\{-\Lambda(\tau)\}}{(C+K)a'(0)\exp\{-\Lambda(\tau)\} - c_s},$$

then $b^* > 0$.

Example 6.4 Consider the failure rate function in Example 6.3. Suppose that $q(t) = 2t$. Then the upper bound for the optimal burn-in time is given by

$$a^{-1}(t^*) = (\Lambda + Q)^{-1}(\Lambda(t^*)) = 3/4.$$

The choice between AL or PH (AH) models actually depends on physical processes that lead to failures of items and on the impact of changing environment on these processes. Many types of electronic items can be described by the corresponding linear PH model (for two environments), whereas mechanical items are more likely to be described by the AL model [20].

6.4.4 Relationships Between the Models

In this subsection, in line with burn-in models considered before, we briefly reformulate some of the obvious but useful relationships for analysis [32]. As it was already mentioned, the PH model (6.22) and the AH model (6.25) imply the AL model (6.15). On the other hand, as γ in (6.22) is a constant, we can write $r_{\mathrm{A}}(t) = \gamma r(t) = r(t) + (\gamma - 1)r(t)$ and the PH model defined in such a way is a specific case of the AH model. Therefore,

$$AL \supset AH \supset PH.$$

As the PH model (6.22) and the specific linear case (6.14) of the AL model are the popular practical tools for modeling the accelerated environment, it makes sense to point out the relationship between these two models.

Using (6.15), $r_A(t)$ can be written as

$$r_A(t) = \left(\frac{\rho'(t)r(\rho(t))}{r(t)}\right) \cdot r(t).$$

Therefore, if

$$\left(\frac{\rho'(t)r(\rho(t))}{r(t)}\right) = \gamma, \tag{6.27}$$

these models are identical. Specifically, for the linear case (6.14), the condition (6.27) becomes

$$\left(\frac{\rho r(\rho t)}{r(t)}\right) = \gamma. \tag{6.28}$$

It is satisfied if the distribution of the component is Weibull (specifically, exponential), which is, of course, is a well-known fact.

The similar reasoning can be used for obtaining formally the conditions for 'identity' between the AL and the AD (the AD and the PH), however, only for the linear case (6.28) these results have a real practical meaning.

References

1. Badía FG, Berrade MD, Campos CA (2003) Why do failure rates decrease? Proceedings of the VIIth Zaragoza-Pau Conference on Applied Mathematics and Statistics 27:97–104
2. Barlow RE, Proschan F (1975) Statistical theory of reliability and life testing, probability models. Holt, Rinehart and Winston, New York
3. Beichelt FE (1993) A unifying treatment of replacement policies with minimal repair. Naval Research, Logistics 40:51–67
4. Beichelt FE, Fischer K (1980) General failure model applied to preventive maintenance policies. IEEE Trans Reliab 29:39–41
5. Birolini A (1985) On the use of stochastic processes in modeling reliability problems. Springer-Verlag, New York
6. Birolini A (1994) Quality and reliability of technical systems. Springer-Verlag, New York
7. Block HW, Li Y, Savits TH (2003) Initial and final behavior of failure rate functions for mixtures and systems. Journal of Applied Probability 40:721–740
8. Block HW, Li Y, Savits TH (2003) Preservation of properties under mixture. Probability in the Engineering and Informational Sciences 17:205–212
9. Block HW, Savits TH (1997) Burn-in. Statistical Science 12:1–19
10. Cha JH (2000) On a better burn-in procedure. Journal of Applied Probability 37:1099–1103
11. Cha JH (2001) Burn-in procedures for a generalized model. Journal of Applied Probability 38:542–553
12. Cha JH (2003) A further extension of the generalized burn-in model. Journal of Applied Probability 40:264–270
13. Cha JH (2006) A stochastic model for burn-in procedures in accelerated environment. Naval Research Logistics 53:226–234
14. Cha JH (2006) An extended model for optimal burn-in procedures. IEEE Trans Reliab 55:189–198
15. Cha JH (2011) A survey of burn-in and maintenance models for repairable systems. In: Tadj L, Ouali M, Yacout S, Ait-Kadi D (eds) Replacement Models with Minimal Repair. Springer, London, pp 179–203
16. Cha JH, Finkelstein M (2011) On stochastic models for accelerated burn-in procedure. International Journal of Applied Mathematics and Statistics 24:51–59
17. Cha JH, Kim JJ (2002) On the existence of the steady-state availability of imperfect repair model. Sankhya, Series B 64:76–81
18. Cha JH, Mi J (2005) Optimal burn-in procedures in a generalized environment. Int J Reliab Qual Saf Eng 12:189–202
19. Clarotti CA, Spizzichino F (1990) Bayes burn-in and decision procedures. Probability in Engineering and Informational Sciences 4:437–445
20. Finkelstein M (2008) Failure Rate modeling for Reliability and Risk. Springer, London

21. Glaser RE (1980) Bathtub and related failure rate characterizations. Journal of the American Statistical Association 75:667–672
22. Høyland A, Rausand M (1994) System reliability theory: Models and statistical methods. Wiley, New York
23. Hui YV, Lu WL (1996) Cost optimization of accelerated burn-in. International Journal of Quality and Reliability Management 13:759–762
24. Jensen F, Petersen NE (1982) Burn-in. John Wiley, New York
25. Kececioglu D, Sun F (1995) Environmental Stress Screening : Its Qualification, Optimization, and Management. Prentice Hall, New Jersey
26. Kececioglu D, Sun F (2003) Burn-in Testing: Its Quantification and Optimization. DEStech Publications, Lancaster
27. Klutke G, Kiessler PC, Wortman MA (2003) A critical look at the bathtub curve. IEEE Trans Reliab 52:125–129
28. Kuo W, Kuo Y (1983) Facing the headaches of early failures: A state-of-the-art review of burn-in decisions. Proc IEEE 71:1257–1266
29. Leemis LM, Beneke M (1990) Burn-in models and methods: A review. IIE Trans 22:172–180
30. Liu X, Mazzuchi TA (2008) The optimal burn-in: state of the art and new advances for cost function formulation. Recent Advances in Reliability and Quality in Design. Springer, London, In
31. Meeker WQ, Escobar LA (1993) A review of recent research and current issues of accelerated testing. International Statistical Review 61:147–168
32. Meeker WQ, Escobar LA (1998) Statistical Methods for Reliability Data. John Wiley, New York
33. Mi J (1991). *Optimal burn-in*. Ph.D. Thesis. Dept. Statistics; Univ. Pittsburgh, Pittsburgh
34. Mi J (1994) Burn-in and maintenance policies. Advances in Applied Probability 26:207–221
35. Mi J (1994) Maximization of survival probability and its application. Journal of Applied Probability 31:1026–1033
36. Mi J (1995) Bathtub failure rate and upside-down bathtub mean residual life. IEEE Trans Reliab 44:388–391
37. Mi J (1996) Minimizing some cost functions related to both burn-in and field use. Operations Research 44:497–500
38. Mi J (2003) Optimal burn-in time and eventually IFR. Journal of the Chinese Institute of Industrial Engineers 20:533–542
39. Nelson W (1990) Accelerated testing : statistical models, test plans, and data analysis. John Wiley & Sons, Inc, New York
40. Nguyen DG, Murthy DNP (1982) Optimal burn-in time to minimize cost for products sold under warranty. IIE Trans 14:167–174
41. Park KS (1985) Effect of burn-in on mean residual life. IEEE Trans Reliab 34:522–523
42. Rajarshi S, Rajarshi MB (1988) Bathtub distributions: A review. Communication in Statistics - Theory and Methods 17:2597–2621
43. Ross SM (1996) Stochastic Processes, 2nd edn. Wiley, New York
44. Usami K, Yoshioka H (2004). Dynamic sleep control for finite-state-machines to reduce active leakage power. *IEICE Transactions on Fundamentals of Electronics, Communications and Computer Sciences*, E87-A, 3116-3123
45. Vaupel JW, Yashin AI (1985) Heterogeneity ruses: some surprising effects of selection on population dynamics. The American Statistician 39:176–185
46. Watson GS, Wells WT (1961) On the possibility of improving the mean useful life of items by eliminating those with short lives. Technometrics 3:281–298
47. Wu CL, Su CT (2002) Determination of the optimal burn-in time and cost using an envitonmental stress approach:a case study in switch mode rectifier. Reliability Engineering and System Safety 76:53–61

Chapter 7
Burn-in for Repairable Systems

In the previous chapter, the emphasis was made on the burn-in procedures for non-repairable items. If a non-repairable item fails during burn-in, then, obviously, it is just scraped and discarded. However, an expensive, complex product or device will not be discarded on account of failure of its part, but rather a repair will be performed. Therefore, in this chapter, we deal mostly with repairable items. Note that the contents of this chapter are rather technical and it can be skipped by a less mathematically oriented reader.

After the survey provided by Block and Savits [3], there has been much research on burn-in procedures, especially for repairable systems. These studies include: (i) various reliability models which jointly deal with burn-in and maintenance policies; (ii) burn-in procedures for general failure model; (iii) a stochastic model for the accelerated burn-in procedure.

7.1 Burn-in and Maintenance Policies: Initial Models

In this section, reliability models that jointly deal with burn-in and *maintenance policies* will be considered. We describe properties of *joint optimal solutions* for burn-in and replacement times for each of these models. Mi [10] was the first to consider the joint optimization problem for determining optimal burn-in and replacement times.

Let $F(t)$ be the distribution function of the absolutely continuous lifetime X. Mi [10] studied an optimal burn-in and maintenance policy under the assumption that $F(t)$ has a bathtub-shaped failure rate function. The following burn-in procedure was considered.

Burn-in Procedure A
Consider a fixed burn-in time b and begin to burn-in a new device. If the device fails before the burn-in time b, then repair it completely with the shop repair cost $c_s > 0$, then burn-in the repaired device again, and so on. If the device survives the burn-in time b, then it is put into field operation [10].

M. Finkelstein and J. H. Cha, *Stochastic Modeling for Reliability*,
Springer Series in Reliability Engineering, DOI: 10.1007/978-1-4471-5028-2_7,

We assume here that the repair is complete, i.e., the repaired device is as good as new. Let the cost of burn-in be proportional to the total burn-in time with proportionality constant $c_0 > 0$.

Let $h(b)$ denote the total cost incurred for obtaining the device which survives the burn-in procedure. Then, similar to Sect. 6.3, the mean cost $E[h(b)]$ can be obtained as

$$E[h(b)] = c_0 \frac{\int_0^b \bar{F}(t)\,\mathrm{d}t}{\bar{F}(b)} + c_s \frac{F(b)}{\bar{F}(b)}.$$

7.1.1 Model 1

For field operation, Mi [10] considered two types of replacement policies, depending on whether the device is repairable or not. For a non-repairable device, the age replacement policy is considered. That is, the device is replaced by a new burned-in device at the time of its failure or 'field-use age' T, whichever occurs first. Let c_f denote the cost incurred for each failure in field operation and $c_a (0 < c_a < c_f)$, the cost incurred for each non-failed item which is replaced by a new burned-in item at its field-use age T. Then, by the theory of renewal reward processes, the long-run average cost rate $c(b, T)$ is given by

$$c(b, T) = \frac{k(b) + c_f F_b(T) + c_a \bar{F}_b(T)}{\int_0^T \bar{F}_b(t)\,\mathrm{d}t},$$

where $\bar{F}_b(t)$ is the conditional survival function, i.e., $\bar{F}_b(t) \equiv \bar{F}(b+t)/\bar{F}(b)$ and $k(b) \equiv E[h(b)]$. Mi [10] have obtained certain results regarding the optimal burn-in time b^* and the optimal age T^* which satisfy

$$c(b^*, T^*) = \min_{b \geq 0,\, T > 0} c(b, T)$$

However, there are several useful 'hidden' properties which can be found in the proof of the corresponding theorem and, therefore, we reformulate the result as follows.

Theorem 7.1 *Suppose that the failure rate function $r(t)$ is bathtub-shaped and differentiable. Let*

$$B_1 \equiv \left\{ b \geq 0 : \mu(b) r(\infty) > \frac{c_f + k(b)}{c_f - c_a} \right\},$$

where $\mu(b) \equiv \int_0^\infty \bar{F}_b(t)\,\mathrm{d}t$, and $B_2 \equiv [0, \infty) \backslash B_1$. Then properties of the optimal burn-in time b^* and of the optimal replacement policy T^* can be stated in detail as follows:

Case 1. $B_1 = [0, \infty), B_2 = \phi$. Let $T^*(b)$ be the unique solution of the equation

$$r(b + T)\int_0^T \frac{\bar{F}(b + t)}{\bar{F}(b)}\mathrm{d}t + \frac{\bar{F}(b + T)}{\bar{F}(b)} = \frac{c_f + k(b)}{c_f - c_a}. \tag{7.1}$$

Then the optimal $(b^*, T^*) = (b^*, T^*(b^*))$, where $0 \le b^* \le t_1$, is the value that satisfies

$$b^* + T^*(b^*) = \min_{0 \le b \le t_1}(b + T^*(b)).$$

Case 2. $B_1 = \phi, B_2 = [0, \infty)$. The optimal $(b^*, T^*) = (b^*, \infty)$, where $0 \le b^* \le t_1$, is the value that satisfies

$$\frac{c_f + k(b^*)}{\mu(b^*)} = \min_{0 \le b \le t_1}\frac{c_f + k(b)}{\mu(b)}.$$

Case 3. $B_1 = \phi$, $B_2 = \phi$. For $b \in B_1$, let $T^*(b)$ be the unique solution of Eq. (7.1). Furthermore, let $b_1^* \in [0, t_1] \cap B_1$ satisfy

$$b_1^* + T^*(b_1^*) = \min_{b \le t_1,\, b \in B_1}(b + T^*(b)),$$

and $b_2^* \in [0, t_1] \cap B_2$ satisfy

$$\frac{c_f + k(b_2^*)}{\mu(b_2^*)} = \min_{b \le t_1,\, b \in B_2}\frac{c_f + k(b)}{\mu(b)}.$$

If

$$(c_f - c_a)r(b_1^* + T^*(b_1^*)) \le \frac{c_f + k(b_2^*)}{\mu(b_2^*)},$$

then $(b^*, T^*) = (b_1^*, T^*(b_1^*))$. Otherwise the optimal (b^*, T^*) is (b_2^*, ∞).

Proof The proof for a more general model is given in the proof of Theorem 7.4 in this chapter and thus it is omitted. □

7.1.2 Model 2

For a repairable device, applying the same burn-in procedure as before, block replacement with minimal repair on failures is performed in field operation. More precisely, fix a $T > 0$ and replace the component at times T, $2T$, $3T$, …, with a new burned-in component. Also, at each intervening failure, a minimal repair is performed. Let $c_m > 0$ be the cost of a minimal repair, and $c_r > 0$ be the cost of replacement. In this case, the long-run average cost rate is given by

$$c(b,T) = \frac{1}{T}\left(k(b) + c_m \int_b^{b+T} r(t)\,dt + c_r\right). \tag{7.2}$$

The following theorem [10] provides the properties of optimal (b^*, T^*) minimizing $c(b,T)$.

Theorem 7.2 *Suppose that the failure rate function $r(t)$ is bathtub-shaped and differentiable. Let*

$$B_1 \equiv \left\{ b \geq 0 : \int_b^{\infty} [r(\infty) - r(t)]\,dt \right.$$
$$\left. > \frac{1}{c_m \bar{F}(b)}\left[(c_r - c_s)\bar{F}(b) + c_s + c_0 \int_0^b \bar{F}(t)\,dt\right]\right\},$$

and $B_2 \equiv [0, \infty)\backslash B_1$. Then the properties of the optimal burn-in time b^* and the replacement policy T^* can be stated in detail as follows:

Case 1. $B_1 = [0, \infty)$, $B_2 = \phi$. Let $T^*(b)$ be the unique solution of the equation

$$Tr(b + T) - \int_b^{b+T} r(t)\,dt = \frac{1}{c_m \bar{F}(b)}\left[(c_r - c_s)\bar{F}(b) + c_s + c_0 \int_0^b \bar{F}(t)\,dt\right]. \tag{7.3}$$

Then, the optimal $(b^*, T^*) = (b^*, T^*(b^*))$, where $0 \leq b^* \leq t_1$, is the value which satisfies

$$b^* + T^*(b^*) = \min_{0 \leq b \leq t_1} (b + T^*(b)).$$

Case 2. $B_1 = \phi$, $B_2 = [0, \infty)$. The optimal $(b^*, T^*) = (b^*, \infty)$, where b^* can be any value in $[0, \infty)$.

Case 3. $B_1 = \phi$, $B_2 = \phi$. For $b \in B_1$, let $T^*(b)$ be the unique solution of the Eq. (7.3). Then, the optimal $(b^*, T^*) = (b^*, T^*(b^*))$, where b^* is the value which satisfies

$$b^* + T^*(b^*) = \min_{b \leq t_1,\, b \in B_1} (b + T^*(b)).$$

Proof The proof for a more general model is given in the proof of Theorem 7.4 in this chapter and thus it is omitted. □

7.1.3 Model 3

In Model 2, Burn-in Procedure A is applied to repairable devices. In many cases, because of practical limitations, products which fail during burn-in are just scraped, regardless of whether the products are repairable or not. In this case, the burn-in procedure A can be applied. However, an expensive, complex product or device will not be discarded on account of failure of its part, but rather a repair will be performed. Cha [4] proposed the following burn-in procedure.

Burn-in Procedure B
Consider the fixed burn-in time b and begin to burn-in a new component. On each component failure, only minimal repair is done with shop minimal repair cost $c_{sm} > 0$. Continue the burn-in procedure for the repaired component. Immediately after the fixed burn-in time b, the component is put into field operation [4].

Note that the total burn-in time for this burn-in procedure is a constant b. For a burned-in component, the block replacement policy with minimal repairs on failures is adopted in field operation as it was in Model 2. Assume $0 < c_{sm} < c_s$, then this means that the cost of a minimal repair during the burn-in process is smaller than that of the complete (perfect) repair, which is a reasonable assumption. Then, the long-run average cost rate is

$$c(b,T) = \frac{1}{T}(c_0 b + c_{sm}\Lambda(b) + c_m(\Lambda(b+T) - \Lambda(b)) + c_r). \qquad (7.4)$$

where $\Lambda(t) \equiv \int_0^t r(u)\,du$. It can be shown that

$$c_B(b,\, T) \leq c_A(b,\, T), \quad \forall 0 < b < \infty,\ 0 < T < \infty,$$

where $c_A(b,\, T)$ and $c_B(b,T)$ are the cost rate functions in Eqs. (7.2) and (7.4), respectively. This implies that

$$c_B(b_B^*,\, T_B^*) \leq c_A(b_A^*,\, T_A^*),$$

where (b_A^*, T_A^*) and $(b_B^*,\, T_B^*)$ are the optimal solutions which minimize $c_A(b,\, T)$ and $c_B(b,\, T)$, respectively. Thus, we can conclude that the burn-in procedure B is always preferable to the burn-in procedure A when the minimal repair policy is applicable.

Let $(b^*,\, T^*)$ be the optimal burn-in time and the optimal replacement time that minimize the cost rate Eq. (7.4). Then the properties of b^* and T^* are given by the following theorem.

Theorem 7.3 *Suppose that the failure rate function $r(t)$ is bathtub-shaped and differentiable. Let*

$$B_1 \equiv \left\{ b \geq 0 : \int_b^\infty [r(\infty) - r(t)]\,dt > \frac{1}{c_m}[c_r + c_0 b + c_{sm}\Lambda(b)] \right\},$$

and $B_2 \equiv [0, \infty)\backslash B_1$. *Then the properties of the optimal burn-in time* b^* *and of the replacement policy* T^* *can be stated in detail as follows:*

Case 1. $B_1 = [0, \infty)$, $B_2 = \phi$. Let $T^*(b)$ be the unique solution of the equation

$$Tr(b + T) - \int_{b}^{b+T} r(t)\,\mathrm{d}t = \frac{1}{c_m}[c_r + c_0 b + c_{sm}\Lambda(b)]. \tag{7.5}$$

Then, the optimal $(b^*, T^*) = (b^*, T^*(b^*))$, where $0 \le b^* \le t_1$, is the value which satisfies

$$b^* + T^*(b^*) = \min_{0 \le b \le t_1} (b + T^*(b)).$$

Case 2. $B_1 = \phi$, $B_2 = [0, \infty)$. The optimal $(b^*, T^*) = (b^*, \infty)$, where b^* can be any value in $[0, \infty)$.

Case 3. $B_1 = \phi$, $B_2 = \phi$. For $b \in B_1$, let $T^*(b)$ be the unique solution of Eq. (7.5). Then, the optimal $(b^*, T^*) = (b^*, T^*(b^*))$, where b^* is the value which satisfies

$$b^* + T^*(b^*) = \min_{b \le t_1,\, b \in B_1} (b + T^*(b)).$$

Proof Clearly, $b_2^* \neq \infty$, since $c_2(\infty, T) = \infty$ for any $0 < T \le \infty$. For any fixed $0 \le b < \infty$,

$$\frac{\partial c_2}{\partial T} = \frac{c_m}{T^2}\left\{\Psi_b(T) - \frac{1}{c_m}[c_r + c_0 b + c_{sm}\Lambda(b)]\right\},$$

where

$$\Psi_b(T) \equiv Tr(b + T) - \int_{b}^{b+T} r(t)\,\mathrm{d}t.$$

Hence, $\partial c_2/\partial T = 0$ if and only if

$$\Psi_b(T) = \frac{1}{c_m}[c_r + c_0 b + c_{sm}\Lambda(b)].$$

Note that, $\Psi_b(0) = 0$ and that $\Psi_b(T)$

$$\begin{cases} \text{strictly decreases} & \text{if } 0 \le T \le t_1 - b \\ \text{is a constant} & \text{if } t_1 - b \le T \le t_2 - b \\ \text{strictly increases} & \text{if } t_2 - b \le T \end{cases}$$

Then define

$$B_1 \equiv \left\{ b \geq 0 : \Psi_b(\infty) = \int_b^\infty [r(\infty) - r(t)]\,\mathrm{d}t > \frac{1}{c_m}[c_r + c_0 b + c_{sm}\Lambda(b)] \right\}$$

and set $B_2 \equiv [0, \infty)\backslash B_1$.

Now, as in the proof of Theorem 2 in [10], the following three separate cases are considered.

Case 1. $B_1 = [0, \infty),\ B_2 = \phi$.

Case 2. $B_1 = \phi,\ B_2 = [0, \infty)$.

Case 3. $B_1 = \phi,\ B_2 = \phi$.

Case 1 is equivalent to the condition that $\Psi(\infty) \equiv \int_b^\infty [r(\infty) - r(t)]\,\mathrm{d}t = \infty$ for at least one $b \geq 0$. In particular, it occurs when $r(\infty) = \infty$ and $r(0) < \infty$. Let $T_2^*(b)$ be the value which satisfies

$$c_2(b, T_2^*(b)) < c_2(b, T),\ \ \forall T \neq T_2^*(b),$$

for all $b \geq 0$. Then for Case 2, it is easy to see that for all $b \geq 0$,

$$c_2(b, T(b)) > c_2(b, \infty),\ \ \forall T > 0,$$

i.e., $T_2^*(b) = \infty$, for $b \geq 0$ and $c_2(b, T_2^*(b)) = c_m r(\infty)$.

For Case 1 and Case 3, it can be shown, as in Case 2, that for every $b' \in B_2$, $T_2^*(b') = \infty$ and $c_2(b', T_2^*(b')) = c_m r(\infty)$. Moreover, for all $b \in B_1$, the following properties can be established:

(i) There exists $T_2^*(b)$, which is the unique solution of Eq. (7.3).
(ii) $t_2 < b + T_2^*(b) < \infty$.
(iii) $c_2(b, T_2^*(b)) = c_m r(b + T_2^*(b))$.
(iv) For all $b' \in B_2, c_2(b, T_2^*(b)) = c_m r(b + T_2^*(b)) < c_m r(\infty) = c_2(b', T_2^*(b'))$.
(v) The optimal burn-in time b_2^* satisfies: $0 \leq b_2^* \leq t_1$.

Therefore, $b_2^* \in \{b : 0 \leq b \leq t_1\} \cap B_1$ and b_2^* is the value that satisfies:

$$b_2^* + T_2^*(b_2^*) = \min_{b \leq t_1,\ b \in B_1} (b + T_2^*(b)).$$

7.2 Burn-in Procedures for General Failure Model

In this section, we discuss the burn-in procedures for a general failure model that was partly studied in the previous chapter. Recall that according to this model, when the unit fails, the Type I failure and the Type II failure may occur with some probabilities. We assume that the Type I failure is a minor one and thus can be removed by a minimal repair, whereas Type II failure is a catastrophic one and thus can be removed only by a complete repair. Such models have been considered in the literature (e.g., [1, 2]).

7.2.1 Constant Probability Model

In this model, when the unit fails, Type I failure occurs with probability $1 - p$ and Type II failure occurs with probability p, $0 \leq p \leq 1$. Cha [5] proposed the following burn-in procedure for this model.

Burn-in Procedure C
Consider the fixed burn-in time b and begin to burn-in a new component. On each component failure, only minimal repair is done for the Type I failure with shop minimal repair cost c_{sm}, $0 \leq c_{sm} \leq c_s$, and a complete repair is performed for the Type II failure with shop complete repair cost c_s. Then continue the burn-in procedure for the repaired component [5].

Cha [5] studied optimal burn-in and replacement policy for the burn-in procedures A and C under the general failure model defined above.

Note that the burn-in procedure A stops when there is no failure during the fixed burn-in time $(0, b]$ for the first time, whereas procedure C stops when there is no Type II failure during the fixed burn-in time $(0, b]$ for the first time.

Note that, in field operation, the component is replaced by a new burned-in component at the 'field-use age' T or at the time of the first Type II failure, whichever occurs first. For each Type I failure occurring during field use, only minimal repair is done.

Let Y_b be the time to the first Type II failure of a burned-in component with the fixed burn-in time b. If we define $G_b(t)$ as the distribution function of Y_b and $\overline{G}_b(t)$ as $1 - G_b(t)$, then $\overline{G}_b(t)$ is given by

$$\begin{aligned} \overline{G}_b(t) &= \mathrm{P}(Y_b > t) \\ &= \exp\{-\int_0^t pr(b+u)\,du\} \\ &= \exp\{-p[\Lambda(b+t) - \Lambda(b)]\}, \quad \forall t \geq 0, \end{aligned} \tag{7.6}$$

where $\Lambda(t) \equiv \int_0^t r(u)\,du$. Let the random variable $N(b; T)$ be the total number of minimal repairs of a burned-in component which occur during field operation after the burn-in time b and in accordance with the replacement policy T. Then, using the results of Beichelt [2], it is easy to see that, when $p \neq 0$, the expectation of $N(b; T)$ is

$$E[N(b;\,T)] = \frac{1}{G_b(t)}\int_0^T\int_0^t (1-p)r(b+u)\mathrm{d}u\mathrm{d}G_b(t)\cdot G_b(t)$$
$$+\int_0^T (1-p)r(b+u)\,\mathrm{d}u\cdot \overline{G}_b(T)$$
$$=\left(\frac{1}{p}-1\right)(1-\exp\{-p[\Lambda(b+T)-\Lambda(b)]\}). \tag{7.7}$$

When $p = 0$ the expectation is given by

$$E[N(b;\,T)] = \Lambda(b+T) - \Lambda(b).$$

Let c_f denote the cost incurred for each Type II failure in field operation and c_a satisfying $0 < c_a < c_f$ be the cost incurred for each non-failed item which is replaced at field use age $T > 0$. Denote also by c_m the cost of a minimal repair which is performed in field operation. When $p = 0$ or $p = 1$, the burn-in and replacement model discussed in this section reduces to that in [10] or [4]. Thus, in the discussion below, we assume that $0 < p < 1$. Then, using the results given by Eqs. (7.6) and (7.7), the long-run average cost rate functions for procedures A and C are given by [5]

$$c_A(b,\,T) = \frac{1}{\int_0^T \overline{G}_b(t)\,\mathrm{d}t}\left(\left[c_0\frac{\int_0^b \overline{F}(t)\,\mathrm{d}t}{\overline{F}(b)} + c_s\frac{F(b)}{\overline{F}(b)}\right]\right.$$
$$\left.+\,c_m\left[\left(\frac{1}{p}-1\right)(1-\exp\{-p[\Lambda(b+T)-\Lambda(b)]\})\right] + c_f G_b(T) + c_a\overline{G}_b(T)\right), \tag{7.8}$$

and

$$c_C(b,T) = \frac{1}{\int_0^T \overline{G}_b(t)\,\mathrm{d}t}\left(\left[c_0\frac{\int_0^b \overline{G}(t)\,\mathrm{d}t}{\overline{G}(b)} + c_s\frac{G(b)}{\overline{G}(b)} + c_{sm}\left(\frac{1}{p}-1\right)(\exp\{pA(b)\}-1)\right]\right.$$
$$+\,c_m\left[\left(\frac{1}{p}-1\right)(1-\exp\{-p[A(b+T)-A(b)]\})\right] + c_f G_b(T)$$
$$\left.+\,c_u\overline{G}_b(T)\right), \tag{7.9}$$

where $c_A(b,\,T)$ and $c_C(b,\,T)$ represent the cost rate for the burn-in procedures A and C, respectively.

Cha [5] showed that

(i) $c_C(0,\,T;\,p) = c_A(0,\,T;\,p),\ \ \forall 0 < T \le \infty,\ 0 < p < 1,$

(ii) $c_C(b,\,T;\,p) < c_A(b,\,T;\,p),\ \ \forall 0 < b < \infty,\ 0 < p < 1,$

where $c_A(b, T; p)$ and $c_C(b, T; p)$ are the cost rate functions $c_A(b, T)$ and $c_C(b, T)$ when the Type II probability is p, $0 < p < 1$. Then, from the above inequalities, it can be concluded that the burn-in procedure C is always (i.e., for all $0 < p < 1$) preferable to the burn-in procedure A when the minimal repair method is applicable.

Now we discuss the properties of optimal burn-in and of optimal replacement times. Note that the cost rate functions in Eqs. (7.8) and (7.9) can be expressed as

$$c(b, T) = \frac{1}{\int_0^T \overline{G}_b(t)\,\mathrm{d}t}\left(k(b) + c_m\left[\left(\frac{1}{p} - 1\right)(1 - \exp\{-p[\Lambda(b + T) - \Lambda(b)]\})\right] + c_f G_b(T) + c_a \overline{G}_b(T)\right), \tag{7.10}$$

where $k(b)$ is the average cost incurred during the burn-in process for each model. The properties of the optimal (b^*, T^*) which minimizes the cost rate Eq. (7.10) are given by the following theorem.

Theorem 7.4 *Suppose that the failure rate function $r(t)$ is bathtub-shaped and differentiable*. Let

$$B_1 \equiv \left\{ b \geq 0 : pr(\infty) \int_b^\infty \exp\{-p[\Lambda(t) - \Lambda(b)]\}\,\mathrm{d}t - 1 > \frac{1}{\left[c_m\left(\frac{1}{p} - 1\right) + (c_f - c_a)\right]}(c_a + k(b)) \right\},$$

and $B_2 \equiv [0, \infty)\backslash B_1$. *Then the properties of the optimal burn-in time b^* and the replacement policy T^* can be stated in detail as follows:*

Case 1. $B_1 = [0, \infty), B_2 = \phi$. Let $T^*(b)$ be the unique solution of the equation

$$pr(b + T) \int_b^{b+T} \exp\{-p[\Lambda(t) - \Lambda(b)]\}\,\mathrm{d}t + \exp\{-p[\Lambda(b + T) - \Lambda(b)]\} - 1 = \frac{1}{\left[c_m\left(\frac{1}{p} - 1\right) + (c_f - c_a)\right]}(c_a + k(b)). \tag{7.11}$$

Then, the optimal $(b^*, T^*) = (b^*, T^*(b^*))$, where $0 \leq b^* \leq t_1$, is the value which satisfies $b^* + T^*(b^*) = \min\limits_{0 \leq b \leq t_1} (b + T^*(b))$.

Case 2. $B_1 = \phi$, $B_2 = [0, \infty)$. The optimal $(b^*, T^*) = (b^*, \infty)$, where $0 \leq b^* \leq t_1$, is the value which satisfies

$$\frac{1}{\mu(b^*)}\left[c_f + c_m\left(\frac{1}{p} - 1\right) + k(b^*)\right] = \min_{0 \le b \le t_1} \frac{1}{\mu(b)}\left[c_f + c_m\left(\frac{1}{p} - 1\right) + k(b)\right].$$

Case 3. $B_1 = \phi$, $B_2 = \phi$. For $b \in B_1$, let $T^*(b)$ be the unique solution of the Eq. (7.11). Furthermore, let $b_1^* \in [0, t_1] \cap B_1$ satisfy

$$b_1^* + T^*(b_1^*) = \min_{b^* \le t_1,\, b \in B_1} (b + T^*(b)),$$

and $b_2^* \in [0, t_1] \cap B_2$ satisfy

$$\frac{1}{\mu(b_2^*)}\left[c_f + c_m\left(\frac{1}{p} - 1\right) + k(b_2^*)\right] = \min_{b \le t_1,\, b \in B_2} \frac{1}{\mu(b)}\left[c_f + c_m\left(\frac{1}{p} - 1\right) + k(b)\right].$$

If

$$\begin{aligned}&\left[c_m\left(\frac{1}{p} - 1\right) + c_f - c_a\right] pr(b_1^*\\ &+ T^*(b_1^*)) \le \frac{1}{\mu(b_2^*)}\left[c_f + c_m\left(\frac{1}{p} - 1\right) + k(b_2^*)\right],\end{aligned}$$

then the optimal $(b^*, T^*) = (b_1^*, T^*(b_1^*))$. Otherwise the optimal (b^*, T^*) is (b_2^*, ∞).

Proof The cost rate $c(b, T)$ in Eq. (7.10) can be rewritten as

$$\begin{aligned}c(b, T) = &\frac{1}{\int_0^T \overline{G}_b(t)\,\mathrm{d}t}\left(h(b) + c_2 + c_m\left(\frac{1}{p} - 1\right)(1 - \exp\{-p[\Lambda(b + T) - \Lambda(b)]\}) \right.\\ &\left. + c_1[1 - \exp\{-p[\Lambda(b + T) - \Lambda(b)]\}]\right),\end{aligned} \tag{7.12}$$

where $c_1 \equiv c_f - c_a$ and $c_2 \equiv c_a$. Clearly, $b^* \ne \infty$ since $c(\infty, T) = \infty$ for any $0 < T \le \infty$.Then, for any fixed $0 \le b < \infty$, $\partial c/\partial T = 0$ if and only if

$$\Psi_b(T) = \frac{1}{c_3}(c_2 + h(b)), \tag{7.13}$$

where $c_3 \equiv [c_m(1/p - 1) + c_1]$ and

$$\begin{aligned}\Psi_b(T) \equiv\ & pr(b + T)\int_b^{b+T} \exp\{-p[\Lambda(t) - \Lambda(b)]\}\,\mathrm{d}t\\ &+ \exp\{-p[\Lambda(b + T) - \Lambda(b)]\} - 1.\end{aligned}$$

Note that $\Psi_b(0) = 0$ and

$$\Psi_b(T)\begin{cases}\text{strictly decreases} & \text{if } 0 \le T \le t_1 - b,\\ \text{is a constant} & \text{if } t_1 - b \le T \le t_2 - b,\\ \text{strictly increases} & \text{if } t_2 - b \le T.\end{cases}$$

Define

$$B_1 \equiv \left\{ b \ge 0 : \Psi_b(\infty) \equiv \lim_{T\to\infty} \Psi_b(T) = pr(\infty)\int_b^\infty \exp\{-p[\Lambda(t) - \Lambda(b)]\}\mathrm{d}t - 1 > \frac{1}{c_3}(c_2 + h(b)) \right\}$$

and set $B_2 \equiv [0, \infty)\backslash B_1$.

We consider now the following three separate cases.

Case 1. $B_1 = [0, \infty)$and $B_2 = \phi$. This is equivalent to the condition that

$$\Psi_b(\infty) = pr(\infty)\int_b^\infty \exp\{-p[\Lambda(t) - \Lambda(b)]\}\mathrm{d}t - 1 = \infty$$

for at least one $b \ge 0$. In particular, it occurs when $r(\infty) = \infty$ and $r(0) < \infty$. In this case, Eq. (7.13) has a unique solution for all $b \ge 0$. which we denote by $T^*(b)$. Furthermore, from the fact that $\Psi_b(0) = 0$ and the monotonicity of Ψ_b, we can immediately see that $\Psi_b(T) < 0$, for all $0 < T \le t_2 - b$. This implies that the unique solution $T^*(b)$ of Eq. (7.13) must satisfy $T^*(b) > t_2 - b$ for any given $b \ge 0$. Thus, we have shown that

$$t_2 < T^*(b) + b \le \infty \tag{7.14}$$

As $T^*(b)$ satisfies Eq. (7.13),

$$\begin{aligned} &pr(b + T^*(b)\,)\int_b^{b+T^*(b)} \exp\{-p[\Lambda(t) - \Lambda(b)]\}\mathrm{d}t \\ &+ \exp\{-p[\Lambda(b + T^*(b)) - \Lambda(b)]\} - 1 = \frac{1}{c_3}(c_2 + h(b)). \end{aligned} \tag{7.15}$$

Combining Eqs. (7.12) and (7.15), we obtain

$$c(b, T^*(b)) = c_3 pr(b + T^*(b)).$$

Thus, minimizing $c(b, T^*(b))$ is equivalent to minimizing $r(b + T^*(b))$ for $0 \le b < \infty$. By Eq. (7.14), $b + T^*(b) > t_2$, so the problem of finding b^* minimizing $c(b, T^*(b))$ is equivalent to finding b^* which satisfies

$$b + T^*(b) = \min_{b\ge 0}(b + T^*(b)).$$

The inequality $b^* \le t_1$ is now verified. To prove this inequality, it is sufficient to show that $\partial(b + T^*(b))/\partial b > 0$ for all $b \ge t_1$. From Eq. (7.15),

$$pr(b + T^*(\mathrm{b})) \int_{\mathrm{b}}^{\mathrm{b}+\mathrm{T}^*(b)} \exp\{-p\Lambda(t)\}\,\mathrm{d}t + \exp\{-p\Lambda(b + T^*(b))\} = \exp\{-p\Lambda(b)\}\left[1 + \frac{c_2}{c_3} + \frac{1}{c_3}h(b)\right]. \tag{7.16}$$

Taking the derivative with respect to b on both sides of Eq. (7.16), we obtain

$$\begin{aligned} pr'(b + T^*(\mathrm{b}))(1 + T^{*\prime}(b)) \int_b^{b+\mathrm{T}^*(b)} \exp\{-pA(t)\}\,\mathrm{d}t - pr(b + T^*(b)) \exp\{-pA(b)\} \\ = \exp\{-pA(b)\}\frac{1}{c_3}h'(b) - \exp\{-pA(b)\}pr(b)\left(1 + \frac{c_2}{c_3} + \frac{1}{c_3}h(b)\right) \\ > -\exp\{-pA(b)\}pr(b)\left(1 + \frac{c_2}{c_3} + \frac{1}{c_3}h(b)\right), \end{aligned} \tag{7.17}$$

since $h'(b) > 0$. Then, from the Inequality Eq. (7.17),

$$\begin{aligned} pr'(b + T^*(b))(1 + T^{*\prime}(b)) \int_b^{b+T^*(b)} \exp\{-p\Lambda(t)\mathrm{d}t\} \\ > pr(b + T^*(b))\exp\{-p\Lambda(b)\} - \exp\left\{-p\Lambda(b)\left(1 + \frac{c_2}{c_2} + \frac{1}{c_3}h(b)\right)\right\}. \end{aligned} \tag{7.18}$$

However, from Eq. (7.15),

$$\begin{aligned} pr(b + T^*(b)) = \frac{1}{\int_b^{b+T^*(b)} \exp\{-p[\Lambda(t) - \Lambda(b)]\}\,\mathrm{d}t} \\ \times \left\{1 - \exp\{-p[\Lambda(b + T^*(b)) - \Lambda(b)]\} + \frac{c_2}{c_3} + \frac{1}{c_3}h(b)\right\}, \end{aligned} \tag{7.19}$$

and by the bathtub-shaped assumption, if $b \ge t_1$, it follows that

$$\begin{aligned} pr(b) \int_b^{b+T^*(b)} \exp\{-p[\Lambda(t) - \Lambda(b)]\}\,\mathrm{d}t &\le \int_b^{b+T^*(b)} pr(t)\exp\{-p[\Lambda(t) - \Lambda(b)]\}\,\mathrm{d}t \\ &= \exp\{p\Lambda(b)\}[-\exp\{-p\Lambda(t)\}]_b^{b+T^*(b)} \\ &= 1 - \exp\{-p[\Lambda(b + T^*(b)) - \Lambda(b)]\} \\ &\le 1. \end{aligned} \tag{7.20}$$

Then, by combining Eqs. (7.18, 7.19 and 7.20), we obtain

$$pr'(b + T^*(b))(1 + T^{*\prime}(b)) \int_b^{b+T^*(b)} \exp\{-p\Lambda(t)\}dt > 0,$$

which implies that $\partial(b + T^*(b))/\partial b > 0$ for all $b \geq t_1$. Therefore, $b^* \leq t_1$ holds.

Case 2. $B_1 = \phi$, $B_2 = [0, \infty)$. In this case, it can easily be shown that

$$\Psi_b(T) < \frac{1}{c_3}(c_2 + h(b)), \ \ \forall T \geq 0,$$

which implies that $\partial c/\partial T < 0$, for every $T > 0$ for all fixed $b \geq 0$. Hence, for all $T > 0$ and $b \geq 0$

$$\begin{aligned} c(b, T) &\geq c(b, \infty) \\ &= \frac{1}{\mu(b)}\left[c_1 + c_2 + \mathrm{c_m}\left(\frac{1}{\mathrm{p}} - 1\right) + \mathrm{h(b)}\right], \end{aligned}$$

where $\mu(b)$ is defined by

$$\begin{aligned} \mu(b) &\equiv \int_b^\infty \exp\{-p[\Lambda(t) - \Lambda(b)]\}dt \\ &= \frac{\int_b^\infty \overline{G}(t)\,dt}{\overline{G}(b)}, \end{aligned}$$

which is the MRL. Then, as follows from [2, 7], it is easy to see that $\mu(b)$ strictly decreases for all $b \geq t_1$, whereas the term

$$\left[c_1 + c_2 + \mathrm{c_m}\left(\frac{1}{\mathrm{p}} - 1\right) + \mathrm{h(b)}\right]$$

strictly increases as b increases. Therefore, the inequalities

$$\begin{aligned} c(b, T) \geq c(b, \infty), &\quad \forall T > 0, \forall b \geq 0, \\ > c(t_1, \infty), &\quad \forall b > t_1, \end{aligned}$$

hold and, consequently, in this case, we have $(b^*, T^*) = (b^*, \infty)$, $0 \leq b^* \leq t_1$ and $b^* + T^* > t_2$. Also, the optimal burn-in time b^* is the value which satisfies

$$c(b^*, \infty) = \min_{0 \leq b \leq t_1} c(b, \infty).$$

Case 3. $B_1 = \phi$, $B_2 = \phi$. In advance, note that $\Psi_b(\infty)$ is strictly decreasing in b for $b \geq t_1$ since

$$\Psi_b(\infty) = pr(\infty)\mu(b) - 1,$$

and the function

$$\frac{1}{c_3}[c_2 + h(b)], \tag{7.21}$$

strictly decreases as $b \uparrow \infty$. Then, by similar arguments to those in [10], it can be shown that ∞ cannot be in the closure B_1 and there exists $0 \le s < \infty$ such that $[s, \infty) \subseteq B_2$. If we set

$$\beta \equiv \inf\{t : [t, \infty) \subseteq B_2\},$$

then, clearly, $[\beta, \infty) \subseteq B_2$.

First suppose that $\beta \le t_1$, therefore, obviously $[t_1, \infty) \subseteq B_2$. In this case, by the arguments of Case 2, the set $[t_1, \infty)$ cannot contain the optimal b^*. Hence $b^* \le t_1$.

Suppose now that $\beta > t_1$. Since $\Psi_b(\infty)$ strictly decreases for $b \ge t_1$ and the function in Eq. (7.21) strictly increases, the fact that $\beta > t_1$ yields that $[t_1, \beta) \subseteq B_1$. Then, by the procedure described in Case 2, the relationship

$$\min_{b \in [\beta, \infty),\, T > 0} c(b, T) = \min_{b \in [\beta, \infty)} c(b, \infty) > c(t_1, \infty)$$

holds, and, therefore, the set $[\beta, \infty)$ cannot contain the optimal b^*. Also, for $b \in [t_1, \beta)$, by the similar arguments to those in Case 1, we can show that $\partial(b + T^*(b))/\partial b > 0$, for all $t_1 \le b < \beta$, and therefore we can conclude that $b^* \le t_1$.

7.2.2 Time-Dependent Probability Model

In [6], the Constant Probability Model was further extended to the case when the corresponding probabilities change with operating time. Assume now that, when the unit fails at its age t, Type I failure occurs with probability $1 - p(t)$ and Type II failure occurs with probability $p(t)$, $0 \le p(t) \le 1$.

In this model, we employ the same notations and random variables used before. Also, note that if $p(t) = p$ *a.e.* (w.r.t. Lebesgue measure), $0 \le p \le 1$, the models under consideration can be reduced to those of Mi [10] and Cha [4, 5]. Thus, we only consider the set of functions P as the set of all of the Type II failure probability functions, which is given by

$$P = \{p(\cdot) : 0 \le p(t) \le 1,\ \forall t \ge 0\} \backslash \{p(\cdot) : p(t) = p \text{ a.e.},\ 0 \le p \le 1\}$$

It can be shown that

$$\overline{G}_b(t) = \exp\{-[\Lambda_p(b + t) - \Lambda_p(b)]\},\ \forall t \ge 0$$

where $\Lambda_p(t) \equiv \int_0^t p(u)r(u)\,du$, and

$$E[N(b;T)] = \int_0^T r(b+t)\bar{G}_b(t)\,\mathrm{d}t - G_b(T).$$

Then, considering both burn-in procedures A and C for this extended model, the long-run average cost rate functions are given by

$$c_A(b,T) = \frac{1}{\int_0^T \bar{G}_b(t)\,\mathrm{d}t}\left(\left[c_0\int_0^b \exp\{-[\Lambda(t)-\Lambda(b)]\}\,\mathrm{d}t + c_s[\exp\{\Lambda(b)\}-1]\right]\right.$$
$$\left. + c_m\left[\int_0^T r(b+t)\bar{G}_b(t)\,\mathrm{d}t - G_b(T)\right] + c_f G_b(T) + c_a\bar{G}_b(T)\right), \tag{7.22}$$

where $\Lambda(t) \equiv \int_0^t r(u)\,\mathrm{d}u$, and

$$c_C(b,T) = \frac{1}{\int_0^T \bar{G}_b(t)\mathrm{d}t}\left(\left[c_0\int_0^b \exp\{-[\Lambda_p(t)-\Lambda_p(b)]\}\,\mathrm{d}t\right.\right.$$
$$\left. + c_s[\exp\{\Lambda_p(b)\}-1] + c_{sm}\int_0^b (1-p(t))r(t)\exp\{-[\Lambda_p(t)-\Lambda_p(b)]\}\,\mathrm{d}t\right]$$
$$\left. + c_m\left[\int_0^T r(b+t)\bar{G}_b(t)\,\mathrm{d}t - G_b(T)\right] + c_f G_b(T) + c_a\bar{G}_b(T)\right). \tag{7.23}$$

As before, it can be shown that

(*i*) $c_C(0,T;p(\cdot)) = c_A(0,T;p(\cdot)),\ \forall 0<T\le\infty,\ p(\cdot)\in P,$
(*ii*) $c_C(b,T;p(\cdot)) \le c_A(b,T;p(\cdot)),\ \forall 0<b<\infty,\ 0<T\le\infty,\ p(\cdot)\in P,$

which ensures the superiority of the burn-in procedure C when the minimal repair method is applicable.

The cost rate functions in Eqs. (7.22) and (7.23) can be rewritten as

$$c(b,T) = \frac{1}{\int_0^T \bar{G}_b(t)\,\mathrm{d}t}\left(k(b) + c_m\left[\int_0^T r(b+t)\bar{G}_b(t)\,\mathrm{d}t - G_b(T)\right].\right.$$
$$\left. + c_f G_b(T) + c_a\bar{G}_b(T)\right),$$

where $k(b)$ denotes the average cost incurred during the burn-in process. Then, under the following assumptions, the properties regarding the optimal burn-in time $b^*$and the optimal replacement policy T^* can be obtained.

Assumptions

1. The failure rate function $r(t)$ is differentiable and bathtub shaped with the first change point s_1 and the second change point s_2.
2. The Type II failure probability function $p(t)$ is differentiable and bathtub shaped with the first change point u_1 and the second change point u_2.
3. Let $t_1 \equiv \max(s_1, u_1)$ and $t_2 \equiv \min(s_2, u_2)$ then $t_1 < t_2$ holds.
4. $(c_f - c_a) > c_m$.

Theorem 7.5 *Suppose that assumptions (1)–(4) hold. Let the set B_1 be*

$$B_1 \equiv \{b \geq 0 : c_m \int_b^\infty [r(\infty) - r(t)] \exp\{-[\Lambda_p(t) - \Lambda_p(b)]\}dt$$

$$+ ((c_f - c_a) - c_m) \left[p(\infty)r(\infty) \int_b^\infty \exp\{-[\Lambda_p(t) - \Lambda_p(b)]\}dt - 1\right]$$

$$> (c_a + k(b))\},$$

and $B_2 \equiv [0, \infty)\backslash B_1$. Then the properties of the optimal burn-in time b^* and replacement policy T^* can be stated in detail as follows:

Case 1. $B_1 = [0, \infty)$, $B_2 = \phi$. Let $T^*(b)$ be the unique solution of the equation,

$$c_m \int_b^{b+T} [r(b + T) - r(t)] \exp\{-[\Lambda_p(t) - \Lambda_p(b)]\} dt + ((c_f - c_a) - c_m)$$

$$\left[p(b + T)r(b + T) \int_b^{b+T} \exp\{-[\Lambda_p(t) - \Lambda_p(b)]\} dt - (1 - \exp\{-[\Lambda_p(b + T) - \Lambda_p(b)]\})\right]$$

$$= (c_a + k(b)), \tag{7.24}$$

then the optimal $(b^*, T^*) = (b^*, T^*(b^*))$, where $0 \leq b^* \leq t_1$ is the value which satisfies $(b^* + T^*(b^*)) = \min_{0 \leq b \leq t_1} (b + T^*(b))$.

Case 2. $B_1 = \phi$, $B_2 = [0, \infty)$. The optimal $(b^*, T^*) = (b^*, \infty)$, where $0 \leq b^* \leq t_1$ is the value which satisfies

$$\frac{1}{\mu(b^*)}\left[(c_f - c_m) + c_m \int_{b^*}^{\infty} r(t)\exp\{-[\Lambda_p(t) - \Lambda_p(b^*)]\}\mathrm{d}t + k(b^*)\right]$$
$$= \min_{0 \le b \le t_1} \frac{1}{\mu(b)}\left[(c_f - c_m) + c_m \int_{b}^{\infty} r(t)\exp\{-[\Lambda_p(t) - \Lambda_p(b)]\}\mathrm{d}t + k(b)\right],$$

where $\mu(b)$ is given by

$$\mu(b) = \int_{b}^{\infty} \exp\{-[\Lambda_p(t) - \Lambda_p(b)]\}\mathrm{d}t. \tag{7.25}$$

Case 3. $B_1 \neq \phi$, $B_2 \neq \phi$. Let $T^*(b)$, $b \in B_1$, be the unique solution of the Eq.(7.24) and $\mu(b)$ be given by Eq. (7.25). Furthermore, let $b_1^* \in [0, t_1] \cap B_1$ be the value which satisfies

$$(b_1^* + T^*(b_1^*)) = \min_{b \le t_1,\, b \in B_1} (b + T^*(b)),$$

and $b_2^* \in [0, t_1] \cap B_2$ be the value which satisfies

$$\frac{1}{\mu(b_2^*)}\left[(c_f - c_m) + c_m \int_{b_2^*}^{\infty} r(t)\exp\{-[\Lambda_p(t) - \Lambda_p(b_2^*)]\}\mathrm{d}t + k(b_2^*)\right]$$
$$= \min_{b \le t_1,\, b \in B_2} \frac{1}{\mu(b)}\left[(c_f - c_m) + c_m \int_{b}^{\infty} r(t)\exp\{-[\Lambda_p(t) - \Lambda_p(b)]\}\mathrm{d}t + k(b)\right].$$

If

$$c_m r(b_1^* + T^*(b_1^*)) + ((c_f - c_a) - c_m)p(b_1^* + T^*(b_1^*))r(b_1^* + T^*(b_1^*))$$
$$\le \frac{1}{\mu(b_2^*)}\left[(c_f - c_m) + c_m \int_{b_2^*}^{\infty} r(t)\exp\{-[\Lambda_p(t) - \Lambda_p(b_2^*)]\}\mathrm{d}t + k(b_2^*)\right],$$

then the optimal $(b^*, T^*) = (b_1^*, T^*(b_1^*))$. Otherwise, optimal $(b^*, T^*) = (b_2^*, \infty)$.

Remark 7.1 In this theorem, we assume that both $r(t)$ and $p(t)$ are bathtub-shaped functions. Cha and Mi [7] investigated how this assumption can practically be satisfied when a device is composed of two statistically independent parts (Part A and Part B) in series. Assume that the failure of Part A causes a catastrophic failure, whereas that of Part B causes a minor failure. The failure rate of the device is

$$r(t) = r_1(t) + r_2(t)$$

and the probability of Type II failure $p(t)$ is given by

$$p(t) = \frac{r(t)}{r_1(t) + r_2(t)}$$

where $r_1(t)$ and $r_2(t)$ are the failure rate functions of Parts A and B, respectively (see [7] for a detailed discussion and several examples when $r(t)$ and $p(t)$ have various shapes).

7.3 Accelerated Burn-in and Maintenance Policy

Burn-in is generally considered to be expensive and its duration is typically limited. Stochastic models for accelerated burn-in were introduced in the previous chapter. In this section, we will discuss reliability models that jointly deal with accelerated burn-in and maintenance policies. In [8], the burn-in and replacement models 1, 2, and 3 of Sect. 7.1 were extended to the case when burn-in is performed in an accelerated environment assuming the failure rate model described in Sect. 6.4 of the previous chapter.

7.3.1 *Model 1*

We consider burn-in and replacement Model 1: the component is burned-in in accordance with the burn-in procedure A under the accelerated environment. The component that had survived burn-in is put into field operation. In field operation, an age replacement policy is applied. We will use the notation of Sects. 6.4 and 7.1.

The corresponding long-run average cost rate is given by (see Sects.6.4 and 7.1)

$$c(b, T) = \frac{1}{\int_0^T \bar{F}_b(t)\,\mathrm{d}t}\left(\left[c_0\frac{\int_0^b \bar{F}_A(t)\,\mathrm{d}t}{\bar{F}_A(b)} + c_s\frac{F_A(b)}{\bar{F}_A(b)}\right] + c_f F_b(T) + c_a\bar{F}_b(T)\right),$$

where

$$\overline{F}_b(t) \equiv \exp\left(-\int_0^t r(a(b) + u)\,\mathrm{d}u\right) = \frac{\overline{F}(a(b) + t)}{\overline{F}(a(b))},$$

and $F_A(t) = F(\rho(t)),\ \forall t \geq 0$.

Let b^* be the optimal accelerated burn-in time and T^* be the optimal replacement policy which satisfy

$$c(b^*, T^*) = \min_{b \geq 0,\, T > 0} c(b, T).$$

Then the properties regarding the optimal accelerated burn-in time b^* and the optimal replacement policy T^* are given by the following theorem [8], which is similar in formulation to Theorem 7.1.

Theorem 7.6 *Suppose that the failure rate function $r(t)$ is bathtub-shaped and differentiable. Let the set B_1 be*

$$B_1 \equiv \left\{ b \geq 0 : r(\infty) \int_{a(b)}^{\infty} \exp\{-[\Lambda(t) - \Lambda(a(b))]\} dt - 1 \right.$$
$$> \frac{1}{c_f - c_a} \left[c_a + c_s[\exp\{\Lambda(\rho(b))\} - 1 \right]$$
$$\left. + c_0 \int_0^b \exp\left\{ -\left[\Lambda(\rho(t)) - \Lambda(\rho(b)) \right] \right\} dt \right] \Bigg\},$$

and $B_2 \equiv [0, \infty) \backslash B_1$. Furthermore, let $a^{-1}(t_1) \geq 0$ be the unique solution of the equation $a(t) = t_1$. Then the properties of the optimal accelerated burn-in time b^* and replacement policy T^* can be stated in detail as follows:

Case 1. $B_1 = [0, \infty)$, $B_2 = \phi$. Let $T^*(b)$ be the unique solution of the equation

$$r(a(b) + T) \int_{a(b)}^{a(b)+T} \exp\{-[\Lambda(t) - \Lambda(a(b))]\} dt + \exp\{-[\Lambda(a(b) + T) - \Lambda(a(b))]\} - 1$$
$$= \frac{1}{c_f - c_a} \left[c_a + c_s[\exp\{\Lambda(\rho(b))\} - 1] + c_0 \int_0^b \exp\{-[\Lambda(\rho(t)) - \Lambda(\rho(b))]\} dt \right]. \tag{7.26}$$

Then the optimal $(b^*, T^*) = (b^*, T^*(b^*))$, where $0 \leq b^* \leq a^{-1}(t_1)$, is the value which satisfies $a(b^*) + T^*(b^*) = \min_{0 \leq b \leq a^{-1}(t_1)} (a(b) + T^*(b))$.

Case 2. $B_1 = \phi$, $B_2 = [0, \infty)$. In this case, the optimal $(b^*, T^*) = (b^*, \infty)$, where $0 \leq b^* \leq a^{-1}(t_1)$ is the value which satisfies

$$\frac{1}{\mu(a(b^*))} \left[c_f + c_s[\exp\{\Lambda(\rho(b^*))\} - 1] + c_0 \int_0^{b^*} \exp\{-[\Lambda(\rho(t)) - \Lambda(\rho(b^*))]\} dt \right]$$
$$= \min_{0 \leq b \leq a^{-1}(t_1)} \frac{1}{\mu(a(b))} \left[c_f + c_s[\exp\{\Lambda(\rho(b))\} - 1] + c_0 \int_0^b \exp\{-[\Lambda(\rho(t)) - \Lambda(\rho(b))]\} dt \right],$$

where $\mu(a(b))$ is given by

$$\mu(a(b)) \equiv \int_{a(b)}^{\infty} \exp\{-[\Lambda(t) - \Lambda(a(b))]\} \mathrm{d}t. \tag{7.27}$$

Case 3. $B_1 \neq \phi$, $B_2 \neq \phi$ For $b \in B_1$, let $T^*(b)$ be the unique solution of the Eq. (7.26) and let $\mu(a(b))$ be given by Eq. (7.27). Furthermore, let $b_1^* \in [0, a^{-1}(t_1)] \cap B_1$ satisfy

$$a(b_1^*) + T^*(b_1^*) = \min_{b \le a^{-1}(t_1),\, b \in B_1} (a(b) + T^*(b)),$$

and

$$b_2^* \in [0, a^{-1}(t_1)] \cap B_2$$

satisfy

$$\frac{1}{\mu(a(b_2^*))}\left[c_f + c_s\left[\exp\{\Lambda(\rho(b_2^*))\} - 1\right] + c_0 \int_0^{b_2^*} \exp\{-[\Lambda(\rho(t)) - \Lambda(\rho(b_2^*))]\} \mathrm{d}t\right]$$

$$= \min_{b \le a^{-1}(t_1),\, b \in B_2} \frac{1}{\mu(a(b))}\left[c_f + c_s[\exp\{\Lambda(\rho(b))\} - 1] + c_0 \int_0^{b} \exp\{-[\Lambda(\rho(t)) - \Lambda(\rho(b))]\} \mathrm{d}t\right].$$

If

$$(c_f - c_a) r(a(b_1^*) + T^*(b_1^*)) \le \frac{1}{\mu(a(b_2^*))}\left[c_f + c_s\left[\exp\{\Lambda(\rho(b_2^*))\} - 1\right] + c_0 \int_0^{b_2^*} \exp\{-[\Lambda(\rho(t)) - \Lambda(\rho(b_2^*))]\} \mathrm{d}t\right],$$

then the optimal (b^*, T^*) is $(b_1^*, T^*(b_1^*))$. Otherwise, the optimal (b^*, T^*) is (b_2^*, ∞).

7.3.2 Model 2

We consider burn-in and replacement model 2: the component is burned-in by the burn-in procedure C and the block replacement with minimal repair at failure is applied to the component in field use.

In this case, the long-run average cost rate is given by

$$c(b, T) = \frac{1}{T}\left(\left[c_0 \frac{\int_0^b \bar{F}_A(t)\,\mathrm{d}t}{\bar{F}_A(b)} + c_s \frac{F_A(b)}{\bar{F}_A(b)}\right] + c_m[\Lambda(a(b) + T) - \Lambda(a(b))] + c_r\right). \tag{7.28}$$

Then properties of the optimal b^* and T^* minimizing $c(b, T)$ in Eq. (7.28) are given by the following theorem [8]

Theorem 7.7 *Suppose that the failure rate function $r(t)$ is bathtub-shaped and differentiable. Let the set B_1 be*

$$B_1 \equiv \left\{ b \geq 0 : \int_{a(b)}^{\infty} [r(\infty) - r(t)]\mathrm{d}t \right.$$
$$\left. > \frac{1}{c_m}\left[c_r + c_s[\exp\{\Lambda(\rho(b))\} - 1] + c_0 \int_0^b \exp\{-[\Lambda(\rho(t)) - \Lambda(\rho(b))]\}\mathrm{d}t\right]\right\},$$

$B_2 \equiv [0, \infty)\backslash B_1$ and $a^{-1}(t_1) \geq 0$ be the unique solution of the equation $a(t) = t_1$. Then the properties of the optimal burn-in time b^* and the replacement policy T^* can be stated in detail as follows:

Case 1. $B_1 = [0, \infty)$, $B_2 = \phi$. Let $T^*(b)$ be the unique solution of the equation

$$Tr(a(b) + T) - \int_{a(b)}^{a(b)+T} r(t)\,\mathrm{d}t$$
$$= \frac{1}{c_m}\left[c_r + c_s[\exp\{\Lambda(\rho(b))\} - 1] + c_0 \int_0^b \exp\{-[\Lambda(\rho(t)) - \Lambda(\rho(b))]\}\mathrm{d}t\right]. \tag{7.29}$$

Then the optimal $(b^*, T^*) = (b^*, T^*(b^*))$, where $0 \leq b^* \leq a^{-1}(t_1)$, is the value which satisfies $a(b^*) + T^*(b^*) = \min_{0 \leq b \leq a^{-1}(t_1)} (a(b) + T^*(b))$.

Case 2. $B_1 = \phi$, $B_2 = [0, \infty)$. The optimal $(b^*, T^*) = (b^*, \infty)$, where b^* can be any value in $[0, \infty)$.

Case 3. $B_1 \neq \phi$, $B_2 \neq \phi$. For $b \in B_1$, let $T^*(b)$ be the unique solution of the Eq. (7.29). Then the optimal $(b^*, T^*) = (b^*, T^*(b^*))$, where b^* is the value which satisfies

$$a(b^*) + T^*(b^*) = \min_{b \leq a^{-1}(t_1),\, b \in B_1} (a(b) + T^*(b)).$$

7.3.3 Model 3

We consider burn-in and replacement Model 3: the component is burned-in by the burn-in procedure B and the block replacement with minimal repair at failure is applied to the component in field use. Then, obviously, the long-run average cost rate is given by

$$c(b, T) = \frac{1}{T}\left([c_0 b + c_{sm}\Lambda(\rho(b))] + c_m[\Lambda(a(b) + T) - \Lambda(a(b))] + c_r \right), \tag{7.30}$$

The properties of the optimal b^* and T^* minimizing $c(b, T)$ in Eq. (7.30) are given by the following theorem.

Theorem 7.8 *Suppose that the failure rate function $r(t)$is bathtub-shaped and differentiable. Let*

$$B_1 \equiv \left\{ b \geq 0 : \int_b^\infty [r(\infty) - r(t)]\,\mathrm{d}t > \frac{1}{c_m}[c_r + c_0 b + c_{sm}\Lambda(b)] \right\},$$

$B_2 \equiv [0, \infty)\backslash B_1$ and $a^{-1}(t_1) \geq 0$ be the unique solution of the equation $a(t) = t_1$. Then the properties of the optimal burn-in time b^* and the replacement policy T^* can be stated in detail as follows:

Case 1. $B_1 = [0, \infty)$, $B_2 = \phi$. Let $T^*(b)$ be the unique solution of the equation

$$Tr(a(b) + T) - \int_{a(b)}^{a(b)+T} r(t)\,\mathrm{d}t = \frac{1}{c_m}[c_r + c_0 b + c_{sm}\Lambda(\rho(b))]. \tag{7.31}$$

Then the optimal $(b^*, T^*) = (b^*, T^*(b^*))$, where $0 \leq b^* \leq a^{-1}(t_1)$, is the value which satisfies

$$a(b^*) + T^*(b^*) = \min_{0 \leq b \leq a^{-1}(t_1)} (a(b) + T^*(b)).$$

Case 2. $B_1 = \phi$, $B_2 = [0, \infty)$. The optimal $(b^*, T^*) = (b^*, \infty)$, where b^* can be any value in $[0, \infty)$.

Case 3. $B_1 \neq \phi$, $B_2 \neq \phi$. For $b \in B_1$, let $T^*(b)$ be the unique solution of the Eq. (7.31). Then the optimal $(b^*, T^*) = (b^*, T^*(b^*))$, where b^* is the value which satisfies

$$a(b^*) + T^*(b^*) = \min_{b \leq a^{-1}(t_1),\, b \in B_1} (a(b) + T^*(b)).$$

References

1. Beichelt FE, Fischer K (1980) General failure model applied to preventive maintenance policies. IEEE Trans Reliab 29:39–41
2. Beichelt FE (1993) A unifying treatment of replacement policies with minimal repair. Naval Res Logist 40:51–67
3. Block HW, Savits TH (1997) Burn-in. Stat Sci 12:1–19
4. Cha JH (2000) On a better burn-in procedure. J Appl Probab 37:1099–1103
5. Cha JH (2001) Burn-in procedures for a generalized model. J Appl Probab 38:542–553
6. Cha JH (2003) A further extension of the generalized burn-in model. J Appl Probab 40:264–270
7. Cha JH, Mi J (2007) Some probability functions in reliability and their applications. Naval Res Logist 54:128–135
8. Cha JH, Na MH (2009) Accelerated burn-in and system maintenance policies. Commun Stat-Theor Methods 38:719–733
9. Mi J (1991) Optimal burn-in. Ph.D. Thesis, Department Statistics, University Pittsburgh, Pittsburgh
10. Mi J (1994) Burn-in and maintenance policies. Adv Appl Probab 26:207–221

Chapter 8
Burn-in for Heterogeneous Populations

In the previous chapters, we discussed the burn-in procedures for homogeneous populations. When the failure rate of a population is decreasing or bathtub-shaped (BT), burn-in can be usually justified. Note that, as mentioned and illustrated earlier, the heterogeneity of populations is often a reason for the decrease in the resulting failure rate, at least, in some time intervals (see [4], for the corresponding discussion and [18], for some general considerations). In this chapter, the optimal burn-in procedures are investigated without assuming that the population failure rate is BT. We consider the mixed population composed of two ordered subpopulations—the subpopulation of strong items (items with 'normal' lifetimes) and that of weak items (items with shorter lifetimes). In practice, weak items may be produced along with strong items due to, for example, defective resources and components, human errors, unstable production environment, etc. In the later part of this section, we will also consider the continuous mixtures model.

The shape of the mixture failure rate (and the shapes of subpopulation failure rates for the heterogeneous case) will play a crucial role in optimal burn-in problems discussed in this chapter. The mixture failure rate for two ordered subpopulations was intensively studied in the literature. For instance, as was mentioned in Sect. 5.1, Gupta and Warren [19] show that the mixture of two gamma distributions with increasing failure rates (IFRs) (with the same scale parameter) can result either in the increasing mixture failure rate or in the modified bathtub (MBT) mixture failure rate (it increases initially and then behaves like a bathtub failure rate). Similar shapes occur for mixtures of two Weibull distributions with IFRs [22]. Navarro and Hernandez [31] state that the mixture failure rate of two truncated normal distributions, depending on the parameters involved, can also be increasing, BT-shaped or MBT-shaped. Block et al. [5] give explicit conditions describing the possible shapes of the mixture failure rate for two increasing linear failure rates, which are: IFR, BT, and MBT (for the noncrossing linear failure rates).

The shape of the mixture failure rate defines the shape of the mean remaining lifetime (MRL) function, which is also very important for various burn-in problems. If, e.g., it increases (decreases), then the MRL decreases (increases). Another useful

M. Finkelstein and J. H. Cha, *Stochastic Modeling for Reliability*,
Springer Series in Reliability Engineering, DOI: 10.1007/978-1-4471-5028-2_8,

result states (see, e.g., [17]) that, if the failure rate is UBT (upside down BT) and the derivative of the MRL function at $t = 0$ is positive (negative), then the corresponding MRL is increasing (decreasing). The 'symmetrical' statement also holds for the BT shape of the failure rate (see also Chap. 2 and the following section for the corresponding discussion).

Our goal of this chapter is to consider optimization of various characteristics of the performance quality of items after burn-in. This will be done for the case when the component's lifetime distribution function is a mixture of two distributions. The case of continuous mixtures will also be considered. It is well known that when the failure rate of a component is increasing, there is no need to perform the burn-in procedure and only when it is decreasing or nonmonotonic (e.g., BT) there is a possibility for burn-in. This reasoning is usually valid only for homogeneous populations. However, when we deal with heterogeneous populations and the subpopulations are described not only by their failure rates but also by different quality of performance, the situation can be dramatically different. For example, burn-in can be justified even for IFRs! Note that, the precise probabilistic analysis of these problems is usually very complex and, therefore, in this chapter, we mainly concentrate on the qualitative analysis with the corresponding examples.

Furthermore, when we are dealing with heterogeneous populations, there exist the risks of selecting the items with poor reliability characteristics (i.e., with large failure rates), and this cannot be described in the framework of the average quality. In this regard, we will also consider the burn-in procedures aiming at the minimization of these risks in this chapter. For dealing with this problem, we introduce the new measures of quality that govern the corresponding optimal burn-in procedures. While presenting the contents of this chapter, we will mostly follow our recent publications: Cha and Finkelstein [10–14].

8.1 Discrete Mixtures

8.1.1 Ordered Subpopulations and the Effect of Burn-in

Denote the lifetime of a component from the strong subpopulation by X_S and its absolutely continuous cumulative distribution function (Cdf), probability density function (pdf), and the failure rate function by $F_1(t), f_1(t)$ and $\lambda_1(t)$, respectively. Similarly, the lifetime, Cdf, pdf, and the failure rate function of a weak component are X_W, $F_2(t), f_2(t)$ and $\lambda_2(t)$, accordingly. Let the lifetimes in these subpopulations be ordered in the sense of the usual stochastic ordering (Shaked and Shantikhumar 2006):

$$\bar{F}_1(t) \geq \bar{F}_2(t), \text{ for all, } t \geq 0. \tag{8.1}$$

where $\bar{F}_i(t) = 1 - F_i(t)$, $i = 1, 2$; or in the sense of the failure rate ordering (Sect. 2.8):

$$\lambda_1(t) \le \lambda_2(t), \quad t \ge 0. \tag{8.2}$$

The composition of our mixed (infinite) population is as follows: the proportion of strong items is π, whereas the proportion of weak items is $1-\pi$. Then the mixture (population) survival function, in accordance with (5.1–5.3), is

$$\bar{F}_m(t) = \pi\bar{F}_1(t) + (1-\pi)\bar{F}_2(t), \tag{8.3}$$

and the mixture failure rate is

$$\lambda_m(t) = \frac{\pi f_1(t) + (1-\pi)f_2(t)}{\pi\bar{F}_1(t) + (1-\pi)\bar{F}_2(t)}.$$

In 'field use', a component that is picked up at random (at time 0) from the population and *that has survived the burn-in time b* has the following survival function:

$$\bar{F}_m(t|b) = \pi(b)\bar{F}_1(t|b) + (1-\pi(b))\bar{F}_2(t|b), \tag{8.4}$$

where $\bar{F}_i(t|b) = \frac{\bar{F}_i(b+t)}{\bar{F}_i(b)}$, $i = 1, 2$, is the corresponding remaining lifetime distribution whereas the proportions of strong and weak components in the survived population are given by [17]

$$\pi(b) = \frac{\pi\bar{F}_1(b)}{\pi\bar{F}_1(b) + (1-\pi)\bar{F}_2(b)}, \quad 1-\pi(b) = \frac{(1-\pi)\bar{F}_2(b)}{\pi\bar{F}_1(b) + (1-\pi)\bar{F}_2(b)}. \tag{8.5}$$

The mixture failure rate that corresponds to (8.4) is

$$\lambda_m(t|b) = \pi(t|b)\lambda_1(b+t) + (1-\pi(t|b))\lambda_2(b+t),$$

where $\pi(t|b)$ and $1-\pi(t|b)$ are the posterior proportions, which are given by

$$\pi(t|b) = \pi(b+t), \quad 1-\pi(t|b) = 1-\pi(b+t).$$

Therefore,

$$\lambda_m(t|b) = \lambda_m(t+b),$$

which is, in fact, intuitively obvious.

It is clear that due to (8.1), for all $b \ge 0$ that

$$\begin{aligned}\pi(b) &= \frac{\pi\bar{F}_1(b)}{\pi\bar{F}_1(b) + (1-\pi)\bar{F}_2(b)} \\ &= \frac{\pi}{\pi + (1-\pi)\bar{F}_2(b)/\bar{F}_1(b)} \ge \pi, \text{and}, 1-\pi(b) \le 1-\pi.\end{aligned} \tag{8.6}$$

This inequality means that the *quality of the sample has improved as more weak items than strong ones have failed in* $[0, b)$. It is clear that when ordering (8.2) holds $\bar{F}_2(b)/\bar{F}_1(b)$ in (8.6) is decreasing in b and $\pi(b)$ is increasing in b. (Note that, ordering (8.1) is not sufficient for the latter statement.) Therefore, burn-in can be justified as an operation improving the quality of the population, as it increases the proportion of strong items. However, the question arises, at what cost? We will address this question later in this section.

It is well known that, if $\lambda_i(t)$ is nonincreasing in t, $i = 1, 2$, then $\lambda_m(t|b) = \lambda_m(t+b)$ is a decreasing function of its argument and therefore, decreases in b for all fixed $t \geq 0$ as well (the mixture failure rate of distributions with decreasing failure rates is also decreasing). Thus, burn-in decreases the failure rate (increases the MRL). The simplest example of this property is given by the following example.

Example 8.1 Let $\lambda_1(t) = \lambda_1, \quad t \geq 0$, and, $\lambda_2(t) = \lambda_2, \quad t \geq 0$, $\lambda_2 > \lambda_1$. Then

$$\begin{aligned}\lambda_m(t) &= \pi(t)\lambda_1(t) + (1-\pi(t))\lambda_2(t) \\ &= \frac{\pi e^{-\lambda_1 t}}{\pi e^{-\lambda_1 t} + (1-\pi)e^{-\lambda_2 t}}\lambda_1 + \frac{(1-\pi)e^{-\lambda_2 t}}{\pi e^{-\lambda_1 t} + (1-\pi)e^{-\lambda_2 t}}\lambda_2,\end{aligned}$$

which is a decreasing function and therefore $\lambda_m(t+b)$ is decreasing in b for all fixed $t \geq 0$.

Example 8.1 shows that burn-in of items from mixed populations (with non-IFRs of subpopulations) not only increases the proportion of strong items, but also decreases the mixture failure rate and therefore, it is obviously justified. However, in the next subsection, we will see that burn-in can be justified even when the failure rates of subpopulations are increasing.

8.1.2 Optimal Burn-in Time for Performance Quality Measures

In this subsection, using general settings and simple illustrative examples, we will briefly describe the burn-in procedures, which maximize several performance quality characteristics of items. However, our main interest will be focused on optimal burn-in that minimizes average costs taking into account possible gains during a mission time. This case is considered in more detail in the last part.

8.1.2.1 Maximization of the Success Probability of a Mission

Let the time required for performing a mission by a component (system) from our heterogeneous population (that had survived burn-in during time b) be a constant and denote it by τ. Then the probability of performing this mission, which is understood as the mixture survival function, is $\bar{F}_m(\tau|b)$ and $\bar{F}_m(\cdot|b)$ is given by (8.4). It is obvious that this probability is strictly decreases in b, if the mixture failure rate function $\lambda_m(t)$ strictly increases.

Assume now that it takes a random time Y_1 (Y_2) for a strong (weak) component to complete a mission, where Y_i is a random variable with the cdf $G_i(y)$ and the pdf $g_i(y)$; $i = 1, 2$. It is also natural to assume that it takes more time for a weak component to perform a mission than for a strong component due to the difference in the performance quality for two subpopulations. This, e.g., can be expressed as the corresponding stochastic ordering:

$$Y_1 \leq_{st} Y_2 \quad (\bar{G}_1(y) \leq \bar{G}_2(y), y \geq 0).$$

Then the probability of performing a mission is defined by

$$P(b) \equiv \pi(b) \int_0^\infty \bar{F}_1(y|b) g_1(y) \mathrm{d}y + (1 - \pi(b)) \int_0^\infty \bar{F}_2(y|b) g_2(y) \mathrm{d}y. \tag{8.7}$$

It is practically impossible to describe monotonicity properties of $P(b)$ analytically in this general form. However, some useful qualitative considerations can be helpful. As it was previously stated, the proportion of strong items $\pi(b)$ increases in b. However, it does not guarantee that $P(b)$ also increases, because the survival functions $\bar{F}_i(y|b), i = 1, 2$ can decrease in b (as, e.g., when both failure rates $\lambda_i(b + t)$ are increasing). In this case (i.e., when both $\bar{F}_i(y|b), i = 1, 2$, decrease in b) there still can be a *finite* $b^* > 0$ that maximizes the probability $P(b)$ (see Example 8.2). On the contrary, it is obvious from the above considerations that formally $b^* = \infty$ when both failure rates are decreasing (and therefore, the mixture failure rate as well, as in Example 8.1).

Example 8.2 Suppose that $\lambda_1(t) = t^{1/2} + 1.0, t \geq 0, \lambda_2(t) = t^{1/2} + 2.6, t \geq 0$, and $\pi = 1 - \pi = 0.5$. Let $g_i(y) = v_i \exp\{-v_i y\}, y \geq 0, i = 1, 2$, where $v_1 = 1.0$, $v_2 = 0.1$. In this case, the mixture failure rate function $\lambda_m(t)$ is given in Fig. 8.1. As illustrated by this graph (and can be proved analytically), the mixture failure rate function is strictly increasing. Therefore, the burn-in procedure is not needed, if we consider only 'ordinary' reliability measures for a homogeneous population described by the same failure rate (e.g., the mean time to failure in field operation

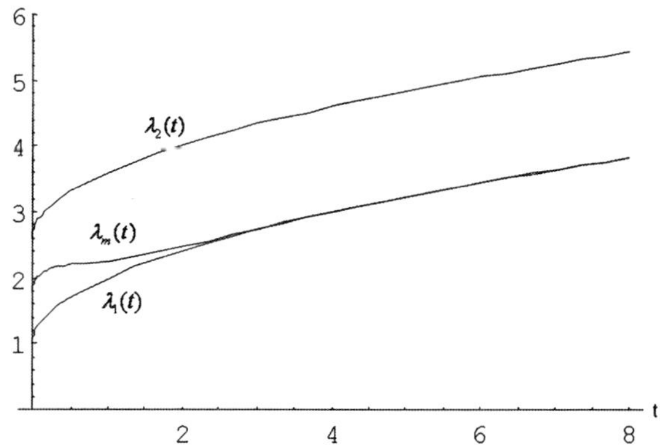

Fig. 8.1 Mixture failure rate function $\lambda_m(t)$ (Example 8.2)

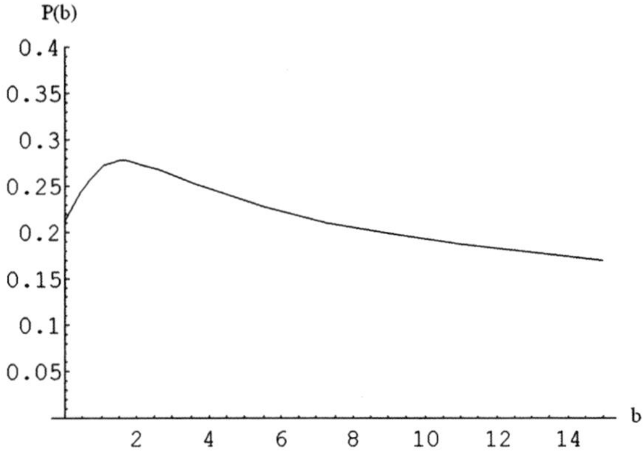

Fig. 8.2 The probability of performing given mission $P(b)$ (Example 8.2)

or success probability of a mission). However, as illustrated by Fig. 8.2, the probability of performing a mission in (8.7) for our heterogeneous population with different quality of performance functions first increases and then monotonically decreases with a maximum at some point. For the considered values of parameters, the optimal burn-in time is $b^* = 1.59$ and the corresponding maximum probability is $P(b^*) = 0.277$.

Note that, the mixture failure rate in this example is contained between the failure rates of subpopulations. Ordering (8.2) holds and Fig. 8.1 also illustrates the well-known fact that the mixture failure rate tends to the failure rate of the strongest subpopulation as the weakest items are 'dying out first' with time [18].

Observe that $P(b)$ in (8.7) can be written in a more explicit way as

$$P(b) \equiv \pi(b) \int_0^\infty \exp\left\{ -\int_b^{b+y} \lambda_1(u)du \right\} g_1(y)dy + (1 - \pi(b)) \int_0^\infty \exp\left\{ -\int_b^{b+y} \lambda_2(u)du \right\} g_2(y)dy.$$

By differentiating,

$$\begin{aligned} P'(b) = {} & \pi'(b) \int_0^\infty \exp\left\{ -\int_b^{b+y} \lambda_1(u)du \right\} g_1(y)dy - \pi'(b) \int_0^\infty \exp\left\{ -\int_b^{b+y} \lambda_2(u)du \right\} g_2(y)dy \\ & - \pi(b) \int_0^\infty (\lambda_1(b+y) - \lambda_1(b)) \exp\left\{ -\int_b^{b+y} \lambda_1(u)du \right\} g_1(y)dy \\ & - (1 - \pi(b)) \int_0^\infty (\lambda_2(b+y) - \lambda_2(b)) \exp\left\{ -\int_b^{b+y} \lambda_2(u)du \right\} g_2(y)dy. \end{aligned} \quad (8.8)$$

It follows from (8.8) that when

$$\pi'(0) > \frac{\pi \int_0^\infty (\lambda_1(y) - \lambda_1(0))\bar{F}_1(y)g_1(y)\mathrm{d}y + (1-\pi)\int_0^\infty (\lambda_2(y) - \lambda_2(0))\bar{F}_2(y)g_2(y)\mathrm{d}y}{\int_0^\infty \bar{F}_1(y)g_1(y)\mathrm{d}y - \int_0^\infty \bar{F}_2(y)g_2(y)\mathrm{d}y}, \tag{8.9}$$

$P(b)$ increases initially, which means that this is a sufficient condition for the existence of the finite ($b^* < \infty$) or the nonfinite ($b^* = \infty$) burn-in time. [Note also that the denominator in (8.9) is positive.] Denote the right-hand side of inequality (8.9) by B. Numerical computation shows that this sufficient condition holds for the setting of Example 8.2, that is,

$$\pi'(0) = 0.4 > B = 0.31.$$

The derivative $\pi'(0)$ is discussed in more detail in the last part of this subsection.

8.1.2.2 Minimization of the Expected Number of Minimal Repairs During the Mission Time

Assume now that the components are minimally repairable and the corresponding quality of performance after burn-in is measured by the expected number of minimal repairs in the fixed interval (mission time) $[0, \tau]$. This setting (including, obviously, the relevant costs) can be of practical interest for manufacturers while assigning the corresponding warranties to their products.

The expected number of minimal repairs during the mission time is given by

$$\pi(b)\int_0^\tau \lambda_1(b+t)\mathrm{d}t + (1-\pi(b))\int_0^\tau \lambda_2(b+t)\mathrm{d}t.$$

As in the previous case, let Y_1 and Y_2 be random mission times for strong and weak components, respectively. The assumptions and notation for Y_1 and Y_2 are the same as in the previous case. Let $M(b)$ be the total number of minimal repairs during the mission time. Then its expectation is given by

$$E(M(b)) \equiv \pi(b)\int_0^\infty\int_0^y \lambda_1(b+t)\mathrm{d}t g_1(y)\mathrm{d}y + (1-\pi(b))\int_0^\infty\int_0^y \lambda_2(b+t)\mathrm{d}t g_2(y)\mathrm{d}y.$$

Given the parameters, the corresponding optimal burn-in time b^* that minimizes $E(M(b))$ can be obtained using numerical procedures. It is clear that our general qualitative considerations of the previous case are also valid and the finite b^* can exist even for populations with IFRs, which was not possible for the homogeneous case.

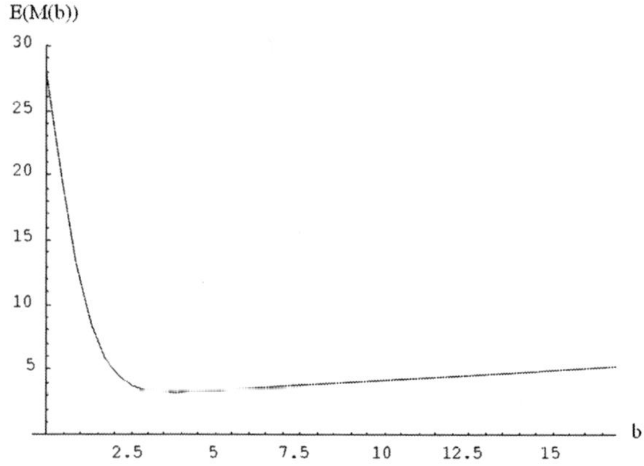

Fig. 8.3 The expected number of minimal repairs $E(M(b))$ (Example 8.3)

Example 8.3 Consider the same setting as in Example 8.2. Then the graphs for the corresponding failure rates are the same as those in Fig. 8.1 and the graph for $E(M(b))$ is given in Fig. 8.3. In this case, the optimal burn-in time is $b^* = 3.73$ and the minimum expected number of minimal repairs is $E(M(b^*)) = 3.31$.

8.1.2.3 Maximization of Expected Total Number of Consecutive Jobs Completed During the Field Operation

Let the components in field operation consecutively perform 'jobs' of the same nature. Assume that the times for completing each job is given by τ_1 (constant) for a strong component and τ_2 (constant) for a weak component, respectively ($\tau_1 < \tau_2$). Therefore, the different quality of performance of our components is described in this way.

Let X_{Sb} be the lifetime of a strong component which has survived burn-in in $[0, b)$ and X_{Wb} be that of a weak component, respectively. Furthermore, let N_{Sb} be the random number of jobs *completed* by a strong component in field operation. Then

$$P(N_{Sb} = k) = P(k\tau_1 < X_{Sb} \leq (k+1)\tau_1) = \bar{F}_1(k\tau_1|b) - \bar{F}_1((k+1)\tau_1|b), k \geq 0$$

and the mean of N_{Sb} is given by

$$E(N_{Sb}) = \sum_{k=0}^{\infty} kP(N_{Sb} = k) = \sum_{k=1}^{\infty} \bar{F}_1(k\tau_1|b).$$

Similarly, the mean number of jobs completed by a weak component in field operation, N_{Wb} is given by

$$E(N_{Wb}) = \sum_{k=0}^{\infty} kP(N_{Wb} = k) = \sum_{k=1}^{\infty} \bar{F}_2(k\tau_2|b).$$

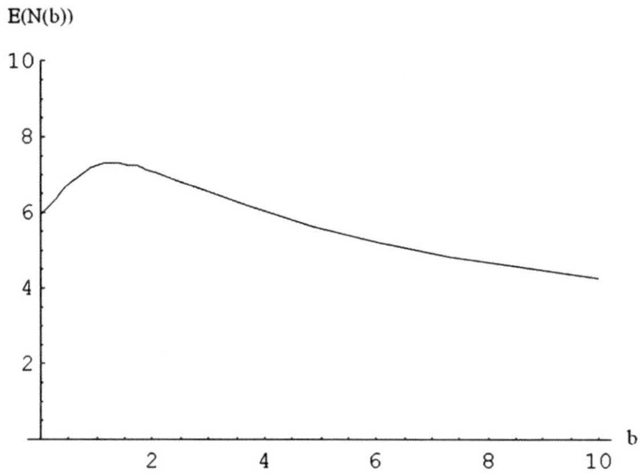

Fig. 8.4 The Average Number of Jobs $E(N(b))$. (Example 8.4)

Let the number of jobs completed during field operation be $N(b)$. Then its expectation is

$$E(N(b)) = \pi(b)\sum_{k=1}^{\infty}\bar{F}_1(k\tau_1|b) + (1-\pi(b))\sum_{k=1}^{\infty}\bar{F}_2(k\tau_2|b).$$

The optimal burn-in time, which maximizes $E(N(b))$ can be obtained numerically and again, unlike the homogeneous case, the finite optimal b^* can exist even when $\lambda_1(t), \lambda_2(t)$ and the mixture failure rate $\lambda_m(t)$ are increasing.

Example 8.4 Consider again the same setting as in Example 8.2 with $\tau_1 = 0.05$ and $\tau_2 = 0.5$. In this case, the graph for $E(N(b))$ is given in Fig. 8.4. It can be seen that the optimal burn-in time is $b^* = 1.32$, and the maximum expected number of jobs is $E(N(b^*)) = 7.31$.

8.1.2.4 Gain Proportional to the Mean Time to Failure

We will describe now a model that already takes into consideration the costs and gains involved. This cost structure (expected costs) accounts for the performance quality after burn-in and defines gains proportional to the MRL. At first, as in the homogeneous case, we do not 'disclose' the composition of our population and deal with the observed mixture distribution function. In accordance with this model, the expected cost function $c(b)$, which accounts for average costs during and after burn-in is:

$$c(b) = c_0(b) - K\frac{\int_b^{\infty}\bar{F}_m(u)\mathrm{d}u}{\bar{F}_m(b)}, \tag{8.10}$$

where $\bar{F}_m(\cdot)$ is given by (8.3), b is the burn-in duration, K is the gain for the unit of time during the mission time and $c_0(b)$ is the average (expected) cost to obtain a component that has passed burn-in. If the first component fails, then the second one is tested, etc., until a component passes burn-in.

Denote the cost of a single item by C and if, for simplicity, we assume that the expected cost of burn-in is just the cost of the failed components then it is easy to show that

$$c_0(b) = C\left(\frac{1}{\bar{F}_m(b)} - 1\right) = \frac{CF_m(b)}{\bar{F}_m(b)},$$

where $1/\bar{F}_m(b)$ corresponds to the expected 'total number of trials' until the first success.

Remark 8.1 As it was mentioned, Eq. (8.10), in fact, formulates the problem exactly like in a homogeneous case, just using the mixture distribution as a governing one. If, e.g., the MRL, $\int_b^\infty \bar{F}_m(u)\mathrm{d}u/\bar{F}_m(b)$ is increasing or initially increasing and is described, e.g., by the UBT-shape, then the problem of obtaining the optimal b^* that minimizes $c(b)$ can be properly formulated. For instance, if the MRL is UBT with the maximum at some $\tilde{b}$, then the optimal duration of the burn-in is obviously smaller (for $c_0(b) > 0) :\ 0 \le b^* < \tilde{b}$. On the other hand, as in the previous cases, if we use the structure of the population described by the time-dependent proportion $\pi(b)$, some other more advanced settings can be considered, e.g., dealing with the quality of performance (gain), which characterizes each subpopulation and not the overall population.

Assume that a component from the strong subpopulation is characterized by the quality (the gain for the unit of time during the mission time) Q_S, whereas the one from the weak subpopulation is characterized by Q_W and $Q_W < Q_S$. Then the expected cost in (8.10) is obviously modified to

$$c(b) = c_0(b) - \left(Q_S\pi(b)\frac{\int_b^\infty \bar{F}_1(u)\mathrm{d}u}{\bar{F}_1(b)} + Q_W(1-\pi(b))\frac{\int_b^\infty \bar{F}_2(u)\mathrm{d}u}{\bar{F}_2(b)}\right). \quad (8.11)$$

When $Q_S = Q_W = K$, as follows from (8.3) and (8.5), Eq. (8.11) reduces to (8.10). Thus, minimization of $c(b)$ can be considered to be a generalization of standard burn-in approaches.

If, for example, distributions of X_S and X_W are exponential with parameters $\lambda_1 \le \lambda_2$, then it is easy to see that, because $Q_W < Q_S$, gains increase with b, as $\pi(b)$ increases with b, whereas $c_0(b)$ also increases. Therefore, under suitable assumptions for parameters there should be a minimum for some b. Similar to the previous cases, the problem becomes much more interesting when both failure rates are increasing (see later).

Example 8.5 Let $\lambda_1 = 0.1, \lambda_2 = 1.0, \pi = 1 - \pi = 0.5, C = 0.1, Q_S = 10.0$ and $Q_W = 1.0$. The corresponding mixture failure rate is given in Example 8.1 and the expected cost function $c(b)$ is plotted in Fig. 8.5. The optimal burn-in time is $b^* = 3.16$, and the minimum expected cost is $c(b^*) = -0.70$.

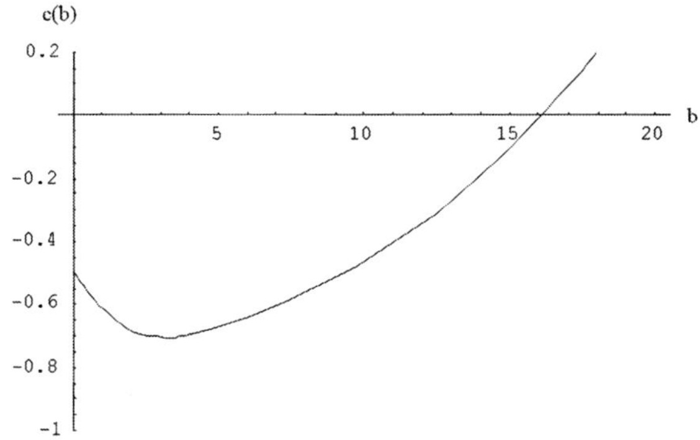

Fig. 8.5 The expected cost $c(b)$ (Example 8.5)

Remark 8.2 There can be other problem formulations, e.g., for missions with high importance (for instance, military). These missions usually need a high level of quality, whereas the costs are not the issue. Assume, e.g., that the corresponding requirement for the unit quality is $Q_R(Q_W < Q_R < Q_S)$. Then we must 'obtain the proportion' that satisfies this requirement via the burn-in procedure, i.e.,

$$\pi(b)Q_S + (1 - \pi(b))Q_W = Q_R. \tag{8.12}$$

This equation can be solved with respect to b and the corresponding solution will define the minimal burn-in time that 'achieves' Q_R.

As in the previous cases, the quality (gains) can change conventional approaches to burn-in problems. To illustrate this, consider the case of increasing, ordered failure rates: $\lambda_1(t) \leq \lambda_2(t)$, such that the mixture failure rate is also increasing (or MBT-shaped) and therefore, the conventional burn-in (without considering different gains for subpopulations) is not needed. However, in our study, when e.g., Q_S is sufficiently larger than Q_W, burn-in can be justified. This is because, in accordance with (8.11), it can decrease the expected cost due to improvement in the population proportion quality that can compensate the effect of the decreasing (in b) remaining lifetime.

First, we present a rather general example with linear failure rates for subpopulations, where the mixture failure rate can be obtained analytically.

Example 8.6 Block et al. [5]:

Let

$$\lambda_1(t) = ct + d_1, \lambda_2(t) = ct + d_2, \quad c > 0, d_2 > d_1$$

The explicit equation for the mixture failure rate is

$$\lambda_m(t) = ct + d_1 + \frac{(1 - \pi)\alpha}{\pi \exp\{\alpha t\} + (1 - \pi)},$$

where $\alpha = d_2 - d_1$. The direct analysis of this function shows that $\lambda_m(t)$ is increasing when $0 < \alpha/\sqrt{c} \leq 2$ and it tends to infinity as t increases approaching $ct + d_1$, the failure rate of the strongest population.

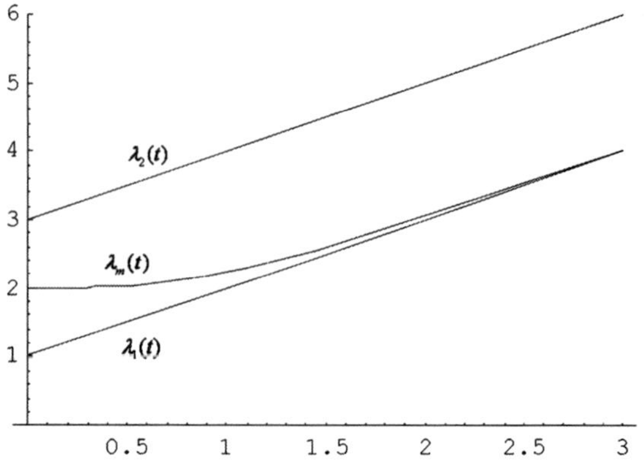

Fig. 8.6 Mixture failure rate function $\lambda_m(t)$ (Example 8.7)

Example 8.7 Consider now the specific case of Example 8.6 with $c = 1.0$, $d_1 = 1.0$ and $d_2 = 3.0$. Let $\pi = 1 - \pi = 0.5, C = 0.1, Q_S = 10.0$ and $Q_W = 1.0$. The graph of the corresponding mixture failure rate function is given in Fig. 8.6 and of the mixture MRL function, in Fig. 8.7. The expected cost function $c(b)$ is given in Fig. 8.8. The optimal burn-in time is $b^* = 0.41$, and the minimum expected cost is $c(b^*) = -3.68$. As we can see from the graph (and can be shown analytically as $\alpha/\sqrt{c} = 2$), the mixture failure rate is not decreasing in $[0, \infty)$ and eventually is converging to the failure rate of the strongest population. In accordance with that, the MRL function is decreasing and therefore, the conventional burn-in is not relevant, whereas in the case under consideration, the optimal burn-in time exists.

The general form of gains in (8.11) can be analyzed further. Taking into account (8.3) and (8.5),

$$q(b) = Q_S \frac{\pi \int_b^\infty \bar{F}_1(u)\mathrm{d}u}{\bar{F}_m(b)} + Q_W \frac{(1-\pi)\int_b^\infty \bar{F}_2(u)\mathrm{d}u}{\bar{F}_m(b)}$$

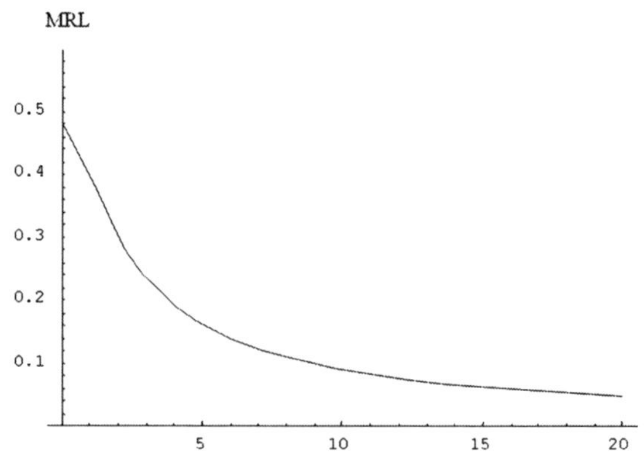

Fig. 8.7 Mean residual lifetime function (Example 8.7)

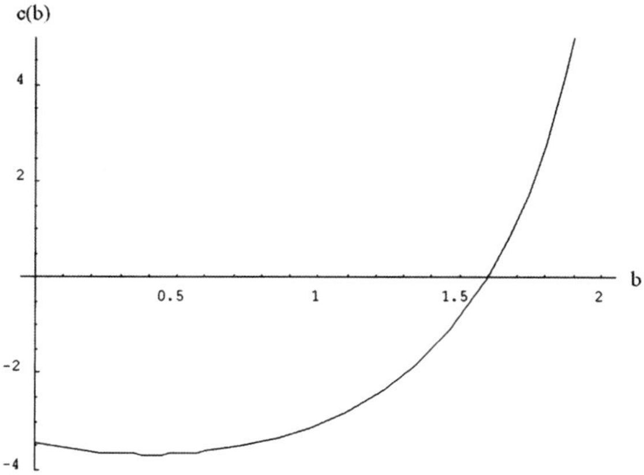

Fig. 8.8 The expected cost $c(b)$ (Example 8.7)

$$= \frac{Q_S\pi \int_b^\infty \bar{F}_1(u)\mathrm{d}u + Q_W(1-\pi)\int_b^\infty \bar{F}_2(u)\mathrm{d}u}{\bar{F}_m(b)}.$$

Therefore, the sign of the derivative $q'(b)$ is defined by the sign of

$$d(b) \equiv -\bar{F}_m(b)(Q_S\pi\bar{F}_1(b) + Q_W(1-\pi)\bar{F}_2(b)) + f_m(b)\left(Q_S\pi\int_b^\infty \bar{F}_1(u)\mathrm{d}u + Q_W(1-\pi)\int_b^\infty \bar{F}_2(u)\mathrm{d}u\right),$$

where $f_m(t) = F'_m(t)$. It is difficult to analyze $d(b)$ for all values of $b \geq 0$, whereas the specific case $b = 0$ can be very helpful for our qualitative analysis:

$$d(0) = -(Q_S\pi + Q_W(1-\pi)) + f_m(0)(Q_S\pi E[X_S] + Q_W(1-\pi)E[X_W]). \quad (8.13)$$

As $f_m(0) = \lambda_m(0)$, (8.13) can be written as

$$d(0) = Q_S\pi(\lambda_m(0)E[X_S] - 1) + Q_W(1-\pi)(\lambda_m(0)E[X_W] - 1). \quad (8.14)$$

When $d(0) > 0$, the gains increase (at least, initially), which is an *important distinction* from the homogeneous case (8.10), where they decrease, as the MRL is decreasing for distributions with IFR $\lambda_m(t)$. This inequality can hold due to the following reasoning: first note that, when both failure rates of subpopulations are ordered, as in (8.2), the mixture failure rate is contained between them. Therefore, obviously, as the failure rates $\lambda_1(t)$ and $\lambda_2(t)$ are increasing, inequality $\lambda_m(0)E[X_W] - 1 < 0$ holds, because $\lambda_m(0) < \lambda_2(t), t \geq 0$, whereas inequality $\lambda_m(0)E[X_S] - 1 > 0$ can still hold (e.g., when $\lambda_m(0) - \lambda_1(0)$ is sufficiently large). Then, if $Q_S - Q_W$ is also sufficiently large, (8.14) is positive and the gains initially increase. This property can constitute the possibility for the optimal burn-in time ($b^* > 0$).

Coming back now to Eq. (8.11), consider obtaining a sufficient condition for the positive optimal burn-in time ($b^* > 0$), which minimizes the expected cost function $c(b)$.

Taking into account that $\bar{F}_i(t) = \exp\{-\int_0^t \lambda_i(x)dx\}$,

$$
\begin{aligned}
c'(b) = & c_0'(b) - Q_S\pi'(b)\int_0^\infty \exp\left\{-\int_b^{b+u}\lambda_1(y)dy\right\}du + Q_W\pi'(b)\int_0^\infty \exp\left\{-\int_b^{b+u}\lambda_2(y)dy\right\}du \\
& + Q_S\pi(b)\int_0^\infty (\lambda_1(b+u)-\lambda_1(b))\exp\left\{-\int_b^{b+u}\lambda_1(y)dy\right\}du \\
& + Q_W(1-\pi(b))\int_0^\infty (\lambda_2(b+u)-\lambda_2(b))\exp\left\{-\int_b^{b+u}\lambda_2(y)dy\right\}du
\end{aligned}
$$

Therefore, if

$$
\pi'(0) > \frac{c_0'(0) + Q_S\pi(1-\lambda_1(0)E[X_S]) + Q_W(1-\pi)(1-\lambda_2(0)E[X_W])}{Q_S E[X_S] - Q_W E[X_W]}, \tag{8.15}
$$

then $c'(0) < 0$, which implies that $c(b)$ is initially decreasing and therefore, the finite or non-finite $b^* > 0$ exists.

Moreover, in accordance with (8.5), the derivative $\pi'(0)$ can be explicitly written as

$$
\pi'(0) = \pi(1-\pi)(\lambda_2(0) - \lambda_1(0)).
$$

This means that increasing $\lambda_2(0) - \lambda_1(0)$ and $Q_S - Q_W$ (for the latter, see the corresponding discussion of Eq. (8.12)) can eventually lead to the desired inequality (8.15). It is also clear that $\pi(1-\pi)$ achieves its maximum at $\pi = 0.5$. Note that, the difference $\lambda_2(0) - \lambda_1(0)$ is important for defining the initial shape of the corresponding mixture failure rate [18]. Note also that the sufficient condition (8.15) is satisfied for Example 8.7 (which should be the case, as the expected cost function is decreasing in the neighborhood of 0 in Fig. 8.8):

$$
\pi'(0) = 0.5 > \tilde{B} = 0.31, \tag{8.16}
$$

where $\tilde{B}$ denotes the right-hand side of inequality (8.15).

As it was mentioned before, mixtures of IFR functions can also result in the modified bathtub-shaped (MTB) failure rate function. Even in this case, as illustrated by the following example, the MRL function can be strictly decreasing and therefore, the conventional burn-in should not be performed. However, burn-in in our setting can be justified even in this case.

Example 8.8 Let $\lambda_1(t) = 0.2t + 0.5, t \geq 0, \lambda_2(t) = t + 1.0, t \geq 0$ and $C = 0.1$, $Q_S = 20.0, Q_W = 1.0$. The graph of the mixture failure rate function is given in Fig. 8.9 and the corresponding MRL and expected cost functions are given in

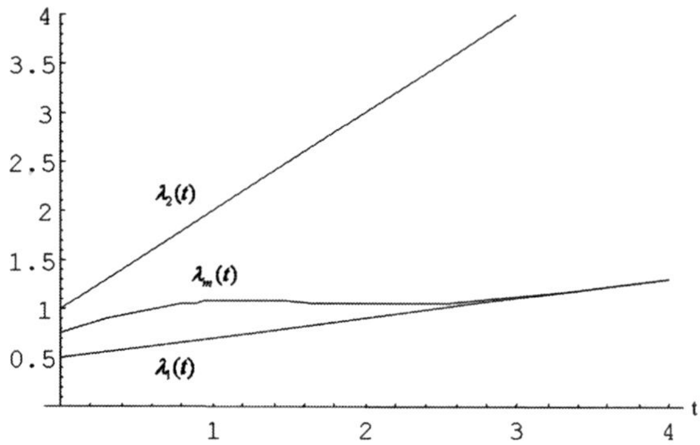

Fig. 8.9 Mixture failure rate function $\lambda_m(t)$ (Example 8.8)

Figs. 8.10 and 8.11, respectively. The optimal burn-in time is $b^* = 1.73$ and the minimal expected cost is $c(b^*) = -16.99$.

As in (8.16), the sufficient condition (8.15) can also be easily verified:

$$\pi'(0) = 0.125 > \tilde{B} = 0.1242.$$

The mixture failure rate in this example has the MBT shape (Fig. 8.9), whereas the MRL function is strictly decreasing (Fig. 8.10), which can be also verified numerically. Thus, this example shows *empirically* that the MBT shape of the failure rate can correspond to the decreasing MRL function. As this fact was not theoretically studied before, we present here some initial findings.

We start with the well-known result for the BT failure rate that was already mentioned in the Introduction (see, e.g., [17]):

Let $\lambda(t)$ be a differentiable BT failure rate in $(0, \infty)$ and $m(t)$ denote the corresponding MRL function. If

$$m'(0) = \lambda(0)m(0) - 1 \leq 0,$$

then $m(t)$ is decreasing (non-increasing).

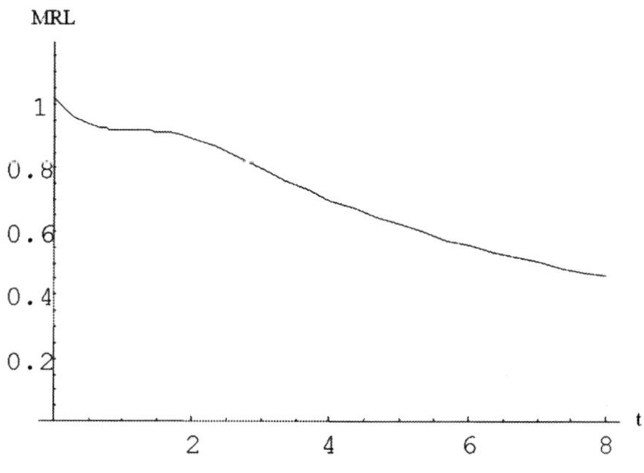

Fig. 8.10 Mean residual lifetime function (Example 8.8)

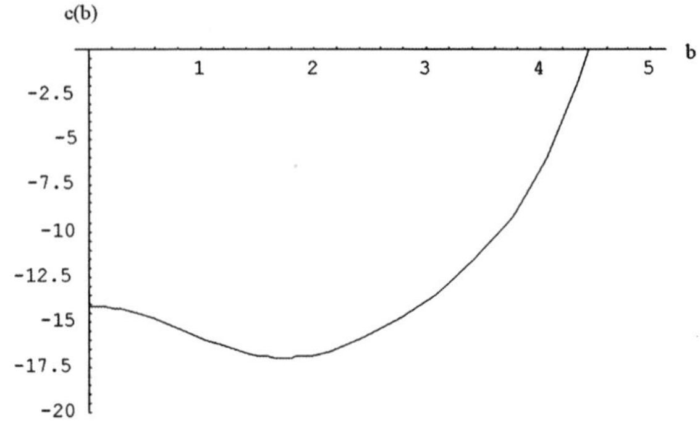

Fig. 8.11 The expected cost $c(b)$ (Example 8.8)

Coming back to the MBT shape of the failure rate, denote the local maximum by t_m (in Fig. 8.9, it is about 1). Assume that $\lambda(t_m)m(t_m) \leq 1$, which means, in accordance with the foregoing result, that $m(t)$ is decreasing for $t \geq t_m$, (it obviously holds for Fig. 8.9). Let us modify the *initial failure rate* to a constant in $0 \leq t < t_m$ and do not change it in $t_m \leq t < \infty$. This means that the resulting failure rate is still the BT and, as $\lambda(t_m) = \lambda(0)$, we can use the condition $\lambda(t_m)\tilde{m}(0) \leq 1$ (where $\tilde{m}(0)$ denotes the corresponding MRL function) as the characterization of the decreasing property of $\tilde{m}(t)$ in $(0, \infty)$. If this condition holds for *the defined BT shape* of the failure rate, the MRL function is decreasing in $(0, \infty)$ for *any MBT-shaped failure rate* that is equal to the given modified failure rate in $t \geq t_m$. Indeed, the initial, increasing in $0 \leq t < t_m$ segment of the failure rate obviously 'additionally contributes to the 'decreasing property', as compared with the flat one. This means that the condition $\lambda(t_m)\tilde{m}(0) \leq 1$ can be considered as a sufficient one, thus expanding the admissible class of failure rates to the class of MBT-shaped failure rates 'constructed' in the described way. On the other hand, this condition is rather crude and the real admissible class of the MBT-shaped failure rates is wider.

8.2 Continuous Mixtures

8.2.1 The Effect of Burn-in

It is well known that continuous mixtures of distributions constitute a useful tool for describing the heterogeneity of population due to random effect. Consider a general 'continuous' mixing model for a heterogeneous population, i.e.,

$$F_m(t) = \int_0^\infty F(t,z)\pi(z)\mathrm{d}z, \; f_m(t) = \int_0^\infty f(t,z)\pi(z)\mathrm{d}z, \tag{8.17}$$

where $F(t,z), f(t,z)$ are the Cdf and the pdf of subpopulations indexed by the frailty parameter Z and $\pi(z)$ is the pdf of Z with support in $(0,\infty)$. The general support $[a,b)$, $0 \le a < b \le \infty$ can be considered as well. Then the mixture (the *observed* or the *population*) failure rate $\lambda_m(t)$, in accordance with (5.11, 5.12), is defined as

$$\lambda_m(t) = \frac{\int_0^\infty f(t,z)\pi(z)\mathrm{d}z}{\int_0^\infty \bar{F}(t,z)\pi(z)\mathrm{d}z} = \int_0^\infty \lambda(t,z)\pi(z|t)\mathrm{d}z, \tag{8.18}$$

where the conditional density (on condition that the item did not fail in $[0,t)$) is

$$\pi(z|t) \equiv \pi(z)\frac{\bar{F}(t,z)}{\int_0^\infty \bar{F}(t,z)\pi(z)\mathrm{d}z}.$$

In the next subsection we will need the following lemma, which defines an expression for the derivative of this density.

Lemma 8.1 *The derivative of the conditional density $\pi(z|t)$ with respect to t is*

$$\pi'(z|t) = \pi(z|t)(\lambda_m(t) - \lambda(t,z)).$$

The *proof* is straightforward as:

$$\begin{aligned}\pi'(z|t) &= -\frac{f(t,z)\pi(z)}{\int_0^\infty \bar{F}(t,z)\pi(z)\mathrm{d}z} + \frac{\bar{F}(t,z)\pi(z)\lambda_m(t)}{\int_0^\infty \bar{F}(t,z)\pi(z)dz} \\ &= \lambda_m(t)\pi(z|t) - \frac{f(t,z)\pi(z)}{\int_0^\infty \bar{F}(t,z)\pi(z)\mathrm{d}z} \\ &= \lambda_m(t)\pi(z|t) - \frac{\lambda(t,z)\bar{F}(t,z)\pi(z)}{\int_0^\infty \bar{F}(t,z)\pi(z)\mathrm{d}z} = \pi(z|t)(\lambda_m(t) - \lambda(t,z)).\end{aligned}$$

□

Denote the Cdfs of $\pi(z)$ and $\pi(z|t)$ by $\Pi(z)$ and $\Pi(z|t)$, respectively, and by $Z|t$ the conditional frailty (on condition that the item did not fail in $[0,t)$). The following theorem describes monotonicity of $\Pi(z|t)$ as a function of t.

Theorem 8.1 *Let our subpopulations be ordered in the sense of the failure rate ordering*

$$\lambda(t,z_1) \le \lambda(t,z_2), \quad z_1 < z_2, \forall z_1, z_2 \in [0,\infty],\ t \ge 0. \tag{8.19}$$

Then $\Pi(z|t)$ is increasing in t for each $z > 0$.

Proof As,

$$\Pi(z|t) = \frac{\int_0^z \bar{F}(t,u)\pi(u)\mathrm{d}u}{\int_0^\infty \bar{F}(t,u)\pi(u)\mathrm{d}u},$$

it is easy to see that the derivative of this function is positive if

$$\frac{\int_0^z \bar{F}'(t,u)\pi(u)\mathrm{d}u}{\int_0^z \bar{F}(t,u)\pi(u)\mathrm{d}u} > \frac{\int_0^\infty \bar{F}'(t,u)\pi(u)\mathrm{d}u}{\int_0^\infty \bar{F}(t,u)\pi(u)\mathrm{d}u}.$$

Taking into account that $\bar{F}'(t,z) = -\lambda(t,z)\bar{F}(t,z)$, it is sufficient to show that

$$\frac{\int_0^z \lambda(t,z)\bar{F}(t,z)\pi(u)\mathrm{d}u}{\int_0^z \bar{F}(t,u)\pi(u)\mathrm{d}u}$$

is increasing in z, or equivalently, that

$$\lambda(t,z)\int_0^z \bar{F}(t,u)\pi(u)\mathrm{d}u > \int_0^z \lambda(t,u)\bar{F}(t,u)\pi(u)\mathrm{d}u,$$

which immediately follows from (8.19). □

A useful and practically relevant model of ordering (8.19) is the multiplicative (proportional hazards):

$$\lambda(t,z) = z\lambda(t).$$

This theorem indicates that the 'proportion' of stronger subpopulations (with smaller failure rates, which corresponds to the smaller values of the frailty parameter Z) increases as time increases. This property can be very *important for justification of the burn-in procedure*, as the 'quality of population' in the described sense (i.e., the proportion of stronger items increases) improves after burn-in of any duration. But along with this effect, the failure rates of subpopulations can decrease, increase or have a nonmonotonic shape (e.g., bathtub) and this should be also taken into account for defining and justifying the corresponding burn-in procedure.

Consider now the burn-in time b. The above relationships should be slightly adjusted. As follows from (8.17), after burn-in during time b, the component that is picked up at random from the population *that has survived burn-in* has the following survival function in 'field use':

$$\bar{F}_m(t|b) = \int_0^\infty \bar{F}(t,z|b)\pi(z|b)\mathrm{d}z,$$

where $\overline{F}(t,z|b) = \frac{\bar{F}(b+t,z)}{\bar{F}(b,z)}$ is the corresponding remaining lifetime distribution. The mixture failure rate function after burn-in is then expressed as

$$\lambda_m(t|b) = \lambda_m(t+b),$$

which is, in fact, intuitively obvious.

It is well known that if $\lambda(t,z)$ is nonincreasing in t, $i=1,2$, then $\lambda_m(t|b) = \lambda_m(t+b)$ is a decreasing function of its argument and therefore decreases in b for all fixed $t \geq 0$ as well (the mixture failure rate of distributions with decreasing failure rates is also decreasing). Thus, in this specific case, burn-in is decreasing the failure rate in field use (increasing the MRL). The simplest example of this property is:

Example 8.9 Suppose that $\lambda(t,z) = z\lambda, t \geq 0$, where λ is a constant and Z is exponentially distributed with parameter θ. Then by direct integration in (8.18):

$$\lambda_m(t) = \frac{\int_0^\infty z\lambda \exp\{-z\lambda t\}\theta \exp\{-\theta z\}dz}{\int_0^\infty \exp\{-z\lambda t\}\theta \exp\{-\theta z\}dz} = \frac{\lambda}{\theta + \lambda t},$$

which is a decreasing function. Thus, substituting $t+b$ instead of t obviously means that $\lambda_m(t+b)$ is decreasing in b for all fixed $t \geq 0$.

Thus, as is shown in Example 8.9, burn-in in the case of mixture of subpopulations with non-IFRs not only increases the proportion of the strong subpopulations, but also decreases the mixture failure rate and thus it is obviously justified. However, in the next subsection, we will see that burn-in may be justified *even when the failure rates of subpopulations and the mixture failure rate are increasing*.

8.2.2 Optimal Burn-in Time for Performance Quality Measures

Now, we will describe a model that already takes into consideration the costs and gains involved. This cost structure (expected costs) accounts for the performance quality after burn-in and defines gains (negative costs) proportional to the MRL. Other types of cost structures considered in the literature (e.g., [25, 27]) can be also discussed in a similar way. At first, as in the homogeneous case, we do not 'reveal' the composition of our population and deal with the observed mixture distribution function. Thus, in accordance with this model:

$$c(b) = c_0(b) - Km_m(b), \tag{8.20}$$

where

- b the burn-in duration.
- K the gain for the unit of time during the mission time, which has a negative sign as the equation is formulated in terms of costs.

- $c_0(b)$ the cost to obtain a component that has passed burn-in. If the first component fails, then the second one is tested, etc., until a component passes burn-in.
- $m_m(b) = \frac{\int_b^\infty \bar{F}_m(u)\mathrm{d}u}{\bar{F}_m(b)}$ is the corresponding mixture MRL after burn-in during the time b.

Denote the cost of a single item by C, and if, for simplicity, we assume that the expected cost of burn-in is just the cost of the failed components, then it is easy to show that

$$c_0(b) = C\left(\frac{1}{\bar{F}_m(b)} - 1\right) = \frac{CF_m(b)}{\bar{F}_m(b)}, \tag{8.21}$$

where $1/\bar{F}_m(b)$ corresponds to the expected 'total number of trials' until the first success. Equation (8.21) can be easily adjusted to the case when there are additional costs proportional to the duration b (see, e.g., [26, 6]). Obviously, $c_0(b)$ is increasing with b.

Remark 8.3 As it was mentioned, Eq. (8.20), in fact, formulates the problem exactly like in a homogeneous case, just using the mixture distribution as a governing one. If, e.g., the MRL, $m_m(b)$ is increasing or initially increasing in b and is described, e.g., by the UBT-shape, then the problem of obtaining the optimal b^* that minimizes $c(b)$ can be properly formulated. On the other hand, if we use the structure of the population described by the time-dependent $\pi(z|b)$, some other more advanced settings can be considered, e.g., dealing with the quality of performance (gain), which characterizes each subpopulation and not the overall population.

Assume now that a component from the strong subpopulation is characterized by the quality (the gain for the unit of time during the mission time), $Q(z)$ also indexed by the frailty parameter Z. Assume also that this function is decreasing: the larger values of Z (weaker items) correspond to the smaller values of gains, which is a realistic assumption at many instances. Then (8.20) is modified:

$$\begin{aligned} c(b) &= c_0(b) - \int_0^\infty Q(z)\frac{\int_b^\infty \bar{F}(u,z)\mathrm{d}u}{\bar{F}(b,z)}\pi(z|b)\mathrm{d}z \\ &= c_0(b) - \frac{\int_0^\infty Q(z)\int_b^\infty \bar{F}(u,z)\mathrm{d}u\pi(z)\mathrm{d}z}{\int_0^\infty \bar{F}(b,z)\pi(z)\mathrm{d}z}. \end{aligned} \tag{8.22}$$

whereas the time-dependent case $Q(z,t)$ corresponds to:

$$\begin{aligned} c(b) &= c_0(b) - \int_0^\infty \frac{\int_0^\infty \int_0^u Q(z,b+t)\mathrm{d}tf(b+u,z)\mathrm{d}u}{\bar{F}(b,z)}\pi(z|b)\mathrm{d}z \\ &= c_0(b) - \frac{\int_0^\infty \int_b^\infty \int_b^{b+u} Q(z,t)\mathrm{d}t\, f(u,z)\mathrm{d}u\pi(z)\mathrm{d}z}{\int_0^\infty \bar{F}(b,z)\pi(z)\mathrm{d}z}. \end{aligned}$$

When $Q(z) = K$, as follows from (8.19), Eq. (8.22) reduces to (8.20). Thus minimization of $c(b)$ can be considered to be a generalization of standard burn-in approaches. For simplicity, we will proceed further with a not 'time-constant case' (8.22).

If the quality lower than some level Q_0 due to some reasons is not acceptable and therefore corresponding realizations should not contribute to the expected quality, then we must set:

$$\tilde{Q}(z) = \begin{cases} Q(z), & z \le z_0 \\ 0, & z > z_0 \end{cases} \tag{8.23}$$

where z_0 is obtained from the equation $Q(z) = Q_0$, which has a unique solution as $Q(z)$ is strictly decreasing in z.

If, e.g., $F(t,z)$ is an exponential family of distributions: $\lambda(t,z) = z\lambda, \quad t \ge 0$, then it is easy to see that gains increase with b as $Q(z)$ is decreasing and the proportion of subpopulations with small values of frailties is increasing with b. Therefore, under suitable assumptions for parameters there should be a minimum for some b for the expected costs function $c(b)$:

$$b^* = \arg \inf_{b \in [0,\infty]} c(b).$$

Obviously, monotonicity properties of $c(b)$ are defined by its derivative. As the costs $c_0(b)$ are increasing, its derivative is positive.

Theorem 8.2 *The derivative of expected costs* (8.22) *is given by the following relationship:*

$$c'(b) = c_0'(b) - \int_0^\infty Q(z)[m(b,z)\lambda_m(t) - 1]\pi(z|b)\mathrm{d}z, \tag{8.24}$$

where $m(t,z) = \int_b^\infty \bar{F}(t,z)\mathrm{d}u / \bar{F}(b,z)$ *is the* MRL *for the subpopulation with frailty* z.

Proof Using this notation the first line in (8.22) can be written as

$$c(b) = c_0(b) - \int_0^\infty Q(z)m(b,z)\pi(z|b)\mathrm{d}z. \tag{8.25}$$

Using Lemma 8.1:

$$c'(b) = c_0'(b) - \int_0^\infty Q(z)[m'(b,z)\pi(z|b) + m(b,z)\pi(z|b)(\lambda_m(b) - \lambda(b,z))]\mathrm{d}z$$

and the well-known equality $m'(t) = \lambda(t)m(t) - 1$ describing the link between the MRL and the failure rate, we obtain (8.24):

$$c'(b) = c_0'(b) - \int_0^\infty Q(z)[\lambda(b,z)m(b,z) - 1 + m(b,z)(\lambda_m(b) - \lambda(b,z))]\pi(z|b)\mathrm{d}z$$
$$= c_0'(b) - \int_0^\infty Q(z)[m(b,z)\lambda_m(b) - 1]\pi(z|b)\mathrm{d}z.$$

□

Using this theorem, we can further analyze the derivative of expected gains. First, note that when $Q(z) \equiv K$, as it should be, we arrive at the derivative of gains (heterogeneous case) that corresponds to the setting defined by Eq. (8.20) for the homogeneous case:

$$d(b) \equiv \int_0^\infty Q(z)[m(b,z)\lambda_m(b) - 1]\pi(z|b)\mathrm{d}z$$
$$= K(\lambda_m(b)m_m(b) - 1) = Km_m'(b),$$

where

$$m_m(t) = \int_0^\infty m(t,z)\pi(z|t)\mathrm{d}z,$$

which, similar to (8.18), defines the mixture (population) MRL. If we assume that the mixture failure rate is increasing, then $m_m'(t) \le 0$ (expected gains are decreasing with time) and the burn-in obviously should not be performed in this case.

What happens now when $Q(z)$ is a decreasing function? In this case, $Q(z)$ can be considered as a kind of weight that gives higher values of performance measure to stronger subpopulations with smaller values of z and therefore, to smaller values of $\lambda(t,z)$ (larger values of $m(t,z)$). Thus, depending on parameters, the inequality $d(b) > 0$ can hold even for the case of increasing mixture failure rates. To illustrate this statement, assume that the mixture failure rate and the subpopulations failure rates are increasing in time and that $\lambda_m(0) \neq 0$. Therefore, $m_m(t)\lambda_m(t) - 1 < 0$. Let, e.g., $b = 0$. Let $Q(z)$ in (8.23) be a step function: $Q(z) = K$ for $0 < z \le z_0$, and z_0 can be chosen 'as small as we wish'. In fact, we must show that

$$d(0) = K\int_0^{z_0} [m(0,z)\lambda_m(0) - 1]\pi(z)\mathrm{d}z$$
$$= K\left[\lambda_m(0)\int_0^{z_0} m(0,z)\pi(z)\mathrm{d}z - P(Z \le z_0)\right] > 0.$$

Note that, as populations are ordered, $m(0,z)$ is decreasing in z and therefore, $m_m(0) < m(0,0)$. The inequality $d(0) > 0$ holds for the sufficiently small z_0, for which inequality $\lambda_m(0)m(0,z_0) - 1 > 0$ (the corresponding lower bound approximation) is satisfied. The sufficient condition for that is $\lambda_m(0)m(0,0) - 1 > 0$ (although $m'(t) = m_m(t)\lambda_m(t) - 1 < 0!$). It is easy to see that this condition is satisfied for the important and widely used proportional hazards model $\lambda(t,z) = z\lambda(t), \quad t \geq 0$, as $m(0,z_0) \to \infty$ for $z_0 \to 0$ (see Example 8.10 of the next subsection).

Remark 8.4 If the lower bound of the support of $\pi(z)$ is not 0, but $a > 0$, the above reasoning is valid, as for sufficiently small z_0, the function $m(0, a + z_0)$ can be as close to $m(0,a)$ as we wish, and, $\lambda_m(t) > \lambda(t,a) \Rightarrow m_m(0) < m(0,a), t \geq 0$. Therefore, similar to the case $a = 0$:

$$d(0) = K \int_{a}^{a+z_0} [m(0,z)\lambda_m(0) - 1]\pi(z)\mathrm{d}z > 0$$

and the sufficient condition for this inequality to hold is $\lambda_m(0)m(0,a) - 1 > 0$

Remark 8.5 It is clear that similar results should hold for the exponentially decreasing quality function $Q(z) = \exp\{-\alpha z\}$ as well (for the sufficiently large α).

The foregoing reasoning can be applied (under stated conditions) to the case $b > 0$. The sufficiently small z_0 (or large α) will result in $d(b) > 0$, but, obviously, this procedure is not uniform, as the larger values of b require the smaller (larger) values of $z_0(\alpha)$.

An obvious sufficient condition for the existence of the finite (or nonfinite) optimal burn-in time is

$$\begin{aligned} c_0'(0) &< \int_0^\infty Q(z)[m(0,z)\lambda_m(0) - 1]\pi(z)\mathrm{d}z \\ &= \lambda_m(0)\int_0^\infty Q(z)m(0,z)\pi(z)\mathrm{d}z - \int_0^\infty Q(z)\pi(z)\mathrm{d}z. \end{aligned} \tag{8.26}$$

Clearly, this condition is rather strong and, e.g., does not hold for the first example in the next subsection, although the optimal burn-in time exists.

8.2.3 Examples

It is tempting to use the setting of Example 8.9 for the simplest illustration, but the mixture MRL in this case is nonfinite. Indeed, the mixture failure rate, as follows from Example 8.9, (when $t \to \infty$) tends to the failure rate of the Pareto distribution of the form $1 - t^{-1}$, which does not have the finite first moment. Other cost structures (defined, e.g., by gains during the fixed interval of mission time) can be considered in this case.

We will describe, first, a meaningful example when the subpopulations failure rates and the mixture failure rate are increasing and therefore, the *conventional burn-in should not be performed,* whereas considering the quality of performance function changes the situation and justifies the necessity of burn-in.

Example 8.10 Consider the truncated extreme value distribution (Gompertz) defined in a following way:

$$\bar{F}(t,z) = \exp\{-zk(\exp\{t\} - 1)\}, t \geq 0,$$
$$\lambda(t,z) = zk\exp\{t\},$$

where $k > 0$ is a constant. As in Example 8.9, let Z be exponentially distributed with parameter θ (proportional hazards model discussed in the previous section) Direct integration [17] gives

$$\int_0^\infty f(t,z)\pi(z)dz = \int_0^\infty zk\exp\{t\}\exp\{-zk(\exp\{t\}-1)\}\theta\exp\{-\theta z\}dz$$
$$= \frac{\theta k\exp\{t\}}{\omega^2}; \ \omega = k\exp\{t\} - k + \theta,$$
$$\int_0^\infty \bar{F}(t,z)\pi(z)dz = \theta\int_0^\infty \exp\{-\omega z\}dz = \frac{\theta}{\omega}.$$

Eventually, using definition (8.18):

$$\lambda_m(t) = \frac{k\exp\{t\}}{\omega} = 1 + \frac{k-\theta}{k\exp\{t\} - k + \theta}.$$

Let $k < \theta$. Then $\lambda_m(t)$ is monotonically increasing asymptotically converging to 1. Thus, the baseline failure rate $k\exp\{t\}$ and the mixture failure rate $\lambda_m(t)$ are increasing, whereas $m_m(t)$ is decreasing. However, the gains in (8.25):

$$\int_0^\infty Q(z)m(b,z)\pi(z|b)dz = \frac{\int_0^\infty Q(z)\int_b^\infty \bar{F}(u,z)du\pi(z)dz}{\int_0^\infty \bar{F}(b,z)\pi(z)dz},$$

as was discussed in the previous section, for the sufficiently rapidly decreasing $Q(z)$ can increase (at least initially) which constitutes the *possibility* of the optimal burn-in time b^*. This is illustrated by the following specific case: $Q(z) = 10\exp\{-10z\}, z \geq 0, C = 0.1, k = 0.1$ and $\theta = 1.0$. Observe that $c_0'(0) = C \cdot f_m(0)$, and

$$c'(0) = c_0'(0) - \lambda_m(0)\int_0^\infty Q(z)m(0,z)\pi(z)dz + \int_0^\infty Q(z)\pi(z)dz = 0.49 > 0.$$

Fig. 8.12 Cost function $c(b)$

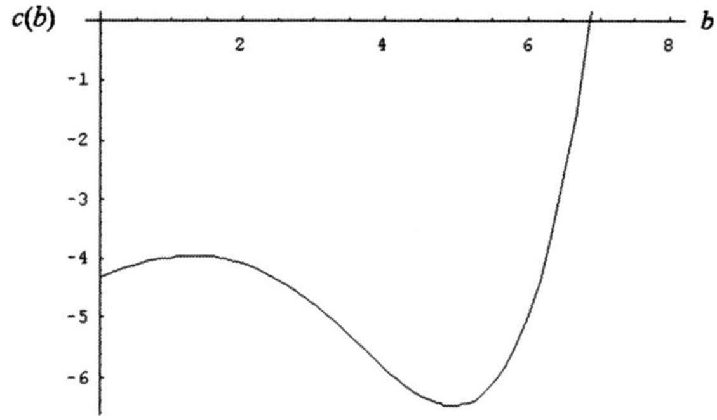

Thus, the condition (8.26) is not satisfied. However, as is shown in Fig. 8.12, there is a positive optimal burn-in time.

By numerical search, the optimal burn-in time and minimum cost is $b^* = 4.95$ and $c(4.95) = -6.48$.

Example 8.11 Assume now that we have time constraints on the duration of burn-in: $b \le b_c$. Consider the case of the UBT shape of the corresponding mixture failure rate in conventional Model (8.20). It is well known (see, e.g., [17]), that if

$$m'_m(0) = \lambda_m(0)m_m(0) - 1 \le 0$$

then the MRL has a bathtub shape and the corresponding gains initially decrease. Therefore, if the interval, where the gains decrease (although they can increase afterward), is larger than b_c then burn-in is not usually performed, as the overall cost function $c(b)$ (monotonic or nonmonotonic in $[0, b_c))$ is initially increasing and has a minimum at $b = 0$. However, considering Model (8.22) with the rapidly decreasing $Q(z)$ can change this decision as a minimum can be achieved at $b = b_c$ (burn-in is justified).

For illustration of the foregoing reasoning, consider the mixture of the Weibull distributions with linearly IFRs: $\lambda(t, z) = 2zt$, then, again assuming that the frailty Z is exponentially distributed with parameter θ, it is easy to show that

$$\lambda_m(t) = \frac{2t}{\theta + t^2}.$$

This function is equal to zero at $t = 0$ and tends to zero as $t \to \infty$ with a single maximum at $t = \sqrt{\theta}$ (BT shaped, as $m'_m(0) = -1 \le 0$).

Let $Q(z) = \exp\{-10z\}, z \ge 0, C = 0.1,$ and $\theta = 1.0, b_c = 1.$ Figure 8.13 shows that $c(b)$ is initially slightly increasing and then decreasing in interval $(0, b_c)$, with a minimum at b_c, and therefore, burn-in is justified: $b^* = b_c$.

On the other hand, Fig. 8.14 shows $c(b)$ for the conventional case ($Q(z) = K = 1$ and all other parameters are the same), with a minimum in $(0, b_c)$

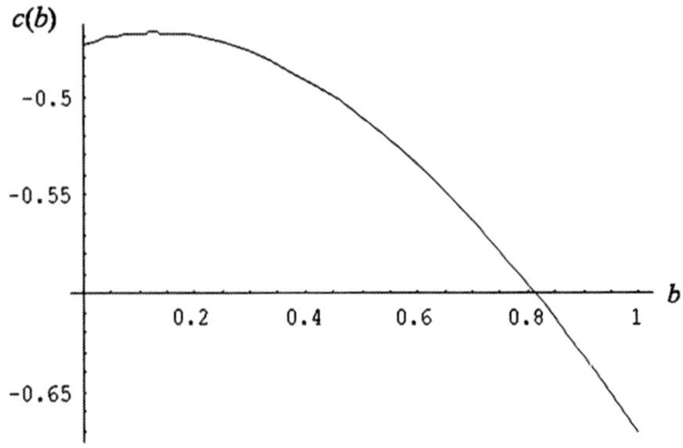

Fig. 8.13 Cost function $c(b)$

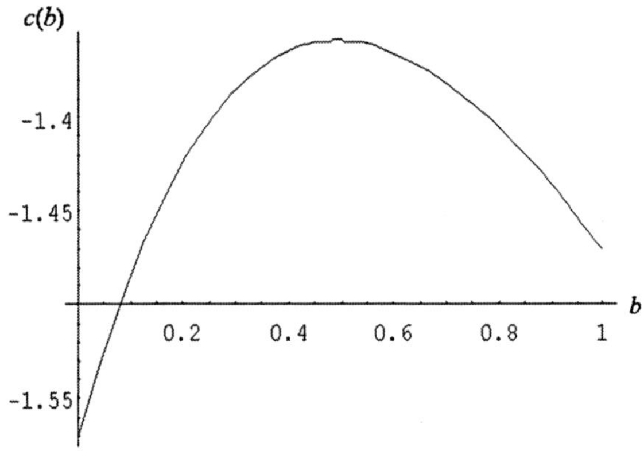

Fig. 8.14 Cost function $c(b)$ for $K = 1$

at $b = 0$. Therefore, burn-in is not justified. Note that, as it should be expected, the function $Q(z)$ has also changed the initial shape of $c(b)$ from 'rapidly increasing' to 'slightly increasing'.

8.3 Burn-in for Minimizing Risks

8.3.1 Burn-in for Avoiding Large Risks: Discrete Mixture

In this subsection, we consider burn-in for avoiding large risks (or losses) that can occur during *important missions*. Most of the references on burn-in consider items from homogeneous populations. Although a few studies on optimal burn for heterogeneous population have been performed (e.g., [2, 3, 10–12, 15]), all of them were considering the mixture failure rate as a characteristic of population quality. However, the mixture (population) failure rate (at each time instant) is the expectation of the failure rates of subpopulations (see later). Therefore, as usual in statistical analysis, dealing with expectations only is not the best approach, especially when substantial risks and losses are involved. In this section, we depart from the conventional reasoning and model the burn-in procedures that minimize the risks that occur due to choosing items with large individual failure rates. As our population is heterogeneous (before and after burn-in), these risks always exist.

Consider the case of $n = 2$ subpopulations (that can be generalized to the arbitrary finite n). Denote the lifetime of a component from the 'strong subpopulation' by T_S and its absolutely continuous Cdf, pdf, and the failure rate function by $F_1(t)$, $f_1(t)$ and $\lambda_1(t)$, respectively. Similarly, the lifetime, the Cdf, pdf, and the failure rate function of the 'weak' component are $T_W, F_2(t), f_2(t)$ and $\lambda_2(t)$, accordingly. Definitions of the strong and weak subpopulations will be given after introducing the corresponding notation. The initial $(t = 0)$ composition of our mixed population is as follows: the proportion of the strong items is π, whereas the proportion of the weak items is $1 - \pi$, which means that the distribution of the discrete frailty Z with realizations z_1 and z_2 in this case is

$$\pi(z) = \begin{cases} \pi, & z = z_1 \\ 1 - \pi, & z = z_2 \end{cases},$$

where the values $z_1, z_2 (z_1 < z_2)$, correspond to the strong and the weak subpopulations, respectively. As previously [see Eq. (8.3)], the mixture (population) survival function is

$$\bar{F}_m(t) = \pi\bar{F}_1(t) + (1 - \pi)\bar{F}_2(t),$$

whereas the mixture failure rate is defined as

$$\lambda_m(t) = \frac{\pi f_1(t) + (1 - \pi)f_2(t)}{\pi\bar{F}_1(t) + (1 - \pi)\bar{F}_2(t)} = \pi_1(t)\lambda_1(t) + \pi_2(t)\lambda_2(t),$$

where the time-dependent probabilities are

$$\pi_1(t) = \frac{\pi\bar{F}_1(t)}{\pi\bar{F}_1(t) + (1 - \pi)\bar{F}_2(t)}, \quad \pi_2(t) = \frac{(1 - \pi)\bar{F}_2(t)}{\pi\bar{F}_1(t) + (1 - \pi)\bar{F}_2(t)}.$$

Assume further that our populations are ordered (and therefore, the weak and the strong subpopulations are defined accordingly) in the sense of the failure rate ordering:

$$\lambda_2(t) \geq \lambda_1(t), \quad t \geq 0.$$

It can be shown [10] that, in this case, the proportion of strong items $\pi_1(t)$ is increasing in t, which is important for our further reasoning.

For illustration and motivation of our further reasoning, consider now the mixture of two distributions with decreasing and IFRs given in Fig. 8.15.

In Fig. 8.15, the proportion of items from the strong subpopulation is 0.80 and that from the weak subpopulation is 0.20 (see also Example 8.12). The mixture

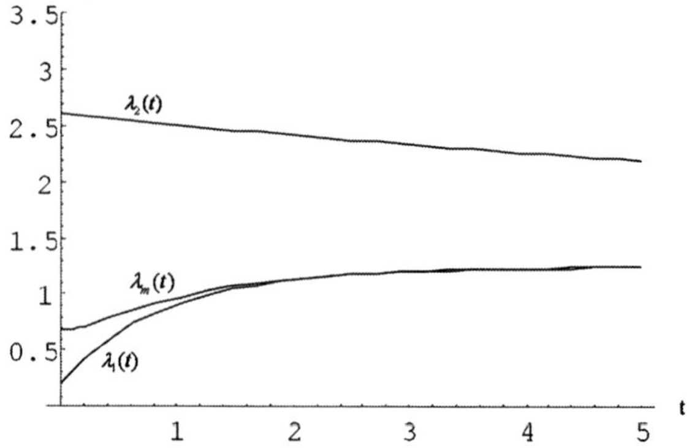

Fig. 8.15 The mixture failure rate for two subpopulations

failure rate in this case *strictly increases* and therefore, there is no need for burn-in from the conventional perspective. However, the situation is more complex when we consider the corresponding risks. Suppose that, at time $t = 0$ (without applying burn-in), we choose an item from the above mixed population for a field usage. If we select a weak item, then its failure rate is $\lambda_2(t)$, which is substantially larger than $\lambda_1(t)$. Therefore, it can result in the unsatisfactory reliability performance. For instance, for the mission time τ, the probability $P(T_W > \tau)$ can be substantially smaller than $P(T_S > \tau)$ (see Example 8.12) and this may cause large risk during usage especially for missions of high importance. We can reduce this risk, if the proportion of items from the weak subpopulation is substantially decreased. It can be achieved via the corresponding burn-in, as $\pi_2(t)$ is decreasing in t. As follows from Fig. 8.15, the population (mixture) failure rate is increasing and therefore, the quality of the population described by this characteristic is decreasing, whereas at the same time the risk of selection of the weak item is decreasing. The joint consideration of the corresponding gains and losses can help to answer to the question: to perform or not to perform burn-in.

Let us formalize now the corresponding measure based on the above reasoning. Suppose that the item is operable at time $t > 0$ (during field operation). For an item from the weak population, the risk of instantaneous failure is obviously larger than that from the strong one. Therefore, a larger penalty (loss) should be imposed to the item with a larger risk. This allows us to define the following "point loss" at time t for the subpopulation i:

$$L_i(t) = g(\lambda_i(t)), i = 1, 2,$$

where $g(\cdot)$ is a strictly increasing function of its argument. Let τ be the usage (mission) time for our components. As the above point loss varies during the mission time, it should be averaged, i.e.,

$$\frac{\int_0^\tau L_i(t)\mathrm{d}t}{\tau} = \frac{\int_0^\tau g(\lambda_i(t))\mathrm{d}t}{\tau}, i = 1, 2.$$

As the selection of a component from a heterogeneous population is made at time $t = 0$ and the corresponding proportions are given by $\pi_i(0), i = 1, 2$, the mean loss for our mixture population (without burn-in) is

$$\sum_{i=1}^{2} \frac{\int_0^\tau g(\lambda_i(t))\mathrm{d}t}{\tau} \cdot \pi_i(0), \text{where } \pi_1(0) = \pi \text{ and } \pi_2(0) = 1 - \pi. \tag{8.27}$$

If the burn-in procedure of duration b is performed, $\lambda_i(t)$ and $\pi_i(0)$ in (8.27) should be replaced by $\lambda_i(b + t)$ and $\pi_i(b)$, respectively, and the mean loss after burn-in is

$$\Psi(b) \equiv \sum_{i=1}^{2} \frac{\int_0^\tau g(\lambda_i(b+t))\mathrm{d}t}{\tau} \cdot \pi_i(b). \tag{8.28}$$

The gains that are already taken into account by this formula are due to the increase of the proportion of strong items.

Based on the measure defined above, we consider the following criterion for obtaining the optimal burn-in time:

Criterion 1 Find b^* which minimizes $\Psi(b)$.

Example 8.12 We describe now in more detail the example that corresponds to Fig. 8.15. Let $\lambda_1(t) = 1.2 - \exp\{-1.2t\} + 0.01t, \lambda_2(t) = 1.4\exp\{-0.08t\} + 1.2 + 0.01t$, with $\pi = \pi_1(0) = 0.80$. Suppose that $\tau = 3.0$ and $g(x) = x^2$. Then $\Psi(b)$ is given by Fig. 8.16.

Therefore, in this case, the optimal burn-in time is $b^* \approx 1.10$. The proportion of strong items after burn-in is now $\pi_1(1.10) = 0.97$ and therefore, about 85 % (!) of weak items have been eliminated. This effect increases our gain. On the other hand, what is the undesirable but inevitable consequence of this operation? Obviously, it is the increase in the failure rate of the strong items after burn-in. By sacrificing the 'quality of the strong subpopulation', the risk that can be caused by the weak subpopulation has been substantially reduced.

Fig. 8.16 $\Psi(b)$

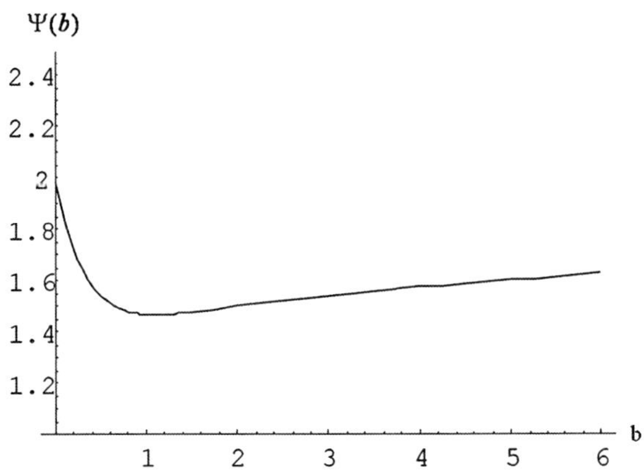

Let $M_i(b), i = 1, 2$, be the mean residual life time of the items in subpopulation i after the burn-in time b:

$$M_i(b) = \int_0^\infty \exp\left\{-\int_0^t \lambda_i(b+u)\mathrm{d}u\right\}\mathrm{d}t, i = 1, 2.$$

Then, similar to (8.28), define the following mean loss after burn-in:

$$\Phi(b) = \sum_{i=1}^{2} g(1/M_i(b))\pi_i(b).$$

Criterion 2 Find b^* which minimizes $\Phi(b)$.

Example 8.12 (Continued) For the setting described above, the corresponding loss function $\Phi(b)$ is given in Fig. 8.17.

Therefore, the optimal burn-in time also exists: $b^* \approx 0.8$.

It is interesting also to see how this risk-based criterion works in the cases when the conventional burn-in approach is applicable (i.e., the mixture failure rate is initially decreasing). We consider an example, where the mixture failure rate has a BT failure rate and therefore, the burn-in is justified and the optimal burn-in time can exist.

Example 8.13 Suppose that $\lambda_1(t) = 1, 0 \le t \le 5.0, \lambda_1(t) = 0.2(t-5)+1, t > 5.0$, and $\lambda_2(t) = \lambda_1(t) + 2, t \ge 0$, with $\pi = \pi_1(0) = 0.7$ and $1 - \pi = 0.3$. Then the mixture failure rate $\lambda_m(t)$ is given in Fig. 8.18.

The failure rate strictly decreases for $0 \le t \le 5.0$ and then it is strictly increasing. Thus it is BT with one change point, $t_1 = 5.0$. Note that, in conventional burn-in, the optimal burn-in time b^*, which optimizes the corresponding criteria (e.g., MRL, the probability of the successful completion of mission, the

Fig. 8.17 $\Phi(b)$

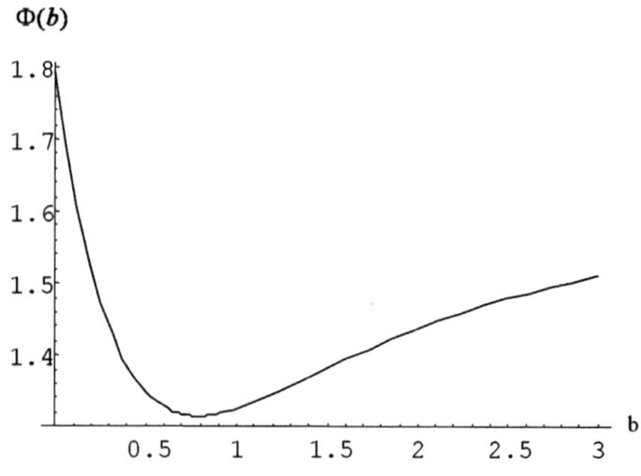

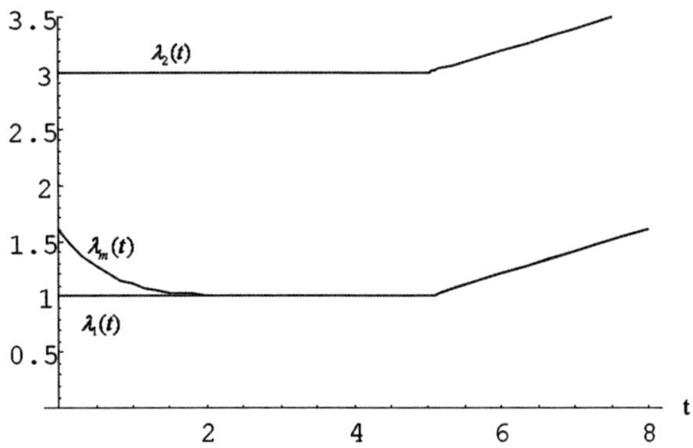

Fig. 8.18 The mixture failure rate for two subpopulations

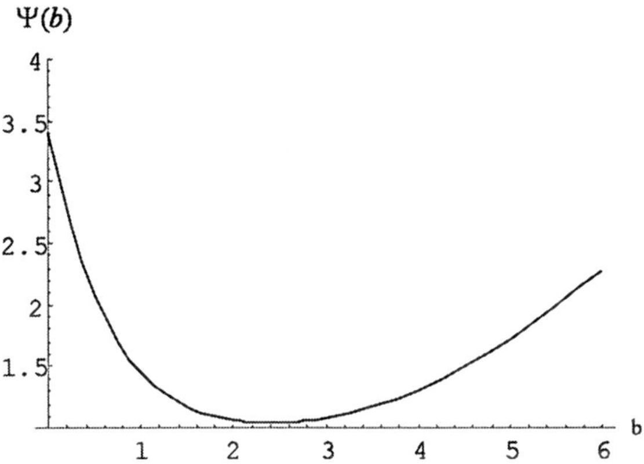

Fig. 8.19 $\Psi(b)$

expected cost, etc.) for this case is positive and $b^* < t_1$ (See [7, 8] and [25–28]). Let $\tau = 3.0$ and $g(x) = x^2$. Then $\Psi(b)$ is given by Fig. 8.19.

In this case, the optimal burn-in time is $b^* \approx 2.58$.

As was already mentioned, a 'sort of sacrifice' takes place for this conventional setting [without implementing average loss (8.28)] as well. Indeed, as the failure rates of both subpopulations are initially constant, burn-in shortens these parts and therefore, makes them 'worse' in terms of the failure rate ordering. On the positive side, the proportion of strong items is increasing and overall the quality of our population is improving with a maximum achieved at b^*.

Remark 8.6 As mentioned before, the mixture setting described in this subsection is often realized in practice, as items belonging to the 'weak distribution' can be produced along with the items of the 'strong (main)' distribution due to the variation in the quality of resources and components, human errors, unstable production environment caused by uncontrolled significant quality factors, etc. The experts' opinions and other prior knowledge can often also be used for identifying the mixture setting: $\bar{F}_m(t) = \pi\bar{F}_1(t) + (1-\pi)\bar{F}_2(t)$. In many practical situations, this identification can be performed using the corresponding density curves (e.g., bimodal) and percentage failures graphs. The next step is the estimation of π and the corresponding parameters for $\bar{F}_1(t)$ and $\bar{F}_2(t)$ from the failure data using various statistical methods. Plenty of examples and detailed procedures for model setup and parameter estimation in relevant settings can be found in Jensen and Petersen [21], Kececioglu and Sun [23] and Klugman et al. [24]. For a specific example, the interested reader could refer to Example 4.2 of Jensen and Petersen [21].

8.3.2 Burn-in for Avoiding Large Risks: Continuous Mixture

Consider now the case of the 'continuous' mixture model for a heterogeneous population, i.e.,

$$F_m(t) = \int_0^\infty F(t,z)\pi(z)\mathrm{d}z,\ f_m(t) = \int_0^\infty f(t,z)\pi(z)\mathrm{d}z,$$

where $F(t,z) \equiv F(t|z), f(t,z) \equiv f(t|z)$ are the Cdf and the pdf of subpopulations indexed (conditioned) by the frailty parameter Z and $\pi(z)$ is the pdf of Z with support in $[0,\infty)$. Then the mixture failure rate $\lambda_m(t)$, as previously [see Eqs. (5.10–5.12)], is defined as

$$\lambda_m(t) = \frac{\int_0^\infty f(t,z)\pi(z)\mathrm{d}z}{\int_0^\infty \bar{F}(t,z)\pi(z)\mathrm{d}z} = \int_0^\infty \lambda(t,z)\pi(z|t)\mathrm{d}z,$$

where the conditional density (on condition that the item did not fail in $[0,t)$) is

$$\pi(z|t) \equiv \pi(z)\frac{\bar{F}(t,z)}{\int_0^\infty \bar{F}(t,z)\pi(z)\mathrm{d}z}. \tag{8.29}$$

As in the discrete case, let our subpopulations be ordered in the sense of the failure rate ordering:

$$\lambda(t,z_1) \le \lambda(t,z_2), \quad z_1 < z_2, \forall z_1, z_2 \in [0,\infty), t \ge 0.$$

Continuous mixtures is an effective tool for modeling population heterogeneity due to randomly changing production environment and other causes of 'random effects' (see also [1]).

For the continuous mixture case, the criteria defined in the discrete case can obviously be generalized as follows:

Criterion 1C Find b^* which minimizes

$$\Psi(b) = \int_0^\infty \frac{\int_0^\tau g(\lambda(b+t,z))\mathrm{d}t}{\tau} \cdot \pi(z|b)\mathrm{d}z.$$

Criterion 2C Find b^* which minimizes

$$\Phi(b) = \int_0^\infty g(1/M(b,z))\pi(z|b)\mathrm{d}z,$$

where $M(b,z)$ is the mean residual life time (for the fixed frailty parameter z) after the burn-in time b:

$$M(b,z) = \int_0^\infty \exp\left\{-\int_0^t \lambda(b+u,z)\mathrm{d}u\right\}\mathrm{d}t, z \geq 0.$$

Example 8.14 Suppose that $\lambda(t,z) = 0.1z\exp\{0.1t\} + 0.02t + 1$, and let Z be exponentially distributed with parameter θ. In this case, the mixture failure rate strictly increases as shown in Fig. 8.20 for the case when $\theta = 0.5$. Let $\tau = 3.0$ and $g(x) = x^2$. Then $\Psi(b)$ is given in Fig. 8.21. It can be seen that the optimal burn-in time is $b^* \approx 1.11$. The frailty distributions before and after burn-in, which are useful for analysis, are given in Fig. 8.22. From the graphs in Fig. 8.22, the following can be obtained:

$$\overline{\Pi}(1|0) \approx 0.61, \overline{\Pi}(1|b^*) \approx 0.6$$

$$\overline{\Pi}(2|0) \approx 0.37, \overline{\Pi}(2|b^*) \approx 0.29$$

$$\overline{\Pi}(3|0) \approx 0.22, \overline{\Pi}(3|b^*) \approx 0.16$$

$$\overline{\Pi}(4|0) \approx 0.14, \overline{\Pi}(4|b^*) \approx 0.08$$

$$\overline{\Pi}(5|0) \approx 0.08, \overline{\Pi}(5|b^*) \approx 0.05$$

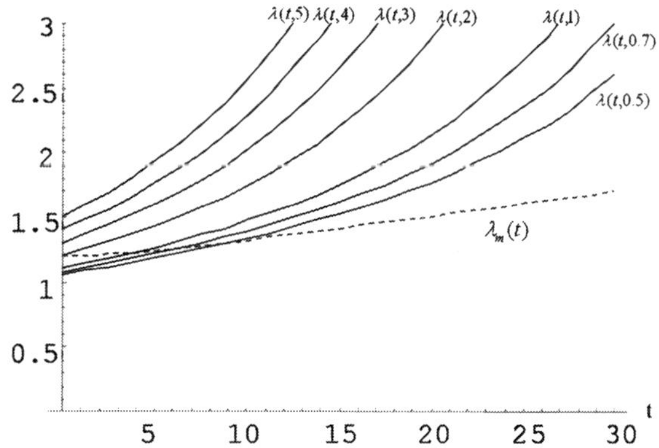

Fig. 8.20 $\lambda_m(t)$ and $\lambda(t,z)$ for $z = 0.5, 0.7, 1, 2, 3, 4$ and 5

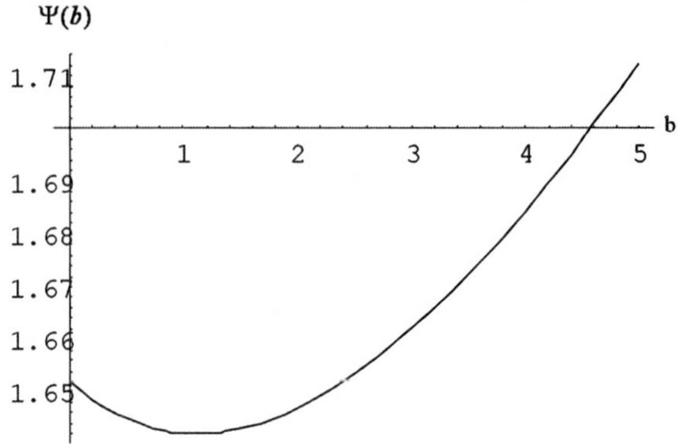

Fig. 8.21 $\Psi(b)$

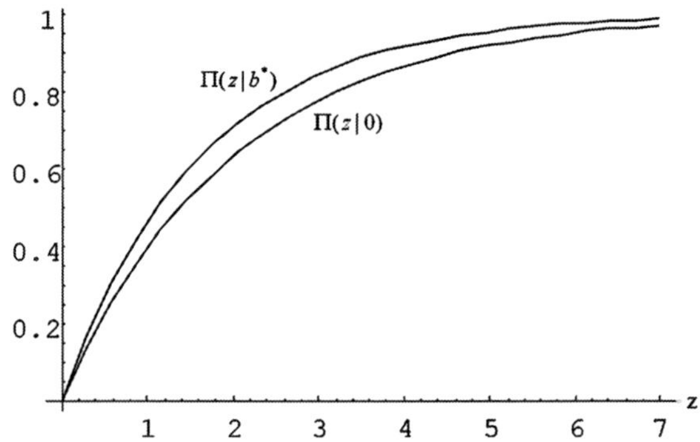

Fig. 8.22 The CDF's of the frailty distribution before and after burn-in

where $\overline{\Pi}(z|t)$ is the conditional *survival function,* which corresponds to $\pi(z|t)$. We can see that the corresponding frailty distributions are stochastically ordered and thus the risks of selecting the 'poor items' have been decreased.

Applying **Criterion 2C**, we can obtain the average loss given in Fig. 8.23. As shown in this figure, there exists the optimal burn-in time $b^* \approx 0.74$.

8.3.3 Optimal Burn-in Based on Conservative Measures

Failures of items may often result in the catastrophic or disastrous events. For example, failures in jet engines of aircrafts or those in gas safety valves may cause fatal consequences. Similarly, failures during important missions can cause huge economic loss. In these cases, rather than the 'average quality' of the heterogeneous population, which (as in the previous sections) we understand as $\lambda_m(t)$, it is reasonable to define the 'marginal' quality in the population that refers in some sense to the "worst scenario". That is, if this marginal quality is still acceptable, then the quality of our population as a whole should also be considered as satisfactory. This reasoning can create an alternative approach to the one discussed in

Fig. 8.23 $\Phi(b)$

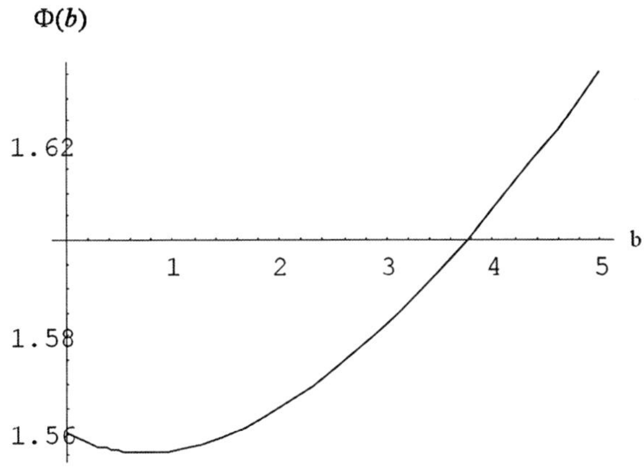

the previous subsections. The marginal quality can be used as a conservative measure (or bound) for the population quality. In this subsection, we consider the optimal burn-in procedures that optimize the conservative measures for continuous mixture models.

After burn-in during time b, the components in the population will have the failure rate $\lambda(t+b,z)$ in accordance with the conditional frailty distribution $\pi(z|b)$ defined in Sect. 8.3.2. Thus, we can define the following 'αth worst' realization of the 'residual' failure rate in the population:

$$\lambda_\alpha(t|b) = \lambda(b+t, z(\alpha|b)), t \geq 0, \tag{8.30}$$

where $z(\alpha|b) \equiv \inf\{z : \prod(z|b) \geq \alpha\}$ and α is usually close to 1 (e.g., 0.9 or 0.95) and, as previously, $\Pi(z|b)$ is the conditional distribution function, which corresponds to the conditional pdf $\pi(z|b)$. Accordingly, $\lambda_\alpha(t|b)$ is the failure rate of an item that has survived burn-in during time b, which corresponds to the αth percentile $z(\alpha|b)$ of the conditional distribution of frailty $\prod(z|b)$. When α is close to 1, this can be interpreted as the αth worst scenario. Based on the above setting, we can define the αth worst MRL of the population after the burn-in time b:

$$M_\alpha(b) \equiv \int_0^\infty \exp\left\{-\int_0^t \lambda_\alpha(u|b)du\right\}dt.$$

Therefore, the following criterion can be applied:

Criterion 3 Determine the optimal burn-in time b^* as the minimal burn-in time b, such that $M_\alpha(b) \geq m_r$, where m_r is the MRL that corresponds to the αth worst scenario.

Example 8.15 Consider the continuous mixture of exponentials. Let the conditional failure rate and the mixing distribution be $\lambda(t,z) = z$ and $\pi(z) = \theta \exp\{-\theta z\}$, respectively. Then

$$\lambda_m(t) = E[Z|t] = 1/(\theta + t),$$

where $Z|t \equiv Z|T > t$. Observe that the conditional mixing pdf and Cdf for this case are

$$\pi(z|t) = (\theta + t)\exp\{-(\theta + t)z\},$$

$$\Pi(z|t) = 1 - \exp\{-(\theta + t)z\},$$

respectively. Therefore,

$$z(\alpha|b) = -\ln(1 - \alpha)/(\theta + b), \tag{8.31}$$

and

$$\lambda_\alpha(t|b) = \lambda(b + t, z(\alpha|b)) = -\ln(1 - \alpha)/(\theta + b), t \geq 0.$$

For obtaining the optimal burn-in time, we will use Criterion 3 defined above. Let, for our example, $\alpha = 0.9$ and $m_r = 1.25$. As $\lambda(t, z) = z$, the corresponding MRL as a function of the burn-in time b is

$$M_\alpha(b) = 1/z(\alpha|b) = -(\theta + b)/\ln(1 - \alpha), \text{ where } \alpha = 0.9.$$

This linear function is given by Fig. 8.24 ($\theta = 1.0$).

It follows from this graph that the corresponding optimal burn-in time is $b^* \approx 1.88$.

The conservative measure (8.30) can be modified (generalized) to account for the average of the lower $(1 - \alpha)$ % quality of items among those that have survived burn-in during time b. Thus, instead of one realization, as previously, we now define the marginal quality as some average for the corresponding 'tail'.

The *initial conditional frailty distribution* after burn-in during time b, [which corresponds to $\pi(z)$ in (8.29)] for the items with the quality lower than $(1 - \alpha)$ % is

Fig. 8.24 $M_\alpha(b)$ for $\alpha = 0.9$, $\theta = 1.0$

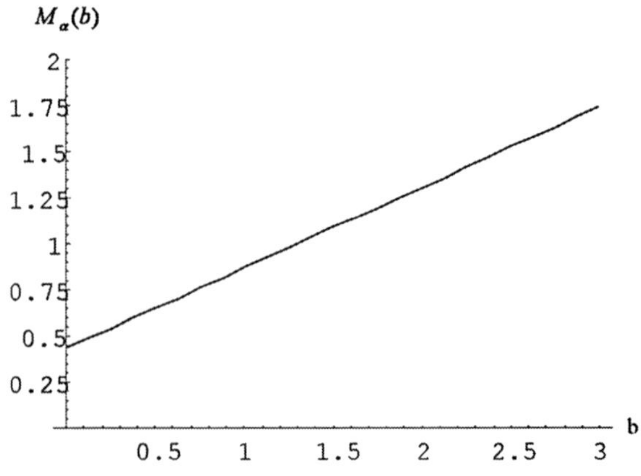

$$\frac{\pi(z|b)}{1-\alpha},\ z(\alpha|b) \le z \le \infty,$$

where, as previously, $z(\alpha|b) \equiv \inf\{z : \prod(z|b) \ge \alpha\}$. Accordingly, the conditional frailty distribution at time t, which corresponds to $\pi(z|t)$ in (8.29), is

$$\pi_\alpha(z|t;b) \equiv \frac{\pi(z|b)}{1-\alpha} \frac{\bar{F}(b+t,z)/\bar{F}(b,z)}{\int_{z(\alpha|b)}^{\infty} \bar{F}(b+t,z)/\bar{F}(b,z)\frac{\pi(z|b)}{1-\alpha}\mathrm{d}z},\ z(\alpha|b) \le z \le \infty.$$

Therefore, after burn-in during time b, the mixture failure rate at time t for the items in the survived population with the quality lower than $(1-\alpha)$ % is

$$\lambda_m(t|b,\alpha) = \int_{z(\alpha|b)}^{\infty} \lambda(b+t,z)\pi_\alpha(z|t;b)\mathrm{d}z.$$

Example 8.15. (Continued) In this case, Eq. (8.31) holds and

$$\int_{z(\alpha|b)}^{\infty} \bar{F}(b+t,z)/\bar{F}(b,z)\frac{\pi(z|b)}{1-\alpha}\mathrm{d}z = \frac{1}{(1-\alpha)} \cdot \frac{\theta+b}{\theta+b+t} \cdot (1-\alpha)^{\frac{\theta+b+t}{\theta+b}}.$$

Thus

$$\begin{aligned}\pi_\alpha(z|t;b) &\equiv \frac{\pi(z|b)}{1-\alpha} \frac{\bar{F}(b+t,z)/\bar{F}(b,z)}{\int_{z(\alpha|b)}^{\infty} \bar{F}(b+t,z)/\bar{F}(b,z)\frac{\pi(z|b)}{1-\alpha}\mathrm{d}z} \\ &= (\theta+b+t) \cdot (1-\alpha)^{-\frac{\theta+b+t}{\theta+b}} \cdot \exp\{-(\theta+b+t)z\},\end{aligned}$$

and

$$\lambda_m(t|b,\alpha) = \int_{z(\alpha|b)}^{\infty} \lambda(b+t,z)\pi_\alpha(z|t;b)\mathrm{d}z = -\frac{\ln(1-\alpha)}{\theta+b} + \frac{1}{\theta+b+t},\ t \ge 0.$$

The criterion for burn-in is practically the same as Criterion 3 with a slight difference that the MRL is calculated not for one realization but for the corresponding "partial" mixture population of items with low quality.

As previously, let $\alpha = 0.9$ and $m_r = 1.25$. Then we have to obtain the MRL of the items with the quality lower than $(1-\alpha)$ % at each b, which is given by

$$\int_0^{\infty} \exp\left\{-\int_0^t \lambda_m(u|b,\alpha)\mathrm{d}u\right\}\mathrm{d}t = \int_0^{\infty} (1-\alpha)^{t/(\theta+b)} \cdot \frac{\theta+b}{\theta+b+t}\mathrm{d}t,$$

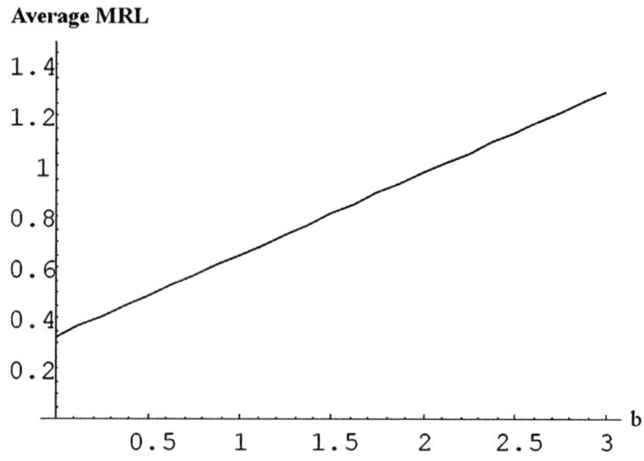

Fig. 8.25 The 'average' MRL as a function of b for $\alpha = 0.9$, $\theta = 1.0$

where $\alpha = 0.9$ and $\theta = 1.0$ This approximately linear function is given in Fig. 8.25.

It follows from this graph that the corresponding optimal burn-in time is $b^* \approx 2.47$ in this case.

8.4 Burn-in for Repairable Items

8.4.1 Basic Setup

In this section, a new burn-in approach for repairable items is proposed and optimal burn-in procedure is investigated. We consider the mixed population composed of two ordered subpopulations—the subpopulation of strong items (items with 'normal' lifetimes) and that of weak items (items with shorter lifetimes). Based on the information obtained during the burn-in procedure, items are classified into two groups: one class of items, which is considered to belong to the strong subpopulation and the other class of items that is believed to belong to the weak subpopulation. Then the items belonging to the second class are eliminated (discarded) and only the remaining items are considered to be suitable for the field operation.

In the first part, we consider two types of risks—(i) the risk that a strong component will be eliminated during burn-in and (ii) the risk that a weak component will pass the burn-in procedure. Optimal burn-in, which minimizes the weighted average of these risks, is investigated. The second part deals with optimal burn-in which minimizes the mean number of failures during the given mission time. It should be emphasized that the obtained optimal burn-in procedure (which minimizes the mean number of repairs during field usage) is suggested mainly for the case when the field mission is very important and the failures (even minimally repaired) during this mission are very undesirable (e.g., military missions). The costs incurred during burn-in are usually not so important in this case.

Let the lifetime of a component from the strong subpopulation be denoted by X_S and its absolutely continuous Cdf be $F_S(t)$. Similarly, the lifetime and the Cdf of a weak component is denoted by X_W and $F_W(t)$, respectively. It is reasonable to assume that these lifetimes are ordered as:

$$X_W \leq_{st} X_S, \tag{8.32}$$

which means that (see Sect. 2.8)

$$F_S(t) \leq F_W(t), t \geq 0. \tag{8.33}$$

These inequalities define a general stochastic ordering between two random variables. Note that, since the Cdf of an absolutely continuous random variable is a continuous function that increases from 0 to 1, the relationship defined in (8.33) is equivalent to the following equation:

$$F_W(t) = F_S(\rho(t)), \forall t \geq 0, \tag{8.34}$$

where $\rho(t)$ is nondecreasing, $\rho(t) \geq t, \forall t \geq 0$, and $\rho(0) = 0$. Throughout this section, we assume the stochastic ordering (8.33–8.34). Let $r_S(t)$ be the failure rate which corresponds to X_S. Then, the failure rate $r_W(t)$ for X_W, as follows from (8.34), is given by

$$r_W(t) = \rho'(t) r_S(\rho(t)). \tag{8.35}$$

Another important ordering in reliability applications is the failure (hazard) rate ordering, which is defined as (see Sect. 2.8)

$$r_S(t) \leq r_W(t), t \geq 0. \tag{8.36}$$

It can be easily seen that Ordering (8.36) implies (8.32), and therefore, Eq. (8.34) also holds. A practical specific case of (8.36) is the proportional hazards model that can be defined in our case as

$$r_W(t) = \rho\, r_S(t), t \geq 0, \tag{8.37}$$

where $\rho > 1$. From a practical point of view, (8.37) constitutes a reasonable model for defining subpopulations of interest. For practical applications, when exponential distribution is assumed, (8.37) turns to:

$$r_W = \rho\, r_S.$$

We assume that the proportion of items from the strong subpopulation in the total population is p. Then the Cdf of the total population is given by the following mixture:

$$G(t) = pF_S(t) + (1-p)F_S(\rho(t)),$$

whereas the proportional hazards model (8.37) results in

$$G(t) = pF_S(t) + (1-p)(1-(\bar{F}_S(t))^\rho),$$

where $\bar{F} \equiv 1 - F$.

Furthermore, assume that items are repairable and undergo minimal repair upon failure (See also [6, 9]).

8.4.2 Optimal Burn-in for Minimizing Weighted Risks

In this subsection, we adopt the following burn-in procedure.

Burn-in Procedure

The item is burned-in during $(0, b]$ and if the number of ***minimally repaired*** *failures during burn-in process $N(b)$ satisfies $N(b) \le n$, then the item is considered as one from the strong subpopulation and put into field operation; otherwise, the item is considered as one from the weak subpopulation and is discarded.*

At $t = 0$ an item from a mixed population is chosen and put into test operation via burn-in. Upon failure at $t = a$ it is minimally repaired, etc. An item that does not meet our burn-in criterion is discarded. Therefore, the main goal is to classify the mixed populations into the weak and strong populations. We assume that the corresponding minimal repair is, in fact, a *physical minimal repair* [16] in the sense that a 'physical operation' of repair (not a replacement) brings an item in the state which is 'statistically identical' to the state it had just prior the failure. Note that, obviously, we do not know whether an item is 'strong' or 'weak'. On the other hand, the described operation in some sense 'keeps a memory of that': if it is, e.g., 'strong', the time to the next failure is distributed as $(F_S(t+a) - F_S(a))/(1 - F_S(a))$, etc. An example of this 'physical operation' is when a small realized defect (fault) is corrected upon failure, whereas the number of the possible inherent defects in the item is large. In practice, physical minimal repair of the described type can be usually performed and, therefore, our assumption is quite realistic.

By various practical reasons, the total burn-in time is generally limited. Therefore, in this section, we assume that the burn-in time is fixed as b. Then the above burn-in procedure can be defined in terms of n and we find an optimal burn-in procedure n^* which minimizes the appropriately defined risk.

For description of related risks, define the following four events:

- Event F_1: the item passes the burn-in process;
- Event F_2: the item is eliminated by the burn-in process;
- Event S: the item is from the strong subpopulation;
- Event W: the item is from the weak subpopulation.

Then

$$P(F_2|S) = 1 - P(F_1|S) \text{ and } P(F_1|W) = 1 - P(F_2|W).$$

Note that $P(F_2|S)$ is, the so-called, the risk of the first order (the probability that the strong component is eliminated) and $P(F_1|W)$ is the risk of the second order (the probability that the weak component had passed the burn-in). Therefore, our goal is to minimize these risks. Basically, we have three options:

First, we minimize the first risk $P(F_2|S)$ not taking into account the second risk. Then this problem is equivalent to maximizing $P(F_1|S)$. In accordance with the well-known property, the process of minimal repairs is the corresponding non-homogeneous Poisson process (NHPP). Therefore, taking into consideration our reasoning with respect to minimal repair:

$$P(F_1|S) = \sum_{i=0}^{n} \frac{(\Lambda_S(b))^i e^{-\Lambda_S(b)}}{i!},$$

where $\Lambda_S(t) \equiv \int_0^t r_S(u)du$ is the corresponding cumulative failure rate. Obviously, the maximum is achieved when $n = \infty$. This is an intuitively clear trivial solution, as we are not concerned about the other risk and 'are free' to minimize $P(F_2|S)$. Therefore, this value can be as close to 0 as we wish. In practice, sometimes this setting can occur but then the optimal n^* should be defined via the corresponding restrictions on the allocated burn-in resources, burn-in costs, etc.

Second, we minimize $P(F_1|W)$ without taking the first risk into account. Then this problem is equivalent to maximizing $P(F_2|W)$. In this case,

$$P(F_2|W) = 1 - \sum_{i=0}^{n} \frac{(\Lambda_S(\rho(b)))^i e^{-\Lambda_S(\rho(b))}}{i!},$$

where, as follows from (8.35),

$$\Lambda_W(t) \equiv \int_0^t r_W(u)du = \int_0^{\rho(t)} r_S(u)du = \Lambda_S(\rho(t)). \tag{8.38}$$

The maximum is achieved when $n = 0$. The corresponding value is

$$P_{n=0}(F_2|W) = 1 - e^{-\Lambda_S(\rho(b))},$$

which means that the second order risk in this case is equal to the probability that an item from the weaker population will survive the burn-in process without any failures, which makes a perfect sense.

The previous two options were illustrative, as they are usually nonrealistic in practice. The appropriate approach should take into account *both* types of risk. Therefore, it is reasonable to consider minimization of the weighted risks:

$$\begin{aligned}\Psi(n) &\equiv w_1 P(F_2|S) + w_2 P(F_1|W) \\ &= 1 - [w_1 P(F_1|S) + w_2 P(F_2|W)],\end{aligned}$$

where w_1 and w_2 are the weights satisfying $w_1 + w_2 = 1$. When $w_1 = 1$, $w_2 = 0$, we arrive at the first considered option, whereas the case $w_1 = 0$, $w_2 = 1$ corresponds to the second one. Furthermore, if $w_1 = w_2 = 1/2$, then we should minimize the sum of two risks $[P(F_2|S) + P(F_1|W)]$ or, equivalently, maximize the sum of the probabilities of correct decisions $[P(F_1|S) + P(F_2|W)]$.

Let n^* be the optimal burn-in procedure that satisfies

$$\Psi(n^*) = \min_{n \geq 0} \Psi(n). \tag{8.39}$$

This value is given by the following theorem:

Theorem 8.3 *Let* $0 < w_i < 1$, $i = 1, 2$, *and* n^* *be the nonnegative integer which satisfies* (8.39). *If*

$$\frac{(\Lambda_S(\rho(b)) - \Lambda_S(b)) + (\ln w_1 - \ln w_2)}{\ln\left(\frac{\Lambda_S(\rho(b))}{\Lambda_S(b)}\right)} < 1,$$

then the optimal n^* *is given by* $n^* = 0$, *otherwise* n^* *is the largest integer which is less than or equal to*

$$\frac{(\Lambda_S(\rho(b)) - \Lambda_S(b)) + (\ln w_1 - \ln w_2)}{\ln\left(\frac{\Lambda_S(\rho(b))}{\Lambda_S(b)}\right)}.$$

Proof Note that the problem is equivalent to the problem of maximizing

$$\Phi(n) \equiv w_1 P(F_1|S) + w_2 P(F_2|W).$$

Substitution gives:

$$\begin{aligned}\Phi(n) &\equiv w_1 P(F_1|S) + w_2 P(F_2|W) \\ &= w_1 P(N(b) \leq n|S) + w_2 P(N(b) > n|W) \\ &= w_1 \sum_{i=0}^{n} \frac{(\Lambda_S(b))^i e^{-\Lambda_S(b)}}{i!} + w_2 \left(1 - \sum_{i=0}^{n} \frac{(\Lambda_S(\rho(b)))^i e^{-\Lambda_S(\rho(b))}}{i!}\right).\end{aligned}$$

Observe that, for $n \geq 1$,

$$\Phi(n)-\Phi(n-1)=w_1\frac{(\Lambda_S(b))^n e^{-\Lambda_S(b)}}{n!}-w_2\frac{(\Lambda_S(\rho(b)))^n e^{-\Lambda_S(\rho(b))}}{n!}\geq 0$$
$$\Leftrightarrow e^{\Lambda_S(\rho(b))-\Lambda_S(b)}\geq\frac{w_2}{w_1}\left(\frac{\Lambda_S(\rho(b))}{\Lambda_S(b)}\right)^n$$
$$\Leftrightarrow n\leq\frac{(\Lambda_S(\rho(b))-\Lambda_S(b))+(\ln w_1-\ln w_2)}{\ln\left(\frac{\Lambda_S(\rho(b))}{\Lambda_S(b)}\right)}.$$

Case 1. Let

$$\frac{(\Lambda_S(\rho(b))-\Lambda_S(b))+(\ln w_1-\ln w_2)}{\ln\left(\frac{\Lambda_S(\rho(b))}{\Lambda_S(b)}\right)}<1.$$

Then, there is no positive integer which satisfies

$$n\leq\frac{(\Lambda_S(\rho(b))-\Lambda_S(b))+(\ln w_1-\ln w_2)}{\ln\left(\frac{\Lambda_S(\rho(b))}{\Lambda_S(b)}\right)}.$$

This implies that

$$\Phi(n)-\Phi(n-1)<0,\forall n\geq 1,$$

and thus we have $n^*=0$.

Case 2. Let

$$\frac{(\Lambda_S(\rho(b))-\Lambda_S(b))+(\ln w_1-\ln w_2)}{\ln\left(\frac{\Lambda_S(\rho(b))}{\Lambda_S(b)}\right)}\geq 1.$$

Then n^* is the largest integer which is less than or equal to

$$\frac{(\Lambda_S(\rho(b))-\Lambda_S(b))+(\ln w_1-\ln w_2)}{\ln\left(\frac{\Lambda_S(\rho(b))}{\Lambda_S(b)}\right)}.$$

□

Corollary 8.1 *When the specific proportional hazard model* (8.37) *holds, the cumulative failure rate in* (8.38) *can be expressed in a more explicit way:*

$$\Lambda_W(t)=\int_0^t r_W(u)\mathrm{d}u=\rho\int_0^t r_S(u)\mathrm{d}u=\rho\Lambda_S(t).$$

In this case, if

$$\frac{(\rho - 1)\Lambda_S(b) + (\ln w_1 - \ln w_2)}{\ln \rho} < 1,$$

then the optimal n^* *is given by* $n^* = 0$*, otherwise* n^* *is the largest integer which is less than or equal to*

$$\frac{(\rho - 1)\Lambda_S(b) + (\ln w_1 - \ln w_2)}{\ln \rho}.$$

Example 8.16 Suppose that the failure rate of the strong subpopulation is given by

$$r_S(t) = 2t, t \geq 0, \text{ (Weibull Distribution)}$$

and $\rho(t)$ in (8.34) is given by $\rho(t) = 3t, t \geq 0$. From (8.35), the corresponding failure rate of the weak subpopulation is then given by

$$r_W(t) = \rho'(t) r_S(\rho(t)) = 18t, t \geq 0, \text{ (Weibull Distribution)}$$

and, therefore, the proportional hazards model in (8.37) holds with $\rho = 9$. Suppose further that the burn-in time for this mixed population is given by $b = 1.0$ and $w_1 = 0.8$, $w_2 = 0.2$. Then, by Corollary 8.1,

$$\frac{(\rho - 1)\Lambda_S(b) + (\ln w_1 - \ln w_2)}{\ln \rho} \approx 4.27.$$

Finally, the optimal burn-in procedure is determined by $n^* = 4$.

8.4.3 *Optimal Burn-in for Minimizing Expected Number of Repairs*

In this subsection, we discuss optimal burn-in that minimizes the mean number of minimal repairs during the mission time τ. We consider the same burn-in procedure as in Sect. 8.4.2, but now it is characterized by both b and n (i.e., b and n are the burn-in parameters).

Observe that

$$\begin{aligned} P(F_1) &= P(F_1|S) \times P(S) + P(F_1|W) \times P(W) \\ &= \left(\sum_{i=0}^{n} \frac{(\Lambda_S(b))^i e^{-\Lambda_S(b)}}{i!}\right) \times p + \left(\sum_{i=0}^{n} \frac{(\Lambda_S(\rho(b)))^i e^{-\Lambda_S(\rho(b))}}{i!}\right) \times (1-p). \\ P(S|F_1) &= \frac{P(S \cap F_1)}{P(F_1)} = P(F_1|S) \times P(S)/P(F_1) \\ &= \frac{\left(\sum_{i=0}^{n} \frac{(\Lambda_S(b))^i e^{-\Lambda_S(b)}}{i!}\right) \times p}{\left(\sum_{i=0}^{n} \frac{(\Lambda_S(b))^i e^{-\Lambda_S(b)}}{i!}\right) \times p + \left(\sum_{i=0}^{n} \frac{(\Lambda_S(\rho(b)))^i e^{-\Lambda_S(\rho(b))}}{i!}\right) \times (1-p)}. \\ P(W|F_1) &= \frac{\left(\sum_{i=0}^{n} \frac{(\Lambda_S(\rho(b)))^i e^{-\Lambda_S(\rho(b))}}{i!}\right) \times (1-p)}{\left(\sum_{i=0}^{n} \frac{(\Lambda_S(b))^i e^{-\Lambda_S(b)}}{i!}\right) \times p + \left(\sum_{i=0}^{n} \frac{(\Lambda_S(\rho(b)))^i e^{-\Lambda_S(\rho(b))}}{i!}\right) \times (1-p)}. \end{aligned}$$

Let $\Psi(b, n)$ be the mean number of minimal repairs during the mission time τ in field operation given that the duration of burn-in is equal to b and that the rejection number is n. Then, in accordance with the above formulas and noting once again that the mean number of minimal repairs is equal to the cumulative intensity function of the corresponding NHPP,

$$\begin{aligned} \Psi(b,n) &= (\Lambda_S(b+\tau) - \Lambda_S(b)) \\ &\quad \times \frac{\left(\sum_{i=0}^{n} \frac{(\Lambda_S(b))^i e^{-\Lambda_S(b)}}{i!}\right) \times p}{\left(\sum_{i=0}^{n} \frac{(\Lambda_S(b))^i e^{-\Lambda_S(b)}}{i!}\right) \times p + \left(\sum_{i=0}^{n} \frac{(\Lambda_S(\rho(b)))^i e^{-\Lambda_S(\rho(b))}}{i!}\right) \times (1-p)} \\ &\quad + (\Lambda_S(\rho(b+\tau)) - \Lambda_S(\rho(b))) \\ &\quad \times \frac{\left(\sum_{i=0}^{n} \frac{(\Lambda_S(\rho(b)))^i e^{-\Lambda_S(\rho(b))}}{i!}\right) \times (1-p)}{\left(\sum_{i=0}^{n} \frac{(\Lambda_S(b))^i e^{-\Lambda_S(b)}}{i!}\right) \times p + \left(\sum_{i=0}^{n} \frac{(\Lambda_S(\rho(b)))^i e^{-\Lambda_S(\rho(b))}}{i!}\right) \times (1-p)}. \end{aligned} \tag{8.40}$$

The objective is to find optimal (b^*, n^*) which satisfies

$$\Psi(b^*, n^*) = \min_{b \geq 0, n \geq 0} \Psi(b, n). \tag{8.41}$$

In order to find the joint optimal solution defined by (8.41), we follow the procedure similar to that given in Mi [26] and Cha [6], where the two-dimensional optimization problems of finding the optimal burn-in time b^* and the age-replacement policy T^* that minimize the long-run average cost rate $c(b, T)$ are considered. At the first stage, we fix the burn-in time b and find optimal $n^*(b)$ that satisfies

$$\Psi(b, n^*(b)) = \min_{n \geq 0} \Psi(b, n). \tag{8.42}$$

At the second stage, we search for b^* that satisfies

$$\Psi(b^*, n^*(b^*)) = \min_{b \geq 0} \Psi(b, n^*(b)).$$

Then the joint optimal solution is given by $(b^*, n^*(b^*))$, since the above procedure implies that

$$\begin{aligned}\Psi(b^*, n^*(b^*)) &\leq \Psi(b, n^*(b)), \text{for all } b \geq 0 \\ &\leq \Psi(b, n), \text{for all } b \geq 0,\ n \geq 0.\end{aligned}$$

As in Mi [26] and Cha [6], in this case, if an uniform upper bound (with respect to n) could be found, then the optimization procedure would be much simpler.

Following the procedure described above, first find optimal $n^*(b)$ satisfying (8.42) for each fixed b. For this purpose, we need to state the following lemma which will be used for obtaining the optimal $n^*(b)$:

Lemma 8.2 [29] *Suppose that $a_i \geq 0, i \geq 1$, and $b_i > 0, i \geq 1$. Then*

$$\min_{1 \leq i \leq n} \frac{a_i}{b_i} \leq \frac{\sum_{i=1}^n a_i}{\sum_{i=1}^n b_i} \leq \max_{1 \leq i \leq n} \frac{a_i}{b_i},$$

where the equality holds if and only if all the $a_i/b_i, i \geq 1$, are equal.

The optimal value $n^*(b)$ is defined by the following theorem.

Theorem 8.4 *For a given fixed $b \geq 0$, let the following inequality:*

$$(\Lambda_S(b + \tau) - \Lambda_S(b)) \leq (\Lambda_S(\rho(b + \tau)) - \Lambda_S(\rho(b))$$

hold. Then the optimal $n^*(b)$ is given by $n^*(b) = 0$, whereas $n^*(b) = \infty$ corresponds to the opposite sign of the inequality.

Proof For the fixed $b \geq 0$, we consider the following two cases:

Case 1. Let

$$(\Lambda_S(b + \tau) - \Lambda_S(b)) \leq (\Lambda_S(\rho(b + \tau)) - \Lambda_S(\rho(b)).$$

As the sum of quotients in Eq. (8.40) is 1 in this case, it can be easily seen that minimizing $\Psi(b, n)$ is equivalent to maximizing

$$P(S|F_1) = \frac{\left(\sum_{i=0}^{n} \frac{(\Lambda_S(b))^i e^{-\Lambda_S(b)}}{i!}\right) \times p}{\left(\sum_{i=0}^{n} \frac{(\Lambda_S(b))^i e^{-\Lambda_S(b)}}{i!}\right) \times p + \left(\sum_{i=0}^{n} \frac{(\Lambda_S(\rho(b)))^i e^{-\Lambda_S(\rho(b))}}{i!}\right) \times (1-p)}.$$

$$\Leftrightarrow \text{Minimize} \frac{p + (1-p) \times \frac{\sum_{i=0}^{n} \frac{(\Lambda_S(\rho(b)))^i e^{-\Lambda_S(\rho(b))}}{i!}}{\sum_{i=0}^{n} \frac{(\Lambda_S(b))^i e^{-\Lambda_S(b)}}{i!}}}{p}$$

$$\Leftrightarrow \text{Minimize} \frac{\sum_{i=0}^{n} \frac{(\Lambda_S(\rho(b)))^i e^{-\Lambda_S(\rho(b))}}{i!}}{\sum_{i=0}^{n} \frac{(\Lambda_S(b))^i e^{-\Lambda_S(b)}}{i!}} \equiv g(b,n)$$

We compare $\Psi(b,n)$ with $\Psi(b,n+1)$, $n = 0,1,2,\ldots$. Observe that $g(b,n) < g(b,n+1)$ if and only if $\Psi(b,n) < \Psi(b,n+1)$. Note that

$$\frac{\frac{(\Lambda_S(\rho(b)))^i e^{-\Lambda_S(\rho(b))}}{i!}}{\frac{(\Lambda_S(b))^i e^{-\Lambda_S(b)}}{i!}}$$

is strictly increasing in $i \geq 0$. This can be easily seen by comparing the values of this function for i and $i+1$, $i \geq 0$. Thus

$$\frac{\frac{(\Lambda_S(\rho(b)))^i e^{-\Lambda_S(\rho(b))}}{i!}}{\frac{(\Lambda_S(b))^i e^{-\Lambda_S(b)}}{i!}} < \frac{\frac{(\Lambda_S(\rho(b)))^{n+1} e^{-\Lambda_S(\rho(b))}}{(n+1)!}}{\frac{(\Lambda_S(b))^{n+1} e^{-\Lambda_S(b)}}{(n+1)!}}, 0 \leq i \leq n.$$

Then using Lemma 8.2:

$$\frac{\sum_{i=0}^{n} \frac{(\Lambda_S(\rho(b)))^i e^{-\Lambda_S(\rho(b))}}{i!}}{\sum_{i=0}^{n} \frac{(\Lambda_S(b))^i e^{-\Lambda_S(b)}}{i!}} < \max_{1 \leq i \leq n} \frac{\frac{(\Lambda_S(\rho(b)))^i e^{-\Lambda_S(\rho(b))}}{i!}}{\frac{(\Lambda_S(b))^i e^{-\Lambda_S(b)}}{i!}} < \frac{\frac{(\Lambda_S(\rho(b)))^{n+1} e^{-\Lambda_S(\rho(b))}}{(n+1)!}}{\frac{(\Lambda_S(b))^{n+1} e^{-\Lambda_S(b)}}{(n+1)!}}.$$

Accordingly, using Lemma 8.2 again:

$$g(b,n) = \min\left\{\frac{\sum_{i=0}^{n} \frac{(\Lambda_S(\rho(b)))^i e^{-\Lambda_S(\rho(b))}}{i!}}{\sum_{i=0}^{n} \frac{(\Lambda_S(b))^i e^{-\Lambda_S(b)}}{i!}}, \frac{\frac{(\Lambda_S(\rho(b)))^{n+1} e^{-\Lambda_S(\rho(b))}}{(n+1)!}}{\frac{(\Lambda_S(b))^{n+1} e^{-\Lambda_S(b)}}{(n+1)!}}\right\}$$

$$< \frac{\sum_{i=0}^{n+1} \frac{(\Lambda_S(\rho(b)))^i e^{-\Lambda_S(\rho(b))}}{i!}}{\sum_{i=0}^{n+1} \frac{(\Lambda_S(b))^i e^{-\Lambda_S(b)}}{i!}} = g(b,n+1),$$

implying that $\Psi(b,n) < \Psi(b,n+1), n = 0,1,2,\ldots,$ Finally, we arrive at $n^*(b) = 0$.

This obviously means that for each fixed duration of the burn-in time b, the failed item is discarded and those that did not fail are put into a field operation. Therefore, the obtained rule is simple and easy for implementation.

Case 2. Let

$$(\Lambda_S(b+\tau)-\Lambda_S(b)) > (\Lambda_S(\rho(b+\tau))-\Lambda_S(\rho(b)).$$

In this case, minimization of $\Psi(b,n)$ is equivalent to minimization of

$$\frac{\left(\sum_{i=0}^{n}\frac{(\Lambda_S(b))^i e^{-\Lambda_S(b)}}{i!}\right)\times p}{\left(\sum_{i=0}^{n}\frac{(\Lambda_S(b))^i e^{-\Lambda(b)}}{i!}\right)\times p+\left(\sum_{i=0}^{n}\frac{(\Lambda_S(\rho(b)))^i e^{-\Lambda_S(\rho(b))}}{i!}\right)\times(1-p)},$$

or, to maximization of $g(b,n)$. Therefore $n^*(b)=\infty$. □

Remark 8.7 When the failure rate ordering (8.36) holds, the first inequality in Theorem 8.4 corresponds to

$$(\Lambda_S(b+\tau)-\Lambda_S(b)) \le (\Lambda_W(b+\tau)-\Lambda_W(b)),$$

which is always obviously satisfied. For the specific case (8.37), it leads to

$$\int_b^{b+\tau} r_S(u)\mathrm{d}u = (\Lambda_S(b+\tau)-\Lambda_S(b)) \le (\Lambda_W(b+\tau)-\Lambda_W(b)) = \rho\int_b^{b+\tau} r_S(u)\mathrm{d}u.$$

Remark 8.8 The result $n^*(b)=\infty$ (Theorem 8.4, Case 2) implies that after the burn-in time b with minimal repair every item is put into field operation regardless of the number of failures during burn-in. This burn-in procedure is the same as that proposed in Cha [6]. Case 2 can obviously occur when the cumulative failure rate in $[0,b)$ for the strong subpopulation is smaller than that for the weak subpopulation, whereas the reverse ordering holds for the interval $[b,b+\tau)$ (e.g., when $r_S(t)$ has a decreasing part). In this case, the 'quality' of items *after burn-in* in the weak subpopulation is better than that in the strong subpopulation. Therefore, the burn-in procedure should leave all weak items in the population, which results in $n^*(b)=\infty$.

Consider now obtaining an uniform upper bound (with respect to n), i.e., we will find an upper bound for b^* denoted by s^*, such that,

$$\min_{0\le b\le s^*}\Psi(b,n) < \min_{b>s^*}\Psi(b,n),$$

for all fixed $n\ge 0$.

The following result gives an uniform upper bound for the optimal burn-in time b^*, but first we need to define the notion of the eventually (ultimately) increasing function [20, 30].

Recall that for the eventually IFR $r(x)$, the first and the second wear-out points t^* and t^{**} are defined as

$$t^* = \inf\{t \geq 0 : r(x) \text{ is nondecreasing in } x \geq t\},$$
$$t^{**} = \inf\{t \geq 0 : r(x) \text{ strictly increases in } x \geq t\}.$$

Observe that the eventually IFR can be constant in parts of the interval (t^*, t^{**}), whereas $t^* = t^{**}$ is obviously a specific case.

Theorem 8.5 *Suppose that*

i. *$r_S(t)$ is eventually increasing with the first wear-out point t^*, the second wear-out point t^{**} and $\lim_{t\to\infty} r_S(t) = \infty$;*
ii. *$\rho(t)$ is a weak (i.e., not necessarily strictly) convex function.*

Then $s^* \in [t^*, \infty)$, defined as

$$s^* = \inf\left\{ b' > t^* \middle| \int_{\rho(t^*)}^{\rho(t^*+\tau)} r_S(u)\mathrm{d}u < \int_b^{b+\tau} r_S(u)\mathrm{d}u, \forall b > b' \right\}, \tag{8.43}$$

is the uniform upper bound for the optimal burn-in time b^*.

Proof Observe that $\Psi(b, n)$ is of the form of weighted average of $(\Lambda_S(b+\tau) - \Lambda_S(b))$ and $(\Lambda_S(\rho(b+\tau)) - \Lambda_S(\rho(b))$, i.e.,

$$\Psi(b, n) = (\Lambda_S(b+\tau) - \Lambda_S(b)) \times p(b) + (\Lambda_S(\rho(b+\tau)) - \Lambda_S(\rho(b))) \times (1 - p(b)),$$

where

$$p(b) = \frac{\left(\sum_{i=0}^{n} \frac{(\Lambda_S(b))^i e^{-\Lambda(b)}}{i!}\right) \times p}{\left(\sum_{i=0}^{n} \frac{(\Lambda_S(b))^i e^{-\Lambda(b)}}{i!}\right) \times p + \left(\sum_{i=0}^{n} \frac{(\Lambda_S(\rho(b)))^i e^{-\Lambda(\rho(b))}}{i!}\right) \times (1-p)}.$$

Also we see that

$$\Lambda_S(b+\tau) - \Lambda_S(b) = \int_b^{b+\tau} r_S(u)\mathrm{d}u \text{ and } \Lambda_S(\rho(b+\tau)) - \Lambda_S(\rho(b)) = \int_{\rho(b)}^{\rho(b+\tau)} r_S(u)\mathrm{d}u.$$

Define $s^* \in [t^*, \infty)$ as

$$s^* = \inf\left\{ b' > t^* \middle| \int_{\rho(t^*)}^{\rho(t^*+\tau)} r_S(u)\mathrm{d}u < \int_b^{b+\tau} r_S(u)\mathrm{d}u, \forall b > b' \right\}.$$

It clear that such s^* exists as $\int_b^{b+\tau} r_S(u)du$ is nondecreasing for $b \in [t^*, \infty)$ and is strictly increasing after some point $t' \in [t^*, t^{**}]$. Observe that $\rho(b+\tau) - \rho(b)$ is nondecreasing in b and

$$\Lambda_S(\rho(b+\tau)) - \Lambda_S(\rho(b)) \geq \Lambda_S(b+\tau) - \Lambda_S(b)$$

for $b \geq t^*$. Then

$$\begin{aligned}\Lambda_S(t^*+\tau) - \Lambda_S(t^*) \leq \Lambda_S(\rho(t^*+\tau)) - \Lambda_S(\rho(t^*)) < \Lambda_S(b+\tau) - \Lambda_S(b) \\ \leq \Lambda_S(\rho(b+\tau)) - \Lambda_S(\rho(b)), \forall b > s^*.\end{aligned}$$

The weighted average of elements in the first group is smaller than that of elements in the second group for any arbitrarily chosen weights in two groups if the maximum element in the first group is smaller than the minimum element in the second group. This fact implies:

$$\Psi(t^*, n) < \Psi(b, n), \forall b > s^*.$$

Then we can conclude that, at least, the optimal burn-in time $b^* \notin (s^*, \infty)$, i.e., $b^* \leq s^*$. This result holds regardless of the value of n. Therefore, s^* is the uniform (with respect to n) upper bound for b^*. □

Example 8.17 As in the previous example, let the failure rate of the strong subpopulation be given by

$$r_S(t) = 2t, t \geq 0. \text{ (Weibull Distribution)}$$

Then $t^* = t^{**} = 0$. Assume that $\rho(t) = 3t, t \geq 0$, $\tau = 2.0$. Then $\int_{\rho(t^*)}^{\rho(t^*+\tau)} r_S(u)du = 36.0$ and it is easy to see that $s^* = 8.0$.

It follows from Theorem 8.4 that, for each b, either $n^*(b) = 0$ or $n^*(b) = \infty$. Moreover, with the uniform upper bound s^* defined by Theorem 8.5, we can search for b^* which minimizes $\Psi(b, n^*(b))$ in the reduced interval $[0, s^*]$. Then Theorems 8.4 and 8.5 imply that the joint optimal solution is given by $(b^*, n^*(b^*))$. Based on these facts, the optimization procedure can be summarized as follows:

< Optimization Procedure (Algorithm)>

(Stage1)

Fix $0 \leq b \leq s^*$. If $\Lambda_S(b+\tau) - \Lambda_S(b) \leq \Lambda_S(\rho(b+\tau)) - \Lambda_S(\rho(b))$ then $n^*(b) = 0$; otherwise $n^*(b) = \infty$.

- (Stage2)

Find b^* which satisfies

$$\Psi(b^*, n^*(b^*)) = \min_{0 \leq b \leq s^*} \Psi(b, n^*(b)).$$

(Joint Optimal Solution)

Then the two-dimensional optimal solution is given by $(b^*, n^*(b^*))$.

Example 8.18 Consider the setting of Example 8.17 and suppose that the proportion of strong subpopulation is $p = 0.9$. Then, as given in Example 8.17, the uniform upper bound s^* is given by $s^* = 8.0$. Thus, in order to find the joint optimal solution (b^*, n^*), we follow the optimization procedure described above. However, in this case, since $\rho(t)$ is a convex function and $r_S(t)$ is a nondecreasing function, the inequality

$$\Lambda_S(b+\tau) - \Lambda_S(b) \le \Lambda_S(\rho(b+\tau)) - \Lambda_S(\rho(b)), \forall b \ge 0 \tag{8.44}$$

always holds. Thus $n^*(b) = 0$, for all $b \ge 0$. Then the optimal solution $(b^*, n^*(b^*))$ is $(b^*, 0)$, where b^* is the value which satisfies

$$\Psi(b^*, 0) = \min_{0 \le b \le 8.0} \Psi(b, 0).$$

By a numerical search, it has been obtained that $b^* = 0.546$ and the minimum value of $\Psi(b, n)$ at the optimal point $(b^*, n^*) = (0.546,\ 0)$ is $\Psi(0.546, 0) = 6.6851112$. Note that, by Theorem 8.4, the minimum value of $\Psi(b, n)$ for each fixed b is $\Psi(b, 0)$ or $\Psi(b, \infty)$. In this specific example, due to Inequality (8.44), $\Psi(b, 0) \le \Psi(b, \infty)$.

The discussion based on the specific setting of Example 8.18 ($\rho(t)$ is a convex function and $r_S(t)$ is a nondecreasing function) can be summarized by the following corollary:

Corollary 8.2. *Suppose that*

i. $r_S(t)$ *is eventually increasing with the first wear-out point* $t^* = 0$, *the second wear-out point* t^{**} *and* $\lim_{t\to\infty} r_S(t) = \infty$;
ii. $\rho(t)$ *is a weak convex function.*

Then the joint optimal solution satisfying Eq. (8.41) *is* $(b^*, 0)$, where b^* is the value which satisfies

$$\Psi(b^*, 0) = \min_{0 \le b \le s^*} \Psi(b, 0),$$

and s^* *is the uniform upper bound given in* (8.43).

References

1. Badía FG, Berrade MD, Clemente AC (2002) Aging properties of the additive and proportional hazard mixing models. Reliab Eng Syst Saf 78:165–172
2. Block HW, Mi J, Savits TH (1993) Burn-in and mixed populations. J Appl Probab 30:692–702
3. Block HW, Mi J, Savits TH (1994) Some results in burn-in. Statistica Sinica 4:525–534

4. Block HW, Savits TH (1997) Burn-in. Stat Sci 12:1–19
5. Block HW, Savits TH, Wondmagegnehu ET (2003) Mixtures of distributions with increasing linear failure rates. J Appl Probab 40:485–504
6. Cha JH (2000) On a better burn-in procedure. J Appl Probab 37:1099–1103
7. Cha JH (2001) Burn-in procedures for a generalized model. J Appl Probab 38:542–553
8. Cha JH (2005) On optimal burn-in procedures—a generalized model. IEEE Trans Reliab 54:198–206
9. Cha JH (2006) An extended model for optimal burn-in procedures. IEEE Trans Reliab 55:189–198
10. Cha JH, Finkelstein M (2010) Burn-in by environmental shocks for two ordered subpopulations. Eur J Oper Res 206:111–117
11. Cha JH, Finkelstein M (2010) Stochastically ordered subpopulations and optimal burn-in procedure. IEEE Trans Reliab 59:635–643
12. Cha JH, Finkelstein M (2011) Burn-in and the performance quality measures in heterogeneous populations. Eur J Oper Res 210:273–280
13. Cha JH, Finkelstein M (2012) Burn-in and the performance quality measures in continuous heterogeenous populations. J Risk Reliab 226:417–425
14. Cha JH, Finkelstein M (2013) Burn-in for heterogeneous populations: how to avoid large risks. Commun Stat—Theory Methods
15. Clarotti CA, Spizzichino F (1990) Bayes burn-in and decision procedures. Probab Eng Inf Sci 4:437–445
16. Finkelstein M (1992) Some notes on two types of minimal repair. Adv Appl Probab 24:226–228
17. Finkelstein M (2008) Failure rate modelling for reliability and risk. Springer, London
18. Finkelstein M (2009) Understanding the shape of the mixture failure rate (with engineering and demographic applications). Appl Stoch Models Bus Ind 25:643–663
19. Gupta RC, Warren R (2001) Determination of change points of nonmonotonic failure rates. Commun Stat—Theory Methods 30:1903–1920
20. Gurland J, Sethuraman J (1995) How pooling failure data may reverse increasing failure rate. J Am Stat Assoc 90:1416–1423
21. Jensen F, Petersen NE (1982) Burn-in. Wiley, New York
22. Jiang R, Murthy DNP (1998) Mixture of weibull distributions—parametric characterization of failure rate function. Appl Stoch Models Data Anal 14:47–65
23. Kececioglu D, Sun F (2003) Burn-in testing: its quantification and optimization. DEStech Publications, Lancaster
24. Klugman SA, Panjer HH, Willmot GE (2004) Loss models: from data to decisions. Wiley, Hoboken
25. Mi J (1991) Optimal burn-in. PhD thesis, Department of Statistics, University of Pittsburgh, Pittsburgh
26. Mi J (1994) Burn-in and maintenance policies. Adv Appl Probab 26:207–221
27. Mi J (1996) Minimizing some cost functions related to both burn-in and field use. Oper Res 44:497–500
28. Mi J (1997) Warranty policies and burn-in. Naval Res Logist 44:199–200
29. Mi J (2002) Age-replacement policy and optimal work size. J Appl Probab 39:296–311
30. Mi J (2003) Optimal burn-in time and eventually IFR. J Chin Inst Ind Eng 20:533–542
31. Navarro J, Hernandez PJ (2004) How to obtain bathtub-shaped failure rate models from normal mixtures. Probab Eng Inf Sci 18:511–531

Chapter 9
Shocks as Burn-in

As described in the previous chapters, in conventional burn-in, the main parameter of the burn-in procedure is its duration. However, in order to shorten the length of this procedure, burn-in is most often performed in an accelerated environment. This indicates that high environmental stress can be more effective in eliminating weak items from a population. In this case, obviously, the larger values of stress should correspond to the shorter duration of burn-in. By letting the stress to increase, we can end up (as some limit) with very short (negligible) durations, in other words, *shocks*. In practice, the most common types of shocks as a method of burn-in are "thermal shock" and "physical drop". In these cases, the item is subjected to a very rapid cold-to-hot, or hot-to-cold, instantaneous thermal change or the item is dropped by a "drop tester" which is specifically designed to drop it without any rotational motion, to ensure the most rigorous impact. In this case, the stress level (to be called shock's *severity*) can be a controllable parameter for the corresponding optimization, which in a loose sense is an analogue of the burn-in duration in accelerated burn-in (see e.g., [1, 9].

This general reasoning suggests that 'electrical' (e.g., the increased voltage for a short period of time for some electronic items), thermal and mechanical shocks can be used for burn-in in heterogeneous populations of items. If the initial population is not 'sufficiently reliable', then the items that have survived a shock might be more suitable for field usage, as their *predicted* reliability characteristics could be better. Therefore, in this chapter, we consider shocks as a method of burn-in and develop the corresponding optimization model. It should be noted that several approaches (such as Environmental Stress Screening to be considered in the next chapter) that exhibit a similar initial reasoning were already implemented in industry as a practical tool (see, for example, [13, 16, 17].

As in the previous chapters, we will also assume that the population is the mixture of stochastically ordered subpopulations. As before, we will consider both discrete and continuous mixture models. Under this and some other natural assumptions, we consider the problem of determining the optimal severity level of a stress. Furthermore, we develop approaches that minimize the risks of selecting items with large levels of individual failure rates for missions of high importance,

M. Finkelstein and J. H. Cha, *Stochastic Modeling for Reliability*, Springer Series in Reliability Engineering, DOI: 10.1007/978-1-4471-5028-2_9,

where failures can result, e.g., in substantial economic losses. We consider some new measures for describing the corresponding optimal burn-in, which boils up in obtaining the optimal severity of shocks. For instance, the losses that are monotonically increasing with the value of the failure rate of items after burn-in are introduced. Furthermore, focusing on the quality of relatively poor (with large failure rates) items in the mixed population, some conservative measures for the population quality are defined and the corresponding optimal burn-in with respect to these measures is also investigated.

We will also consider burn-in for items that will operate (after burn-in) in the environment with shocks. We assume that there are two competing risk causes of failure—the 'usual' one (in accordance with aging processes in a system) and environmental shocks. A new type of burn-in via the controlled (laboratory) test shocks is considered and the problem of obtaining the optimal level (severity) of these shocks is investigated.

9.1 Discrete Mixtures

9.1.1 General Setting

We assume in this section that a population is a mixture of two ordered subpopulations—the strong subpopulation and the weak subpopulation. Let the lifetime of a component from the strong subpopulation be denoted by X_S and its absolutely continuous cumulative distribution function (Cdf), probability density function (pdf) and the failure rate function be $F_1(t), f_1(t)$ and $\lambda_1(t)$, respectively. Similarly, the lifetime, the Cdf, pdf, and the failure rate function of a weak component are denoted by X_W, $F_2(t), f_2(t)$ and $\lambda_2(t)$, respectively. Let the lifetimes in these subpopulations be ordered either in the sense of the failure rate ordering:

$$\lambda_1(t) \le \lambda_2(t), \text{for all } t \ge 0$$

or in the sense of the usual stochastic ordering

$$\overline{F}_1(t) \ge \overline{F}_2(t), \text{for all } t \ge 0,$$

where $\overline{F}_i(t) = 1 - F_i(t)$, $i = 1, 2$. Assume that the mixing proportion (distribution) is given by

$$\pi(z) = \begin{cases} \pi, & z = z_1 \\ 1 - \pi, & z = z_2 \end{cases},$$

where z_1 and z_2, $z_1 < z_2$, are variables that represent the strong and the weak subpopulations, respectively. Therefore, $Z = (z_1, z_2)$ can be considered as the discrete frailty in this case. Then the corresponding mixture distribution and the density functions are defined as in the previous chapters:

$$F_m(t) = \pi F_1(t) + (1-\pi)F_2(t),$$
$$f_m(t) = \pi f_1(t) + (1-\pi)f_2(t),$$

respectively, and the mixture failure rate is

$$\lambda_m(t) = \frac{\pi f_1(t) + (1-\pi)f_2(t)}{\pi \bar{F}_1(t) + (1-\pi)\bar{F}_2(t)} = \pi(z_1|t)\lambda_1(t) + \pi(z_2|t)\lambda_2(t),$$

where the time-dependent probabilities are

$$\pi(z_1|t) = \frac{\pi \bar{F}_1(t)}{\pi \bar{F}_1(t) + (1-\pi)\bar{F}_2(t)},$$
$$\pi(z_2|t) = 1 - \pi(z_1|t) = \frac{(1-\pi)\bar{F}_2(t)}{\pi \bar{F}_1(t) + (1-\pi)\bar{F}_2(t)}.$$

Assume that at time $t = 0$ an instantaneous shock has occurred and with complementary probabilities it either 'kills' an item (i.e., a failure occurs), or 'leaves it unchanged'. The following is the basic assumption in our reasoning:

Basic Assumption
The more frail (e.g., with the larger failure rate) the items are, the more susceptible they are to be 'killed' by a shock.

Let $\pi_s(z)$ denote the frailty distribution *after a shock* and let T_s and $\lambda_{ms}(t)$ be the corresponding lifetime and the mixture (observed) failure rate, respectively. Denote the probabilities of failures caused by each shock for two subpopulations as:

$$p(z) = \begin{cases} p_1, & z = z_1, \\ p_2, & z = z_2. \end{cases} \tag{9.1}$$

Here, in accordance with our Basic Assumption, $p_1 \le p_2$. It is easy to show that

$$\pi_s(z) = \begin{cases} \frac{(1-p_1)\pi}{(1-p_1)\pi + (1-p_2)(1-\pi)} \equiv \pi_s, & z = z_1, \\ \frac{(1-p_2)(1-\pi)}{(1-p_1)\pi + (1-p_2)(1-\pi)} \equiv 1 - \pi_s, & z = z_2. \end{cases}$$

and

$$\lambda_{ms}(t) = \frac{\pi_s f_1(t) + (1-\pi_s)f_2(t)}{\pi_s \bar{F}_1(t) + (1-\pi_s)\bar{F}_2(t)} = \pi_s(z_1|t)\lambda_1(t) + \pi_s(z_2|t)\lambda_2(t), \tag{9.2}$$

where

$$\pi_s(z_1|t) = \frac{\pi_s \bar{F}_1(t)}{\pi_s \bar{F}_1(t) + (1-\pi_s)\bar{F}_2(t)},$$
$$\pi_s(z_2|t) = 1 - \pi_s(z_1|t) = \frac{(1-\pi_s)\bar{F}_2(t)}{\pi_s \bar{F}_1(t) + (1-\pi_s)\bar{F}_2(t)}.$$

The corresponding survival function is given by

$$\overline{F}_{ms}(t) = \pi_s\overline{F}_1(t) + (1-\pi_s)\overline{F}_2(t).$$

The following initial result justifies the fact that a shock can be considered as the burn-in procedure.

Theorem 9.1 *Let* $p_1 \leq p_2$.

(i) *If* $\lambda_1(t) \leq \lambda_2(t)$, *for all* $t \geq 0$, *then* $\lambda_{ms}(t) \leq \lambda_m(t), \forall t \in [0,\infty)$.
(ii) *If* $\overline{F}_1(t) \geq \overline{F}_2(t)$, *for all* $t \geq 0$, *then* $\overline{F}_{ms}(t) \geq \overline{F}_m(t), \forall t \in [0,\infty)$.

Proof Observe that $\lambda_m(t)$ and $\lambda_{ms}(t)$ are weighted averages of $\lambda_1(t)$ and $\lambda_2(t)$. Then it is sufficient to show that $\pi_s(z_1|t) \geq \pi(z_1|t)$. Note that

$$\pi(z_1|t) = \frac{\pi\bar{F}_1(t)}{\pi\bar{F}_1(t) + (1-\pi)\bar{F}_2(t)} = \frac{\bar{F}_1(t)}{\bar{F}_1(t) + (1/\pi - 1)\bar{F}_2(t)}$$

is increasing in π, and

$$\pi_s - \pi = \frac{(1-p_1)\pi}{(1-p_1)\pi + (1-p_2)(1-\pi)} - \pi = \frac{\pi(1-\pi)(p_2-p_1)}{(1-p_1)\pi + (1-p_2)(1-\pi)} \geq 0.$$

Therefore, $\pi_s(z_1|t) \geq \pi(z_1|t)$ and we can conclude that $\lambda_{ms}(t) \leq \lambda_m(t), \forall t \in [0,\infty)$.

On the other hand, $\overline{F}_m(t)$ and $\overline{F}_{ms}(t)$ are also weighted averages of $\overline{F}_1(t)$ and $\overline{F}_2(t)$. Then the second result is obvious from the fact that $\pi_s \geq \pi$. □

Remark 9.1 The above result implies that reliability characteristics of a population of items that have survived a shock have improved. This justifies the described burn-in procedure as a measure of improving the 'quality' of a heterogeneous population. Depending on assumptions, Theorem 9.1 states that the population lifetime random variable after a shock is larger than that before the shock either in the sense of the failure rate ordering, or in the sense of the usual stochastic ordering. Note that individual characteristics of an item that has survived a shock, due to our assumption, are same as before.

9.1.2 Optimal Severity for Population Quality Measures

The optimal burn-in time is the main characteristic of interest in conventional burn-in procedures. In our model, the 'severity' of a shock in a way corresponds to this burn-in time. Therefore, we will suggest now an approach for determining an optimal magnitude of a shock that maximizes the 'quality' of our population after burn-in.

Denote the magnitude of a shock by $s \in [0, \infty]$. Assume that the 'strength' of the component in a strong subpopulation is a continuous random variable, which is denoted by U. By 'strength' we understand here the corresponding measure of resistance to a single shock, i.e., if $s > U$, then the failure occurs. Let the Cdf, the survival function, and the failure rate function of U are denoted by $G(s)$, $\overline{G}(s)$, and $r(s)$, respectively. Similarly, let the strength of the component in a weak subpopulation be denoted by U_w. Then, in accordance with our Basic Assumption, let

$$U \geq_{st} U_w.$$

It is easy to see that this inequality is equivalent to

$$G_w(s) = G(\rho(s)), \text{for all } s \geq 0, \tag{9.3}$$

where $G_w(s)$ is the Cdf of U_w, $\rho(s)$ is an increasing function, $\rho(s) \geq s$ for all $s \geq 0$, and $\rho(0) = 0$. It follows from (9.1) that the probabilities of failure for this case are given by

$$p(z, s) = \begin{cases} p_1 = G(s), & z = z_1, \\ p_2 = G(\rho(s)), & z = z_2. \end{cases} \tag{9.4}$$

Then $p_1 \leq p_2$ holds for all $s \in [0, \infty)$. Under the above setting, $\lambda_{ms}(t)$ is also a function of s and therefore will be denoted as $\lambda_{ms}(t; s)$:

$$\lambda_{ms}(t; s) = \frac{\pi_s \bar{F}_1(t)}{\pi_s \bar{F}_1(t) + (1 - \pi_s)\bar{F}_2(t)} \cdot \lambda_1(t) + \frac{(1 - \pi_s)\bar{F}_2(t)}{\pi_s \bar{F}_1(t) + (1 - \pi_s)\bar{F}_2(t)} \cdot \lambda_2(t),$$

where

$$\pi_s = \frac{(1 - G(s))\pi}{(1 - G(s))\pi + (1 - G(\rho(s)))(1 - \pi)},$$
$$1 - \pi_s = \frac{(1 - G(\rho(s)))(1 - \pi)}{(1 - G(s))\pi + (1 - G(\rho(s)))(1 - \pi)}.$$

Denote the expected lifetime (as a function of s) of an item that has survived a shock by $m(s)$ and, by $P(\tau, s)$, the probability of success (survival probability) for a mission time τ. We are interested in 'pure' maximization of these functions without considering any costs or gains. Thus we want to maximize (with respect to s) the following functions:

$$m(s) = \int_0^\infty \exp\left\{ -\int_0^t \lambda_{ms}(u; s) \mathrm{d}u \right\} \mathrm{d}t, \tag{9.5}$$

$$P(\tau, s) = \exp\left\{ -\int_0^\tau \lambda_{ms}(u; s) \mathrm{d}u \right\}. \tag{9.6}$$

Intuitively, the first guess would be: the larger is the level of severity s, the larger are the functions of interest, which means that *formally* $s^* = \infty$ and we understand this notation here and in the rest of the chapter only in the described sense. However, as the strength of the item is given by distributions in (9.3), there can be the other non-intuitively evident possibility.

In order to investigate the maximizations of (9.5) and (9.6), consider a more general problem—the uniform minimization of $\lambda_{ms}(t;s)$, for all fixed $t \ge 0$, with respect to $s \in [0, \infty]$. That is, find s^* which satisfies

$$s^* = \arg \inf_{s \in [0,\infty]} \lambda_{ms}(t;s), \text{for all fixed } t \ge 0.$$

Denote by $R(s) \equiv \int_0^s r(u)du$ the cumulative failure rate that corresponds to the Cdf $G(s)$. Then the following result describes the optimal severity s^*.

Theorem 9.2 *Let $\lambda_1(t) \le \lambda_2(t)$, for all $t \ge 0$. Then the optimal s^* is the value which maximizes $R(\rho(s)) - R(s)$. In particular,*

(i) *If $r(s)$ is increasing and $\rho'(s) > 1$, then $s^* = \infty$.*
(ii) *If $\frac{\rho'(s)r(\rho(s))}{r(s)} > 1$, for $s < s_0$, and $\frac{\rho'(s)r(\rho(s))}{r(s)} < 1$, for $s > s_0$, then $s^* = s_0$.*

Proof Note again that in accordance with (9.2), $\lambda_{ms}(t;s)$ is the weighted average of $\lambda_1(t)$ and $\lambda_2(t)$ with the corresponding weights $\pi_s(z_1|t)$ and $\pi_s(z_2|t) = 1 - \pi_s(z_1|t)$, respectively, and

$$\pi_s(z_1|t) = \frac{\pi_s \bar{F}_1(t)}{\pi_s \bar{F}_1(t) + (1-\pi_s)\bar{F}_2(t)} = \frac{\bar{F}_1(t)}{\bar{F}_1(t) + (1/\pi_s - 1)\bar{F}_2(t)}$$

is increasing in π_s. Thus, for each fixed $t \ge 0$, as $\lambda_1(t) \le \lambda_2(t)$, the minimum of $\lambda_{ms}(t;s)$ is obtained by maximizing

$$\pi_s = \frac{(1-G(s))\pi}{(1-G(s))\pi + (1-G(\rho(s)))(1-\pi)}. \tag{9.7}$$

This problem is equivalent to minimizing

$$\frac{1-G(\rho(s))}{1-G(s)} = \exp\{-[R(\rho(s)) - R(s)]\}.$$

Therefore, the minimum can now be attained by maximizing $R(\rho(s)) - R(s)$.

(i) Denote $\phi(s) \equiv R(\rho(s)) - R(s)$. Then $\phi'(s) \equiv \rho'(s)r(\rho(s)) - r(s)$. As $\rho'(s) > 1$ and $r(x)$ is increasing,

$$\phi'(s) = \rho'(s)r(\rho(s)) - r(s) > r(\rho(s)) - r(s) \ge 0,$$

where assumption $\rho(s) \ge s$ is used. Thus, in this case, $s^* = \infty$.

Fig. 9.1 Graph for $\eta(s)$

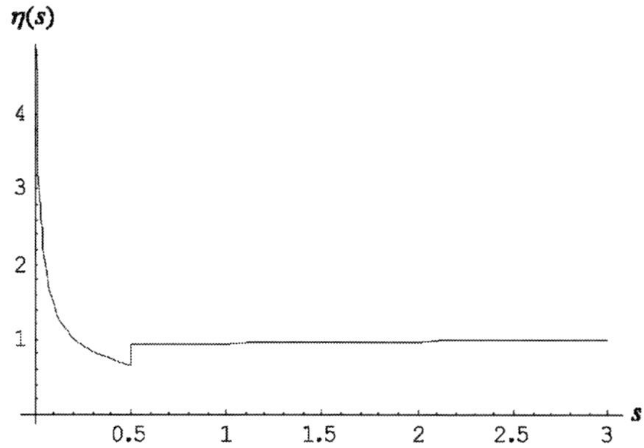

(ii) Assume now that $\frac{\rho'(s)r(\rho(s))}{r(s)} > 1$, for $s < s_0$, and $\frac{\rho'(s)r(\rho(s))}{r(s)} < 1$, for $s > s_0$. Then $\phi'(s) > 0$, for $s < s_0$, and $\phi'(s) < 0$, for $s > s_0$, which implies $s^* = s_0$.

□

Example 9.1 Let $r(s) = e^{-s} + 1, s \geq 0$, and $\rho(s) = \sqrt{s}, 0 \leq s \leq 1/2$; $\rho(s) = s + (1/\sqrt{2} - 1/2,)s \geq 1/2$. The graph for $\eta(s) \equiv \rho'(s)r(\rho(s))/r(s)$ is given in Fig. 9.1. Then it can be seen that there exists some $0 < s_0 < \infty$ which satisfies

$$\frac{\rho'(s)r(\rho(s))}{r(s)} > 1, \text{for } s < s_0 \text{ and } \frac{\rho'(s)r(\rho(s))}{r(s)} < 1, \text{for } s > s_0.$$

Thus, obtaining this value numerically: $s^* = s_0 = 0.204$.

Remark 9.2 In practice, obviously, there exists a maximum level of stress $s_a < \infty$ that can be applied to items without destroying the whole population or without the non-negligible damage in the survived items. In this case, the first part of Theorem 9.2 is modified to $s^* = s_a$, whereas, for the second part of Theorem 9.2, if $s_0 \leq s_a$ then $s^* = s_0$; otherwise $s^* = s_a$.

Let s^* be the optimal severity level which satisfies

$$s^* = \arg \sup_{s \in [0,\infty]} \overline{F}_{ms}(t;s), \text{for all fixed } t \geq 0.$$

Corollary 9.1 *Suppose that $\overline{F}_1(t) \geq \overline{F}_2(t)$, for all $t \geq 0$. Then the optimal s^* is the same as the value which minimizes $\lambda_{ms}(t;s)$, for all fixed $t \geq 0$.*

Proof Observe that $\overline{F}_{ms}(t;s)$ is the weighted average of $\overline{F}_1(t)$ and $\overline{F}_2(t)$:

$$\overline{F}_{ms}(t;s) = \pi_s\overline{F}_1(t) + (1 - \pi_s)\overline{F}_2(t).$$

As $\overline{F}_1(t) \geq \overline{F}_2(t)$ and s^*, in accordance with Theorem 9.2, maximizes π_s, the result follows immediately. □

Note that maximizations of $m(s)$ and $P(\tau; s)$, which can be expressed as [see Eqs. (9.5) and (9.6)]

$$m(s) = \int_0^\infty \overline{F}_{ms}(t; s)\mathrm{d}t,$$

$$P(\tau; s) = \overline{F}_{ms}(\tau; s),$$

is equivalent to uniform maximization of $\overline{F}_{ms}(t; s)$. Therefore, optimal s^* is the same as given in Corollary 9.1.

In the framework of our burn-in model, consider now the corresponding gains and penalties defined for four mutually exclusive events. Denote:

- g_1: gain due to the survival of a strong component
- c_1: penalty incurred by the elimination of a strong component
- g_2: gain due to the elimination of a weak component
- c_2: penalty incurred by the survival of a weak component.

In accordance with this notation and relationship (9.4), the expected gain resulting from the burn-in procedure performed by a shock is given by the following function of severity s:

$$\begin{aligned}\varphi(s) &= g_1\pi\overline{G}(s) + g_2(1-\pi)G(\rho(s)) - c_1\pi G(s) - c_2(1-\pi)\overline{G}(\rho(s)) \\ &= -(\pi g_1 + \pi c_1)G(s) + ((1-\pi)g_2 + (1-\pi)c_2)G(\rho(s)) + g_1\pi - c_2(1-\pi).\end{aligned} \tag{9.8}$$

It is clear that maximization of $\varphi(s)$ is equivalent to minimization of

$$\pi(g_1 + c_1)G(s) + (1-\pi)(g_2 + c_2)(1 - G(\rho(s)))$$

or to minimization of

$$\psi(s) \equiv w_1 G(s) + w_2(1 - G(\rho(s))), \tag{9.9}$$

where the weights w_1 and w_2 are

$$w_1 = \frac{\pi(g_1 + c_1)}{\pi(g_1 + c_1) + (1-\pi)(g_2 + c_2)}, \quad w_2 = 1 - w_1.$$

Note that the probability of failure of a strong component $G(s)$ can be interpreted as the risk that the strong component will be eliminated by a shock. On the other hand, $(1 - G(\rho(s))$ can be regarded as the risk that a weak component will survive a shock. Expressions (9.8) and (9.9) imply that maximization of expected gain is equivalent to minimization of the weighted risk. Observe that when $s = 0$, $\psi(0) = w_2$ and when $s \to \infty$, $\psi(\infty) = w_1$.

The optimal severity s^* should be obtained numerically, however, we can define an upper bound for s^* under some additional conditions.

Theorem 9.3 *Let* $w_1 > w_2$, $\rho'(s) < w_1/w_2$, *for all* $s > s_0$, *and* $r(s)$ *is decreasing for* $s > s_1$. *Then the upper bound for optimal severity level* s^* *is given by* $\max\{s_0, s_1\}$, *that is,* $s^* \le \max\{s_0, s_1\}$.

Proof Observe that

$$\psi'(s) \equiv w_1 r(s) \exp\{-R(s)\} - w_2 \rho'(s) r(\rho(s)) \exp\{-R(\rho(s))\},$$

where $R(s) \equiv \int_0^s r(u)du$. If $\rho'(s) < w_1/w_2$, for all $s > s_0$, and $r(s)$ is decreasing for $s > s_1$, then $\psi'(s) > 0$, for all $s > \max\{s_0, s_1\}$. This implies that $\psi(s)$ is strictly increasing for $s > \max\{s_0, s_1\}$. Thus the upper bound for s^* is given by $\max\{s_0, s_1\}$. □

Example 9.2 Suppose that $w_1 = 0.6, w_2 = 0.4, r(s) = 1, 0 \le s < 2$; $r(s) = e^{s-2}, s \ge 2$, and $\rho(s) = 5s, 0 \le s < 1$; $\rho(s) = s + 4, s \ge 1$. Then, in this case, $s_0 = 1.0$ and $s_1 = 2.0$. Therefore, $s^* \le \max\{s_0, s_1\} = 2.0$. The graph for $\psi(s)$ is given in Fig. 9.2.

It can be obtained numerically that $s^* = 0.302$.

9.1.3 Optimal Severity for Minimizing Expected Costs

In this section, we consider two models of determining the optimal severity minimizing the expected cost function, which takes into account burn-in and field operation.

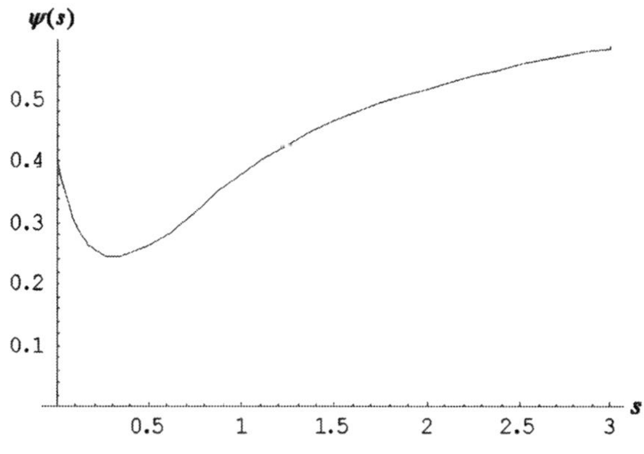

Fig. 9.2 Graph for $\psi(s)$

9.1.3.1 Model 1: Minimization Without Replacement During Field Operation

An item is chosen at random from our heterogeneous population and is exposed to a shock. If it survives, then it is considered to be ready for usage, otherwise the failed item is discarded and the new one is chosen from the population, etc. This procedure is repeated until the first survived item is obtained.

Let c_{sr} be the shop replacement cost and c_s be the cost for conducting a single shock. Let $c_1(s)$, as a function of s, be the expected cost for eventually obtaining a component which has survived a shock. Conditioning on the event that the component survives (or fails) a shock, the following equation can be obtained:

$$c_1(s) = (1 - P)c_s + ((c_s + c_{sr}) + c_1(s))P, \tag{9.10}$$

where $P = G(s)\pi + G(\rho(s))(1 - \pi)$ is the probability that an item from the mixture population does not survive the shock. Then, from Eq. (9.10):

$$c_1(s) = \frac{c_s + c_{sr}P}{1 - P} = -c_{sr} + \frac{c_s + c_{sr}}{1 - P}. \tag{9.11}$$

Let:

The cost c_m is incurred by the event $\{T_s \le \tau\}$ (Failure of the Mission);
The gain g_m results from the event $\{T_s > \tau\}$ (Success of the Mission).

Then the expected costs during field operation, $c_2(s)$, is given by

$$c_2(s) = -g_m\left(\pi_s\overline{F}_1(\tau) + (1 - \pi_s)\overline{F}_2(\tau)\right) + c_m(\pi_s F_1(\tau) + (1 - \pi_s)F_2(\tau)),$$

where π_s is defined by Eq. (9.7). Then the total expected cost $c(s)$ is

$$\begin{aligned} c(s) &= c_1(s) + c_2(s) = -c_{sr} + \frac{c_s + c_{sr}}{\overline{G}(s)\pi + \overline{G}(\rho(s))(1 - \pi)} \\ &\quad - (g_m + c_m)\left(\frac{\overline{G}(s)\pi}{\overline{G}(s)\pi + \overline{G}(\rho(s))(1 - \pi)}\overline{F}_1(\tau) + \frac{\overline{G}(\rho(s))(1 - \pi)}{\overline{G}(s)\pi + \overline{G}(\rho(s))(1 - \pi)}\overline{F}_2(\tau)\right) + c_m. \end{aligned}$$

Let s^* be the optimal severity level that satisfies

$$s^* = \arg \inf_{s \in [0,\infty]} c(s).$$

The following theorem defines properties of optimal s^*.

Theorem 9.4 *Let $\overline{F}_1(t) \ge \overline{F}_2(t)$, for all $t \ge 0$. If $R(\rho(s)) - R(s)$ strictly decreases for $s > s_0$, then $s^* \le s_0$. In particular,*

(i) *If $\rho'(s) > 1$ and $r(x)$ is increasing, then $s^* < \infty$.*
(ii) *If $\frac{\rho'(s)r(\rho(s))}{r(s)} < 1$, for $s > s_0$, then $s^* \le s_0$.*

Proof Note that $c_1(s)$ strictly increases from $c_1(0) = c_s$ to $c_1(\infty) = \infty$. Also observe that $c_2(s) = -(g_m + c_m)\overline{F}_{ms}(\tau; s) + c_m$, where $\overline{F}_{ms}(t; s)$ is the weighted average of $\overline{F}_1(t)$ and $\overline{F}_2(t)$ with the corresponding weights π_s and $1 - \pi_s$, respectively. If $R(\rho(s)) - R(s)$ strictly decreases for $s > s_0$, then, by similar arguments as those described in the proof of Theorem 9.2, $c_2(s)$ strictly increases for $s > s_0$. This imply that $c(s)$ strictly increases for $s > s_0$ and thus we can conclude that optimal $s^* \leq s_0$.

(i) From the proof of Theorem 9.2, it can be seen that if $\rho'(s) > 1$ and $r(x)$ is increasing, then $c_2(s)$ strictly decreases for $s > 0$. But $c(\infty) = \infty$ and thus $s^* < \infty$.
(ii) If $\rho'(s)r(\rho(s))/r(s) < 1$, for $s > s_0$ then, from the proof of Theorem 9.2, it is easy to see that $c_2(s)$ strictly increases for $s > s_0$, and thus $s^* \leq s_0$. □

Assume now that the expected gain during field operation is proportional to the mean lifetime. Then the expected cost (i.e., the negative gain) during field operation is

$$c_2(s) = -k\left(\pi_s \int_0^\infty \overline{F}_1(t)\mathrm{d}t + (1 - \pi_s) \int_0^\infty \overline{F}_2(t)\mathrm{d}t\right),$$

and the total expected cost is given by

$$\begin{aligned} c(s) = & -c_{sr} + \frac{c_s + c_{sr}}{\overline{G}(s)\pi + \overline{G}(\rho(s))(1 - \pi)} \\ & - k\left(\frac{\overline{G}(s)\pi}{\overline{G}(s)\pi + \overline{G}(\rho(s))(1 - \pi)} \int_0^\infty \overline{F}_1(t)\mathrm{d}t + \frac{\overline{G}(\rho(s))(1 - \pi)}{\overline{G}(s)\pi + \overline{G}(\rho(s))(1 - \pi)} \int_0^\infty \overline{F}_2(t)\mathrm{d}t\right), \end{aligned} \tag{9.12}$$

where k is a constant of proportionality. Then the following corollary holds:

Corollary 9.2 *Let $\overline{F}_1(t) \geq \overline{F}_2(t)$, for all $t \geq 0$. Then the properties of optimal s^* for the total expected cost function* (9.12) *are the same as those described in Theorem 9.4.*

The *proof* is similar to that of Theorem 9.4. □

9.1.3.2 Model 2: Minimization with Replacement During Field Operation

Assume that if an item fails during field operation, it is replaced by another item which has survived a shock at a cost $c_f > c_{sr}$. The time intervals between two consecutive replacements constitute a renewal process. Therefore, in accordance with $\overline{F}_{ms}(t) = \pi_s \overline{F}_1(t) + (1 - \pi_s)\overline{F}_2(t)$ and Eq. (9.7), the mean time between two consecutive replacements is equal to

$$\frac{\overline{G}(s)\pi}{\overline{G}(s)\pi + \overline{G}(\rho(s))(1-\pi)} \int_0^\infty \overline{F}_1(t)\mathrm{d}t + \frac{\overline{G}(\rho(s))(1-\pi)}{\overline{G}(s)\pi + \overline{G}(\rho(s))(1-\pi)} \int_0^\infty \overline{F}_2(t)\mathrm{d}t. \tag{9.13}$$

Then, by the renewal reward theory argument, the expected cost rate $\tilde{c}(s)$ is given by

$$\begin{aligned}\tilde{c}(s) = &\frac{1}{\frac{\overline{G}(s)\pi}{\overline{G}(s)\pi + \overline{G}(\rho(s))(1-\pi)} \int_0^\infty \overline{F}_1(t)\mathrm{d}t + \frac{\overline{G}(\rho(s))(1-\pi)}{\overline{G}(s)\pi + \overline{G}(\rho(s))(1-\pi)} \int_0^\infty \overline{F}_2(t)\mathrm{d}t} \\ &\times \left(\frac{c_s + c_{sr}}{\overline{G}(s)\pi + \overline{G}(\rho(s))(1-\pi)} + (c_f - c_{sr}) \right),\end{aligned} \tag{9.14}$$

where the denominator is just an expected duration of a renewal cycle given by Eq. (9.13) and the numerator defines the expected cost incurred during this cycle (taking into account that the expected cost during burn-in is given by (9.11) and the replacement cost during field operation is given by c_f).

Let s^* denote the optimal severity which satisfies

$$s^* = \arg \inf_{s \in [0,\infty]} \tilde{c}(s).$$

Then, similar to Theorem 9.4, the following result is also true:

Theorem 9.5 *Let $\overline{F}_1(t) \geq \overline{F}_2(t)$, for all $t \geq 0$. If $R(\rho(s)) - R(s)$ strictly decreases for $s > s_0$, then optimal $s^* \leq s_0$. In particular,*

(i) *If $\rho'(s) > 1$ and $r(s)$ is increasing, then $s^* < \infty$.*
(ii) *If $\frac{\rho'(s) r(\rho(s))}{r(s)} < 1$, for $s > s_0$, then the optimal $s^* \leq s_0$.*

Proof Rearranging terms in (9.14):

$$\begin{aligned}\tilde{c}(s) = &\frac{c_s + c_{sr}}{\overline{G}(s)\pi \int_0^\infty \overline{F}_1(t)\mathrm{d}t + \overline{G}(\rho(s))(1-\pi) \int_0^\infty \overline{F}_2(t)\mathrm{d}t} \\ &+ \frac{c_f - c_{sr}}{\frac{\overline{G}(s)\pi}{\overline{G}(s)\pi + \overline{G}(\rho(s))(1-\pi)} \int_0^\infty \overline{F}_1(t)\mathrm{d}t + \frac{\overline{G}(\rho(s))(1-\pi)}{\overline{G}(s)\pi + \overline{G}(\rho(s))(1-\pi)} \int_0^\infty \overline{F}_2(t)\mathrm{d}t}.\end{aligned}$$

The first term in the right-hand side strictly increases for $s > 0$. Note that the denominator of the second term is the weighted average of $\int_0^\infty \overline{F}_1(t)\mathrm{d}t$ and $\int_0^\infty \overline{F}_2(t)\mathrm{d}t$ $\left(\int_0^\infty \overline{F}_1(t)\mathrm{d}t \geq \int_0^\infty \overline{F}_2(t)\mathrm{d}t\right)$ with the corresponding weights π_s and

$1 - \pi_s$, respectively. Then, following the procedures described in the proof of Theorem 9.4, we can obtain the desired result. □

Remark 9.3 In 'ordinary' burn-in, as discussed in the previous chapters, when the lifetimes of items are described by the distributions with the bathtub-shaped failure rate, the following property holds: the optimal burn-in time should be smaller than the first change point (see, e.g., [5, 12]). In our reasoning, optimal stress levels, in accordance with Theorems 9.2, 9.4, and 9.5, in a similar way also depend on the properties of the distribution of strength.

Remark 9.4 In practice, the cost parameters $(c_s, c_{sr}, c_f, c_m, g_m)$ might not be estimated precisely, which could make the optimization procedure difficult. In this case, the Receiver Operating Characteristic (ROC) analysis can be adopted and effectively used to determine the optimal burn-in time which minimizes the corresponding cost functions. A reference for this approach can be found in Wu and Xie [15], where the application of ROC analysis is used to remove the weak subpopulation in burn-in problems.

9.2 Continuous Mixtures

9.2.1 The Impact of Shocks on Mixed Populations

Consider a population of identically distributed items with lifetimes $T_i, i = 1, 2, \ldots$. Each item 'is affected' by a non-observable univariate frailty parameter Z_i and the lifetimes T_i are conditionally independent given the values of parameters $Z_i = z_i$. Assume that these parameters are i.i.d with a common pdf $\pi(z)$ and with support in $[0, \infty)$. (The general support $[a, b)$, $0 \le a < b \le \infty$ can be considered as well.) Then, obviously $T_i, i = 1, 2, \ldots$ are also i.i.d. For convenience, the sub index "i" will be omitted and, therefore, the lifetimes and frailties for all items will be denoted by T and Z, respectively. Thus, obviously, T is described by the *mixture* Cdf and pdf

$$F_m(t) = \int_0^\infty F(t, z)\pi(z)\mathrm{d}z,$$

$$f_m(t) = \int_0^\infty f(t, z)\pi(z)\mathrm{d}z,$$

respectively, where $F(t, z) \equiv F(t|z) = \Pr[T \le t|Z = z]$, $f(t, z) = F'(t, z)$ are the corresponding conditional characteristics for realization $Z = z$.

Then the mixture (*observed*) failure rate $\lambda_m(t)$, similar to (5.11, 5.12) is

$$\lambda_m(t) = \frac{f_m(t)}{\bar{F}_m(t)} = \frac{\int_0^\infty f(t,z)\pi(z)\mathrm{d}z}{\int_0^\infty \bar{F}(t,z)\pi(z)\mathrm{d}z} = \int_0^\infty \lambda(t,z)\pi(z|t)\mathrm{d}z, \tag{9.15}$$

where

$$\pi(z|t) \equiv \pi(z)\frac{\bar{F}(t,z)}{\int_0^\infty \bar{F}(t,z)\pi(z)\mathrm{d}z}. \tag{9.16}$$

In the framework of the model described above, we will consider mixed populations of stochastically ordered subpopulations.

Remark 9.5 The foregoing definitions and properties describe a standard statistical mixture (or frailty) model for an item and for the collection of items (population) as well. However, the following *interpretation* can be also useful, as frailty models were initially developed in demographic and actuarial studies as a method of describing heterogeneity in large populations (see, e.g., [3, 11, 14]; and references therein). Thus, we assume that heterogeneity, described by the unobserved frailty, is a property of an infinite population. It usually means that, due to different environments, conditions, different manufacturers, etc., the population consists of subpopulations of items with different statistical properties. Pooling at random items from this population results in the described mixture model.

Assume that an item is put into operation for the mission time τ with the required survival probability $P_r(\tau)$. If

$$\exp\left\{-\int_0^\tau \lambda_m(u)\mathrm{d}u\right\} \geq P_r(\tau), \tag{9.17}$$

then everything is fine and we do not need to improve the performance of our items. On the contrary, if this inequality does not hold, the burn-in procedure can be performed. There are different types of these procedures and we will consider here the burn-in that is performed via shocks that can eliminate the weak items.

Throughout this section, the impact of a shock is described by the following *general assumption:*

Assumption An instantaneous shock either 'kills' an item with a given probability or does not change its stochastic properties with the complementary probability. The more 'frail' (e.g., with larger failure rate or with smaller survival function) an item is, the larger is the probability that a shock will 'kill' it.

The following burn-in procedure is employed:

- *Burn-in procedure by means of shocks.* An item is exposed to a shock. If it survives, it is considered to be ready for usage, otherwise the failed item is discarded and a new one is exposed to a shock, etc.

This setting can be defined probabilistically in the following way: Let $\pi_s(z)$ denote the pdf of the frailty Z_s (with support in $[0,\infty)$) after a shock and let $\lambda_{ms}(t)$ be the corresponding mixture failure rate. In accordance with (9.15):

$$\lambda_{ms}(t)=\int_0^\infty \lambda(t,z)\pi_s(z|t)\mathrm{d}z,$$

where, similar to (9.16), $\pi_s(z|t)$ is defined by the right-hand side of (9.16) with $\pi(z)$ substituted by $\pi_s(z)$.

First, assume formally that population frailties before and after a shock are ordered in the sense of the likelihood ratio (see Sect. 2.8):

$$Z\geq_{LR} Z_s, \tag{9.18}$$

which in our terms is defined as

$$\pi_s(z)=\frac{g(z)\pi(z)}{\int_0^\infty g(z)\pi(z)\mathrm{d}z}, \tag{9.19}$$

where $g(z)$ is a decreasing function and therefore $\pi_s(z)/\pi(z)$ is decreasing. As it will be discussed in the next subsection, the function $g(z)$ can be interpreted as the survival probability of an item with frailty z after the shock. Therefore, the assumption that $g(z)$ is a decreasing function of z corresponds to our general "Assumption". Note that the 'likelihood ratio ordering' for mixing (frailty) distributions was used by Block et al. [4] for ordering optimal burn-in times in 'ordinary' burn-in settings (without shocks): the larger frailty corresponds to the larger optimal burn-in time for some specified cost functions. It seems that this ordering is natural for stochastic modeling in heterogeneous populations. The following important theorem shows that depending on assumptions, the likelihood ratio ordering of frailties leads either to the failure rate or to the usual stochastic ordering of population lifetimes.

Theorem 9.6 *Let relationship* (9.19), *defining the mixing density after a shock, where* $g(z)$ *is a decreasing function, hold.*

(i) *Assume that*

$$\lambda(t,z_1)\leq\lambda(t,z_2),\quad z_1<z_2,\forall z_1,z_2\in[0,\infty],t\geq 0. \tag{9.20}$$

Then

$$\lambda_{ms}(t)\leq\lambda_m(t);\ \forall t\geq 0. \tag{9.21}$$

(ii) *Assume that*

$$\overline{F}(t,z_1) \geq \overline{F}(t,z_2), \quad z_1 < z_2, \forall z_1, z_2 \in [0,\infty], t \geq 0. \tag{9.22}$$

Then

$$\overline{F}_{ms}(t) \geq \overline{F}_m(t), \forall t \geq 0, \tag{9.23}$$

where $\lambda_{ms}(t), \bar{F}_{ms}(t)$ are the population (mixture) failure rate and the survival function after a shock, respectively.

Proof Note that, inequalities (9.20) and (9.22) define two types of stochastic orderings for subpopulations, i.e., the failure rate ordering and the usual stochastic ordering, respectively.

(i) It can be shown [10: p. 164] that:

$$\begin{aligned} & sign[\lambda_{ms}(t) - \lambda_m(t)] \\ & = sign \int\limits_{\substack{0 \\ u>s}}^{\infty} \int\limits_0^{\infty} \bar{F}(t,u)\bar{F}(t,s)(\lambda(t,u) - \lambda(t,s))(\pi_s(u)\pi(s) - \pi_s(s)\pi(u))duds, \end{aligned} \tag{9.24}$$

which is negative due to definition (9.19) and assumptions of this theorem.

(ii) As $g(z)$ is a decreasing function, and the survival function $\bar{F}(t,z)$ is also decreasing in z, it can be easily shown using the mean value theorem that

$$\overline{F}_{ms}(t) - \overline{F}_m(t) = \frac{\int_0^\infty \overline{F}(t,z)g(z)\pi(z)\mathrm{d}z}{\int_0^\infty g(z)\pi(z)\mathrm{d}u} - \int\limits_0^\infty \overline{F}(t,z)\pi(z)\mathrm{d}z \geq 0. \tag{9.25}$$

Indeed

$$\int\limits_0^\infty g(z)\pi(z)\mathrm{d}z = g(z^*)$$

and

$$\int\limits_0^\infty \overline{F}(t,z)g(z)\pi(z)\mathrm{d}z = g(z^{**}) \int\limits_0^\infty \overline{F}(t,z)\pi(z)\mathrm{d}z,$$

where $g(z^*)$ and $g(z^{**})$ are the corresponding mean values. As $\bar{F}(t,z)$ is decreasing in z, $z^{**} \leq z^*$. Therefore, taking into account that $g(z)$ is a decreasing function, (9.25) follows. Note that the usage of the mean value theorem relies on the continuity of $g(z)$. Alternatively, the general case (without this assumption) can be proved similar to the proof in (i) (see also Theorem 9.7). □

Remark 9.6 Inequality (9.20) is a natural ordering in the family of failure rates $\lambda(t,z), z \in [0,\infty)$ and trivially holds, e.g., for the specific multiplicative model:

$$\lambda(t,z) = z\lambda(t). \tag{9.26}$$

Remark 9.7 Theorem 9.6 means that the *population quality* (in terms of the failure rate or the survival function) has improved after a shock. Thus, in accordance with our statistical 'frequentistic' *interpretation* (see Remark 9.5) when 'the whole population' is exposed to a shock, the items that have passed this test form a new population with better stochastic characteristics. On the other hand, following our formal initial setting, it turns out that the benefit of a non-destructive shock is of 'informational' type, i.e., surviving a shock has the 'Bayesian' effect of modifying the posterior distribution of Z, which is Z_s in our notation.

Remark 9.8 In accordance with (9.21) and (9.23), inequality (9.17) can be already achieved after one shock, otherwise new shocks should be applied or the "severity" of a single shock (see later) should be increased. It is also worth noting that the replacement of condition (9.18) by the usual stochastic ordering: $Z \geq_{st} Z_s$ will not guarantee orderings (9.21) and (9.23) for all t.

9.2.2 The Impact of Shocks on an Item

Now we must consider a more specific mechanism of a shock's impact on an item. Let each item fail with probability $p(z)$ and survive (as good as new) with probability $q(z) = 1 - p(z)$. Here, the condition that corresponds to the general "Assumption" in Sect. 9.2.1 is that $p(z)$ is an increasing function of $z\,(0 \leq p(z) \leq 1)$. This assumption makes sense as, in accordance with (9.20), larger values of frailty correspond to larger values of the failure rate. Therefore, items with larger values of frailty are more susceptible to failures. Equation (9.19) reads now

$$\pi_s(z) = \frac{q(z)\pi(z)}{\int_0^\infty q(z)\pi(z)\mathrm{d}z}, \tag{9.27}$$

where $\pi_s(z)$ is the pdf of Z_s (*predictive*, or *posterior* pdf, as it has been called in Bayesian terminology). As $q(z)$ is decreasing with z, it follows from Theorem 9.6 that the failure rate ordering (9.21) and the usual stochastic ordering (9.23) hold.

If we are not concerned about the costs (e.g., when the mission is very important) and inequality

$$\exp\left\{-\int_0^\tau \lambda_{ms}(u)\mathrm{d}u\right\} \geq P_r(\tau) \tag{9.28}$$

holds, then the burn-in is over and the item that has survived a shock can be put into field operation. Otherwise, a shock with the higher level of severity or several shocks should be performed for each item in order to achieve this inequality.

On the other hand, in most practical situations the costs are involved. In order to consider the corresponding optimization, we must define the costs and probabilities of interest. A convenient and useful model for $p(z)$ (although oversimplified for practical usage) is the step function:

$$p(z) = \begin{cases} 0, & 0 \le z \le z_b \\ 1, & z > z_b \end{cases}. \tag{9.29}$$

It means that all 'weak' items with $z > z_b$ will be eliminated and only 'strong' items will remain in the population. In accordance with (9.29), the probability of not surviving the shock in this case is

$$P_{z_b} \equiv \bar{\Pi}(z_b) = \int_{z_b}^{\infty} \pi(z)\mathrm{d}z, \tag{9.30}$$

where $\Pi(z)$ is the Cdf that corresponds to the pdf $\pi(z)$. Obviously, for a general form of $p(z)$, this probability is defined by the following mixture

$$P = \int_{0}^{\infty} p(z)\pi(z)\mathrm{d}z. \tag{9.31}$$

9.2.3 Shock's Severity

It is clear that the parameter z_b in the specific model (9.29) can be considered as a parameter of severity: the larger values of z_b correspond to a smaller severity. Now we can deal with the issue of severity in a more general context, that is, when $p(z)$ is not a simple step function but a continuous function of z.

For this discussion, define the functions $p(z)$ and $q(z)$ as functions of the frailty variable z and the severity parameter $s \in [0, \infty)$, $p(z, s)$ and $q(z, s)$. Assume that $q(z, s)$ is decreasing in z for each fixed s and is decreasing in s for each z. The assumption that $q(z, s)$ is decreasing in z for each fixed s is just what was assumed in our general "Assumption" in Sect. 9.2.1. The assumption that $q(z, s)$ is decreasing in s for each fixed z is also quite natural and implies that items characterized by the same value of frailty have larger failure probabilities under larger severity levels.

Denote the corresponding failure rate and the survival function by $\lambda_{ms}(t; s)$ and $\overline{F}_{ms}(t; s)$, respectively. Similar to (9.19) and (9.16):

$$\pi_s(z,s) = \frac{q(z,s)\pi(z)}{\int_0^\infty q(u,s)\pi(u)\mathrm{d}u}, \pi_s(z,s|t) \equiv \pi_s(z,s)\frac{\bar{F}(t,z)}{\int_0^\infty \bar{F}(t,u)\pi_s(u,s)\mathrm{d}u}.$$

In order to compare two severity levels, we need the following definition.

Definition 9.1

(i) The severity (stress) level s is said to be *dominated* under the failure rate criterion if there exists another level s' such that

$$\lambda_{ms}(t;s) \geq \lambda_{ms}(t;s'), \text{for all } t \geq 0.$$

(ii) The severity (stress) level s is said to be *dominated* under the survival probability criterion if there exists another level s' such that

$$\overline{F}_{ms}(t;s') \geq \overline{F}_{ms}(t;s), \text{for all } t \geq 0.$$

Otherwise, the severity (stress) level s is called *non-dominated.*

Theorem 9.7 *Assume that $q(z,s)$ is decreasing in z for each fixed s and is decreasing in s for each z. Consider two stress levels s and s'. Let*

$$q(u,s')q(v,s) - q(v,s')q(u,s) \leq 0, \text{for all } u > v, \tag{9.32}$$

which means that $q(z,s')/q(z,s)$ is decreasing in z.

(i) *If $\lambda(t,z_1) \leq \lambda(t,z_2), \quad z_1 < z_2, \forall z_1, z_2 \in [0,\infty], t \geq 0,$ then the severity level s is dominated under the failure rate criterion.*

(ii) *If $\overline{F}(t,z_1) \geq \overline{F}(t,z_2), \quad z_1 < z_2, \forall z_1, z_2 \in [0,\infty], t \geq 0,$ then the severity level s is dominated under the survival probability criterion.*

Proof

(i) Similar to (9.24):

$$sign[\lambda_{ms}(t;s') - \lambda_{ms}(t;s)]$$
$$= sign \underset{u>v}{\int_0^\infty \int_0^\infty} \bar{F}(t,u)\bar{F}(t,v)(\lambda(t,u) - \lambda(t,v))(\pi_s(u,s')\pi_s(v,s) - \pi_s(v,s')\pi_s(u,s))dudv.$$

Thus, if (9.32) holds, then

$$\pi_s(u,s')\pi_s(v,s) - \pi_s(v,s')\pi_s(u,s) \leq 0,$$

which implies the result in (i).

(ii)

$$\overline{F}_{ms}(t;s') - \overline{F}_{ms}(t;s) = \frac{\int_0^\infty \overline{F}(t,z)q(z,s')\pi(z)\mathrm{d}z}{\int_0^\infty q(u,s')\pi(u)\mathrm{d}u} - \frac{\int_0^\infty \overline{F}(t,z)q(z,s)\pi(z)\mathrm{d}z}{\int_0^\infty q(u,s)\pi(u)\mathrm{d}u},$$

and the corresponding numerator can be transformed to

$$\int_{0}^{\infty}\int_{0}^{\infty}_{u>v} \pi(u)\pi(v)(\overline{F}(t,u) - \overline{F}(t,v))(q(u,s')q(v,s) - q(v,s')q(u,s))\mathrm{d}u\mathrm{d}v.$$

Therefore, if (9.32) holds, then

$$\overline{F}_{ms}(t;s') - \overline{F}_{ms}(t;s) \geq 0, \text{for all } t \geq 0.$$

□

Remark 9.9 Note that although the assumption that $q(z,s)$ is decreasing in z for each fixed s and is decreasing in s for each z is not used formally in the foregoing proof, it represents some basic 'physical properties' of the model and should be checked in applications.

Remark 9.10 In accordance with Remark 9.7, Theorem 9.7 means that the *population quality* (in terms of the failure rate or the survival function) is better after the shock with severity s' than after the shock with severity s.

Example 9.3 Consider the following illustrative discrete example. Suppose that there are only three stress levels: s_1, s_2, and $s_3 (s_1 < s_2 < s_3)$. Let $q(z,s_1) = 0.2e^{-z} + 0.6, q(z,s_2) = 0.6e^{-z} + 0.2$, and $q(z,s_3) = 0.2e^{-z} + 0.2$. Then $q(z,s_i)$ is decreasing in z, for each $i = 1,2,3$. Furthermore, for each fixed z, $q(z,s_1) \geq q(z,s_2) \geq q(z,s_3)$ and in this way the condition for ordering the stress levels $(s_1 < s_2 < s_3)$ is justified. Observe that

$$\frac{q(z,s_2)}{q(z,s_1)} \text{ and } \frac{q(z,s_2)}{q(z,s_3)}$$

strictly decrease in z. Therefore, as follows from Theorem 9.7, the stress levels s_1 and s_3 are dominated and, in this case, the stress level s_2 minimizes the failure rate and maximizes the survival function after a shock. Thus s_2 is the optimal level.

Remark 9.11 Intuitively, it can be believed that a higher level of severity results in a 'better population' but it is not always true as shown in this example. A similar observation is true for the conventional burn-in in homogeneous populations when the larger time of burn-in does not necessarily lead to a 'better population'. In this case, the shape of the failure rate (e.g., bathtub) plays a crucial role in the corresponding analysis.

Consider again the specific case (9.29). For convenience, and in accordance with our reasoning, let us change the notation in the following way:

$$q(z,s) = \begin{cases} 1, & 0 \leq z \leq z_s \\ 0, & z > z_s \end{cases}, \tag{9.33}$$

where $z_s > z_{s'}$ if $s' > s,\ s, s' \in [0, \infty)$. Then we have the following corollary.

Corollary 9.3 *Let the model* (9.33) *hold and fix* $s' > 0$.

(i) *If* $\lambda(t, z_1) \le \lambda(t, z_2),\quad z_1 < z_2, \forall z_1, z_2 \in [0, \infty], t \ge 0$, *then the severity level* s *for* $\forall s \le s'$ *is dominated under the failure rate criterion. That is,*

$$\lambda_{ms}(t; s) \ge \lambda_{ms}(t; s'), \text{for all } t \ge 0, \text{for all } s \le s'.$$

(ii) *If* $\overline{F}(t, z_1) \ge \overline{F}(t, z_2),\quad z_1 < z_2, \forall z_1, z_2 \in [0, \infty], t \ge 0$, *then the severity level* s *for* $\forall s \le s'$ *is dominated under the survival probability criterion. That is,*

$$\overline{F}_{ms}(t; s') \ge \overline{F}_{ms}(t; s), \text{for all } t \ge 0, \text{for all } s \le s'.$$

Proof It is easy to check that condition

$$q(u, s')q(v, s) - q(v, s')q(u, s) \le 0, \text{for all } u > v,$$

is always satisfied for $q(z, s)$ given by Relationship (9.33) for all $s' > s$. □

It follows from this corollary that the better population quality (see Remark 9.7) can be obtained by increasing s (formally, $s \to \infty$, but the level of severity is always bounded in practice).

Remark 9.12 In Theorem 9.7, considering general form of $q(z, s)$, it was assumed that $q(z, s')/q(z, s)$ decreases in z for some fixed s' and s. If we now assume that this quotient decreases in z for all $s' > s$, then, similar to the specific case of Corollary 9.3, the better population quality can be obtained by increasing $s\ (s \to \infty)$.

Remark 9.13 It should be noted that there is a certain analogy between describing the usual burn-in for heterogeneous populations during a given time period and the burn-in via shocks. It was shown in Finkelstein [10] that, if two different frailty distributions are ordered in the sense of the likelihood ratio and inequality (9.20) holds, then the smaller frailty implies the smaller mixture failure rate (the better population quality after burn-in). In the case under consideration, Inequality (9.32) can be also interpreted as the corresponding likelihood ordering of frailties after the shocks with two stress levels s and s', respectively.

9.2.4 The Cost of Burn-in and Optimal Problem

In field operation, items are frequently required to survive a pre-specified time period, which is called the mission time, τ. In this subsection, optimal severity of a shock, which minimizes the average cost incurred during the burn-in and the operation phase will be considered.

As previously, a new component randomly selected from the heterogeneous population is burned-in by means of a shock. If the first one did not survive then we take another one from infinite heterogeneous population and burn-in again. This procedure is repeated until we obtain the first component which survives burn-in. Then this component is put into the field operation. Assume, first, for simplicity, that the cost of conducting a single shock $c_s = 0$. Denote by c_1 the expected cost of the burn-in until obtaining the first item that has survived shocks. It is clear that

$$\begin{aligned} c_1 &= 0 \times (1-P) + c_{sr}P(1-P) + 2c_{sr}P^2(1-P) + 3c_{sr}P^3(1-P) + \cdots \\ &= c_{sr}P(1-P)(1+2P+3P^2+\cdots) = \frac{c_{sr}P}{1-P}, \end{aligned} \tag{9.34}$$

where c_{sr} is the shop replacement cost. Similarly, when $c_s \neq 0$

$$c_1 = \frac{c_{sr}P + c_s}{1-P}. \tag{9.35}$$

Obviously, this function increases when P increases in $[0, 1)$. Note that P is now a function of the stress level s, that is, $P(s)$ [see definition (9.31), where $p(z)$ should be substituted by $p(z, s)$] and thus, in the following, c_1 in (9.34) and (9.35) should be also understood as a function of $s, c_1(s)$.

Let:

The cost c_m is incurred by the event $\{T_s \leq \tau\}$ (Failure of the Mission);
The gain g_m results from the event $\{T_s > \tau\}$ (Success of the Mission).

Obviously, the expected cost during field operation is:

$$\begin{aligned} c_2(s) &= -g_m\overline{F}_{ms}(\tau; s) + c_m(1 - \overline{F}_{ms}(\tau; s)) \\ &= -(g_m + c_m)\overline{F}_{ms}(\tau; s) + c_m. \end{aligned}$$

Therefore, the total expected cost function (as a function of the stress level s) for the burn-in and the field operation phases is given by

$$c(s) = c_1(s) + c_2(s), \tag{9.36}$$

where $c_1(s)$ is defined in (9.35). The values c_{sr}, c_s, g_m, c_m are assumed to be known. Thus the corresponding optimization problem can be formalized as

$$s^* = \arg\ \min c(s). \tag{9.37}$$

It is worth noting that condition (9.28) can also be imposed as an additional requirement for obtaining minimum of the total costs function.

Theorem 9.8 *Suppose that*

$$\overline{F}(t, z_1) \geq \overline{F}(t, z_2), \quad z_1 < z_2, \forall z_1, z_2 \in [0, \infty], t \geq 0.$$

(i) *If, for any* $s_2 > s_1$, $q(u,s_2)q(v,s_1) - q(v,s_2)q(u,s_1) \leq 0$, *for all* $u > v$, i.e., $q(z,s_2)/q(z,s_1)$ *decreases in* z *for all* $s_2 > s_1$, *then there exists the finite optimal level* $s^* < \infty$ *for the optimization problem* (9.37).
(ii) *If there exists* $s_0 < \infty$ *such that for all levels* $s > s_0$, *the level s is dominated by* s_0 *under the survival probability criterion, then* $s^* < s_0$.

Proof
(i) Observe that $c_1(s)$ strictly increases in s with $c_1(0) = c_s$ to $c_1(\infty) = \infty$ and $c_2(s)$ can be minimized by maximizing $\overline{F}_{ms}(\tau; s)$. If $q(z,s_2)/q(z,s_1)$ decreases in z for all $s_2 > s_1$, then $c_2(s)$ strictly decreases for $s > 0$ since $\overline{F}_{ms}(\tau; s)$. strictly increases for $s > 0$ by Theorem 9.7. But $c(\infty) = \infty$ and thus, $s^* < \infty$.
(ii) If there exists $s_0 < \infty$ such that for all stress levels $s > s_0$, the level s is dominated by s_0 then it is obvious that $c(s_0) \leq c(s)$, for all $s > s_0$. Therefore, $s^* < s_0$. □

Assume now that the expected gain during field operation is proportional to the mean lifetime of an item, which is also a reasonable assumption that is often used in practice. Then the expected cost during the field operation, $c_2(s)$, is given by

$$c_2(s) = -k\int_0^\infty \overline{F}_{ms}(t;s)\mathrm{d}t = -k\frac{\int_0^\infty \{\int_0^\infty \overline{F}(t,z)\mathrm{d}t\}q(z;s)\pi(z)\mathrm{d}z}{\int_0^\infty q(u;s)\pi(u)\mathrm{d}u},$$

where k is the proportionality constant. It is obvious that if

$$\overline{F}(t,z_1) \geq \overline{F}(t,z_2), \quad z_1 < z_2, \forall z_1, z_2 \in [0,\infty], t \geq 0,$$

then

$$\int_0^\infty \overline{F}(t,z_1)\mathrm{d}t \geq \int_0^\infty \overline{F}(t,z_2)\mathrm{d}t, \quad z_1 < z_2, \forall z_1, z_2 \in [0,\infty]$$

and, as in Theorem 9.8, the same result for optimal severity level s^* can be obtained (See also the proof of Theorem 9.7-(ii)).

If our goal is only to achieve minimum of $c(s)$ and a shock can be made as severe as we wish, then no further shocks are needed. However, if the shock's severity beyond certain level (that is usually defined by the physical processes in the item subject to a shock) results in a non-negligible damage in the 'survived' item, then we cannot go above this level of severity and should consider an option of performing additional shocks. Note that additional shocks in the framework of the specific model (9.29) do not improve the quality of a population. This can be easily seen by deriving $P_{z_b}^{(2)}$- the probability of not surviving the second shock with the same level of z_b. Using (9.27) and (9.30),

$$P_{z_b}^{(2)} = \int_{z_b}^{\infty} \pi_s(z)\mathrm{d}z = \frac{\int_{z_b}^{\infty} q(z)\pi(z)\mathrm{d}z}{\int_0^{\infty} q(z)\pi(z)\mathrm{d}z} = 0.$$

On the other hand, the general model (9.31) gives a *positive probability* of not surviving the second shock (with the same level of severity $p(z,s)$) after an item had survived the first shock:

$$P^{(2)}(s) = \int_0^{\infty} p(z,s)\pi_s(z)\mathrm{d}z = \frac{\int_0^{\infty} p(z,s)q(z,s)\pi(z)\mathrm{d}z}{\int_0^{\infty} q(z,s)\pi(z)\mathrm{d}z} > 0.$$

Therefore, when the high level of stress can negatively affect even those items that had formally passed it (did not fail), we can perform a more 'friendly' burn-in with a lower level of stress by increasing the number of shocks as opposed to the option of one shock.

Denote the posterior density after the nth shock by $\pi_s^{(n)}(z)$, where $\pi_s^{(1)}(z) = \pi_s(z)$. Then, (9.27) is generalized to:

$$\pi_s^{(n)}(z) = \frac{q^n(z,s)\pi(z)}{\int_0^{\infty} q^n(z,s)\pi(z)\mathrm{d}z}, \tag{9.38}$$

meaning that for the given $q(z,s)$, this density tends (in the sense of generalized functions) to the 'one-sided' δ-function (in the positive neighborhood of 0). Therefore, if we assume that there is no penalty (cost) for additional shocks, then obviously, we can reach the desired level of severity (the same as with one 'unfriendly' shock) with a finite number of shocks. This 'multi-shock reasoning' can be generalized to an extended model considering the relevant costs and the corresponding optimal problem. In essence, as all shocks are applied in a relatively short period of time, we are treating the sequence of shocks as one 'aggregated' shock.

In this case, the number of shocks can be considered as a measure of severity. Let s_i denote the level of severity with i shocks, $i = 1, 2, \ldots$, that is, for example, at level s_1 only one shock with severity level s is applied; at level s_2 two consecutive shocks with severity level s are applied, and so on. Let $\tilde{q}(z,s_i)(\tilde{q}(z,s_1) \equiv q(z,s))$ be the item's survival probability for this 'multi-shock structure'. Obviously, from (9.38), we have $\tilde{q}(z,s_i) = q^i(z,s)$. As

$$\frac{\tilde{q}(z,s_{i+1})}{\tilde{q}(z,s_i)} = q(z,s)$$

is decreasing in z, by Remark 9.12, we can conclude that the better quality of a population can be obtained by monotonically increasing the number of shocks. Using this property, similar results as in Theorem 9.8 can be obtained when the corresponding cost structure is considered.

Example 9.4 Consider the multiplicative model (9.26) with the constant baseline failure rate $\lambda(t,z)=z\lambda$. This is a real-life example as, e.g., many electronic components have a constant failure rate which is varying from component to component due to production instability, etc. Note that 'traditional' burn-in (i.e., for the specified time) for these heterogeneous populations was usually executed by the manufacturers especially when the items had to meet high reliability requirements (e.g., for military field usage).

Assume for simplicity that Z is also exponentially distributed (it can easily be generalized to the gamma distribution): $\Pr(Z\leq z)=1-\exp\{-\alpha z\}$. It is well known that the mixture failure rate in this case is

$$\lambda_m(t)=\frac{\int_0^\infty z\lambda\exp\{-z\lambda t\}\pi(z)\mathrm{d}z}{\int_0^\infty \exp\{-z\lambda t\}\pi(z)\mathrm{d}z}=\frac{\lambda}{\lambda t+\alpha}. \tag{9.39}$$

Consider a single shock defined by the specific $p(z)$ given by Eq. (9.29) [it is just more convenient for this particular example to use this parameterization rather than the equivalent parameterization (9.33)]. In accordance with (9.27):

$$\pi_s(z)=\frac{q(z)\pi(z)}{\int_0^\infty q(z)\pi(z)\mathrm{d}z}=\frac{1}{\int_0^{z_b}\pi(z)\mathrm{d}z}\begin{cases}\pi(z), & 0\leq z\leq z_b\\ 0, & z>z_b\end{cases}$$

$$=\frac{1}{\Pi(z_b)}\begin{cases}\pi(z), & 0\leq z_b\\ 0, & z>z_b\end{cases}.$$

Therefore, simple integration results in

$$\begin{aligned}\lambda_{ms}(t,z_b)&=\frac{\int_0^{z_b} z\lambda\exp\{-z\lambda t\}\pi(z)\mathrm{d}z}{\int_0^{z_b}\exp\{-z\lambda t\}\pi(z)\mathrm{d}z}\\ &=\frac{\lambda}{\lambda t+\alpha}\left(1-\frac{z_b(\lambda t+\alpha)}{\exp\{z_b(\lambda t+\alpha)\}-1}\right).\end{aligned} \tag{9.40}$$

It can be easily seen that $1-z_b(\lambda t+\alpha)/(\exp\{z_b(\lambda t+\alpha)\}-1)$ is increasing in z_b from 0 at $z_b=0$ to 1 at $z_b=\infty$, for all fixed $t>0$. Note that the value at $z_b=0$ should be considered only like a limit (which obviously does not belong to admissible failure rates). Thus, when $z_b\to\infty$, (9.40) tends to the value defined by Eq. (9.39). It is also clear that the general inequality (9.21) holds in this specific case. It follows from (9.30) that the probability of not surviving a shock in this specific case is:

$$P(z_b)=\int_{z_b}^{\infty}\pi(z)\mathrm{d}z=\exp\{-\alpha z_b\}.$$

In accordance with (9.36), the corresponding total expected cost function is

$$c(z_b)=c_1(z_b)+c_2(z_b),$$

where

$$c_1(z_b) = \frac{c_{sr}\exp\{-\alpha z_b\} + c_s}{1 - \exp\{-\alpha z_b\}},$$

and

$$c_2(z_b) = -(g_m + c_m)\exp\left\{-\int_0^{\tau} \frac{\lambda}{\lambda u + \alpha}\left(1 - \frac{z_b(\lambda u + \alpha)}{\exp\{z_b(\lambda u + \alpha)\} - 1}\right)\mathrm{d}u\right\} + c_m.$$

It is obvious that $c_1(z_b)$ is decreasing in z_b and its limits are ∞ and c_s at $z_b = 0$ and $z_b = \infty$, respectively. On the other hand, as $1 - z_b(\lambda t + \alpha)/(\exp\{z_b(\lambda t + \alpha)\} - 1)$ is increasing in z_b from 0 at $z_b = 0$ to 1 at $z_b = \infty$ (for all fixed $t > 0$), $c_2(z_b)$ is increasing in z_b and its limits are $-g_m$ and $-(g_m + c_m)\exp\{-\int_0^{\tau} \lambda/(\lambda u + \alpha)\mathrm{d}u\} + c_m$, at $z_b = 0$ and $z_b = \infty$, respectively.

Thus, in this case, $c(z_b)$ has its limit

$$c_s - (g_m + c_m)\exp\left\{-\int_0^{\tau} \lambda/(\lambda u + \alpha)\mathrm{d}u\right\} + c_m.$$

Consider the following illustrative numerical values: $\lambda = 1.0, \alpha = 0.1, c_{sr} = 1.0, c_s = 1.0, g_m = 300, c_m = 200$, and $\tau = 5.0$. The corresponding graph is given in Fig. 9.3.

It follows from Theorem 9.8 that there exists a finite optimal stress level $s^* < \infty$, which implies that in our example there exists a positive optimal $z_b{}^* > 0$. For the chosen numerical values, we have: $z_b{}^* = 0.165$ and $c(z_b{}^*) \approx -19.63$. This result shows that for the given values of parameters the optimal stress level is relatively large ($z_b{}^*$is small).

9.3 Burn-in for Minimizing Risks

9.3.1 Discrete Mixtures

In the previous sections, it was shown that under reasonable assumptions, shocks will eliminate weaker items with larger probabilities than strong items, and in this way the burn-in can be performed. The optimal severity of shocks for some population quality measures was also studied. In this section, we will apply this methodology to the shock burn-in that minimizes the risks of selecting items (from heterogeneous populations) with poor reliability characteristics for important missions or missions, where failures can result, e.g., in a substantial economic loss. This type of burn-in can be beneficial when the 'ordinary' time burn-in does not

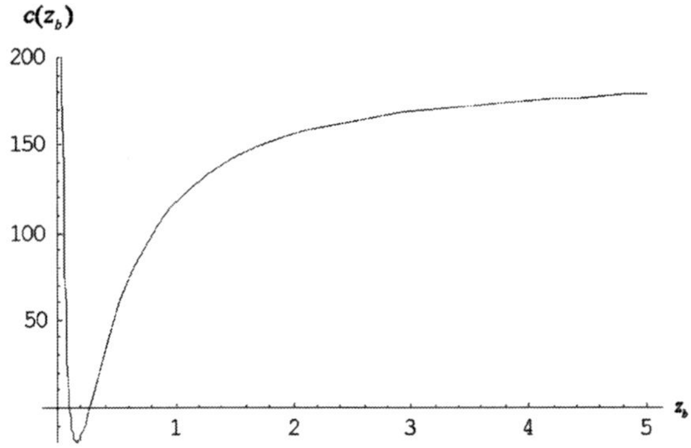

Fig. 9.3 The function $c(z_b)$

make sense (e.g., when the population failure rate is increasing), which will be illustrated by relevant examples (see also [8]). In what follows, we implicitly assume that shocks randomly occurring during 'normal' operation constitute one of the main causes of failure. Therefore, a single shock of a larger magnitude under the assumptions to be discussed can act as a method of burn-in.

Consider now the case of $n = 2$ subpopulations. For convenience, we repeat the initial setting of Sect. 9.1. First, we describe the composition of our population. Denote the lifetime of a component from the 'strong subpopulation' by T_s and its absolutely continuous Cdf, pdf, and the failure rate function by $F_1(t), f_1(t)$ and $\lambda_1(t)$, respectively. Similarly, the lifetime, the Cdf, pdf, and the failure rate function of a 'weak' component are $T_W, F_2(t), f_2(t)$ and $\lambda_2(t)$, accordingly. We define strong and weak subpopulations in the sense of the following failure rate ordering:

$$\lambda_2(t) \geq \lambda_1(t), \quad t \geq 0. \tag{9.41}$$

The initial $(t = 0)$ composition of our mixed population is as follows: the proportion of the strong items is π, whereas the proportion of the weak items is $1 - \pi$, which means that the distribution of the discrete *frailty* Z with realizations z_1 and z_2 in this case is

$$\pi(z) = \begin{cases} \pi, & z = z_1 \\ 1 - \pi, & z = z_2 \end{cases}$$

and $z_1, z_2\ (z_1 < z_2)$, correspond to the strong and the weak subpopulations, respectively. The mixture (population) survival function is

$$\overline{F}_m(t) = \pi\overline{F}_1(t) + (1 - \pi)\overline{F}_2(t).$$

Then the mixture (the *observed* or the *population*) failure rate is

$$\lambda_m(t) = \frac{\pi f_1(t) + (1 - \pi)f_2(t)}{\pi\bar{F}_1(t) + (1 - \pi)\bar{F}_2(t)} = \pi_1(t)\lambda_1(t) + \pi_2(t)\lambda_2(t), \tag{9.42}$$

where the time-dependent probabilities are

$$\pi_1(t) = \frac{\pi \bar{F}_1(t)}{\pi \bar{F}_1(t) + (1-\pi)\bar{F}_2(t)}, \quad \pi_2(t) = \frac{(1-\pi)\bar{F}_2(t)}{\pi \bar{F}_1(t) + (1-\pi)\bar{F}_2(t)}. \tag{9.43}$$

We adopt the same assumption as in Sect. 9.1:

Basic Assumption 1 The more frail (e.g., with the larger failure rate during 'normal' operation) the items are, the more susceptible they are to be 'killed' by a single shock of a larger magnitude (burn-in).

Burn-in is applied in the following way:

- *Burn-in procedure by means of shocks.* An item from our heterogeneous population is exposed to a shock. If it survives, it is considered to be ready for usage, otherwise the failed item is discarded and a new one is exposed to a shock, etc.

Let $\pi_s(z)$ denote the frailty distribution *after the (burn-in) shock* and let $\lambda_{ms}(t)$ be the corresponding mixture (observed) failure rate. Denote the probabilities of failures caused by each shock for two subpopulations as:

$$p(z) = \begin{cases} p_1, & z = z_1, \\ p_2, & z = z_2. \end{cases} \tag{9.44}$$

Then $\pi_s(z), \lambda_{ms}(t)$ and $\overline{F}_{ms}(t)$ are defined as in Sect. 9.1 [see, e.g., Eq. (9.2)].

Consider now a simple motivating example, where the shock burn-in can be effective, whereas the ordinary time burn-in will only decrease reliability characteristics of items.

Example 9.5 Let $\lambda_1(t) = 0.3t + 1$, $\lambda_2(t) = 0.6t + 2$ and $\pi = 0.60$. Then, obviously, $\lambda_2(t) \geq \lambda_1(t), \quad t \geq 0$, and the mixture failure rate $\lambda_m(t)$ given in Fig. 9.4 is strictly increasing. Therefore, the time burn-in *should not be applied* for this heterogeneous population.

Let $p_1 = 0.1$ and $p_2 = 0.8$ [see Eq. (9.44)]. Then the mixture failure rate functions before and after (lower) the shock burn-in are given in Fig. 9.5.

Therefore, the shock burn-in improves the quality (reliability) characteristics of this population.

In the following, we consider the problem of determining the optimal severity of the shock burn-in for suitable measures of risk in operation. Denote the magnitude of a shock by $s \in [0, \infty]$. Assume that the 'strength' of the component in a strong subpopulation is a continuous random variable, which is denoted by U. By 'strength' we understand here the corresponding measure of resistance to a single shock, i.e., if $s > U$, then the failure occurs. Let the Cdf, the survival function, and the failure rate function of U are denoted by $G(s), \overline{G}(s)$, and $r(s)$, respectively. Similarly, let the strength of the component in a weak subpopulation be denoted by U_W. Then, in accordance with our Basic Assumption 1, let

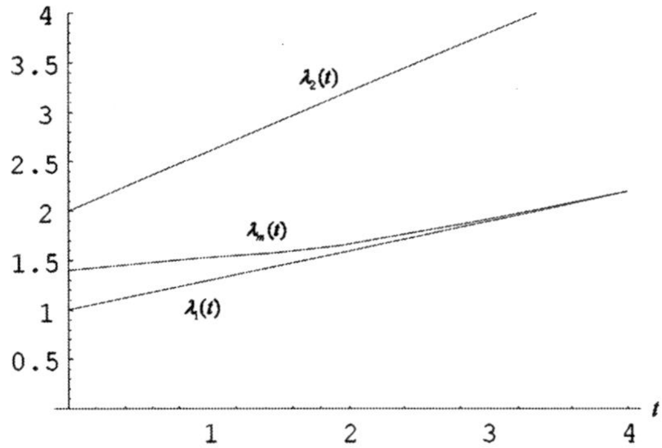

Fig. 9.4 Mixture failure rate

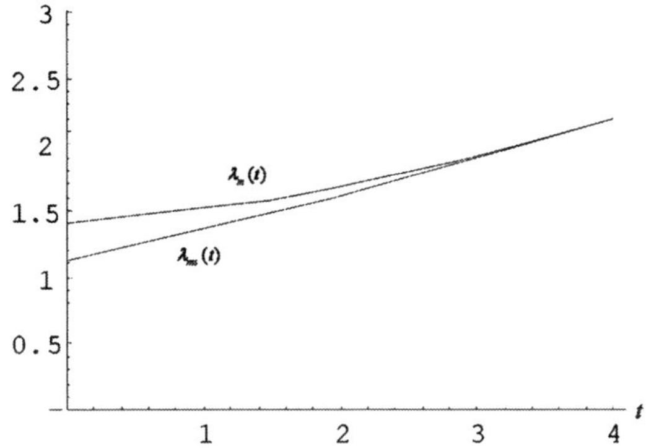

Fig. 9.5 The mixture failure rate functions before and after Shock Burn-in

$$U \geq_{st} U_W. \tag{9.45}$$

Then Eqs. (9.3) and (9.4) and the corresponding reasoning employed while deriving these equations hold.

Let an item from our population be operable at time $t > 0$ (in field operation). Then, if this is a weak item, the 'risk of instantaneous failure' is larger than that for a strong one. Therefore, a larger penalty (loss) should be imposed to the item with a larger risk. This allows us to define the following "point loss" at time t for the subpopulation i:

$$L_i(t) = g((\lambda_i(t)),\ i = 1, 2, \tag{9.46}$$

where $g(\cdot)$ is a strictly increasing function of its argument.

The following criterion of optimization of shock's severity level stems from definition (9.46):

Criterion 1 Find s^* which minimizes

$$\overline{L}(t|s) = \sum_{i=1}^{2} L(\lambda_i(t), 0)\pi_s(z_i|0) = \sum_{i=1}^{2} g(\lambda_i(t))\pi_s(z_i|0), \text{for all } t \geq 0. \tag{9.47}$$

Observe that $\overline{L}(t|s)$ in (9.47) corresponds to the mean loss at time t of an item which has experienced the shock burn-in with the corresponding magnitude s. Suppose that the subpopulations are ordered as in (9.41). Then, it is easy to see that maximization of the proportion of the strong components, $\pi_s(z_1|0) \equiv \pi_s$ minimizes (9.47) for all $t \geq 0$. Therefore, as follows from (9.45), the problem is the same as maximizing

$$\pi_s = \frac{(1-G(s))\pi}{(1-G(s))\pi + (1-G(\rho(s)))(1-\pi)},$$

which is the same as finding s^* that satisfies

$$s^* = \arg \inf_{s \in [0,\infty]} \lambda_{ms}(t;s), \text{for all fixed } t \geq 0.$$

The corresponding result can be found in Cha and Finkelstein [6]:

Theorem 9.9 [6] *Let $\lambda_1(t) \leq \lambda_2(t)$, for all $t \geq 0$. Then the optimal s^* is the value which maximizes $R(\rho(s)) - R(s)$, where $R(s) \equiv \int_0^s r(u)\mathrm{d}u$. In particular,*

(i) *If $r(s)$ is increasing and $\rho'(s) > 1$, then $s^* = \infty$.*
(ii) *If $\frac{\rho'(s)r(\rho(s))}{r(s)} > 1$, for $s < s_0$, and $\frac{\rho'(s)r(\rho(s))}{r(s)} < 1$, for $s > s_0$, then $s^* = s_0$.*

Consider now the second criterion. Let τ be the usage (mission) time for our components. Then, as the point loss varies during mission time, it should be averaged, i.e., it should be integrated for the mission interval (and then divided by the length of the interval) to measure the 'overall risk' during the mission. Thus, the average loss during the operational interval for subpopulation i can be defined as

$$\frac{\int_0^\tau L_i(t)\mathrm{d}t}{\tau} = \frac{\int_0^\tau g(\lambda_i(t))\mathrm{d}t}{\tau}, i = 1,2.$$

As the selection of a component from a heterogeneous population is made just after the shock burn-in and the corresponding proportions after the burn-in are given by $\pi_s(z_i|0), i = 1,2$, the mean loss for our mixture population (after burn-in) is

$$\Psi(s) = \sum_{i=1}^{2} \frac{\int_0^\tau L(\lambda_i(t),0)\mathrm{d}t}{\tau} \cdot \pi_s(z_i|0) = \sum_{i=1}^{2} \frac{\int_0^\tau g(\lambda_i(t))\mathrm{d}t}{\tau} \cdot \pi_s(z_i|0). \tag{9.48}$$

Criterion 2 Find s^* which minimizes $\Psi(s)$.

Similar to the optimization based on **Criterion 1**, as the subpopulations are ordered in the sense of failure rate ordering, Theorem 9.9 could be also applied, which is illustrated by the following example.

Example 9 Let $\lambda_1(t) = 1.2 - \exp\{-1.2t\} + 0.01t$, $\lambda_2(t) = 1.4\exp\{-0.08t\} + 1.2 + 0.01t$, with $\pi = \pi_1(0) = 0.80$. Then $\lambda_2(t) \geq \lambda_1(t)$, $t \geq 0$ and the

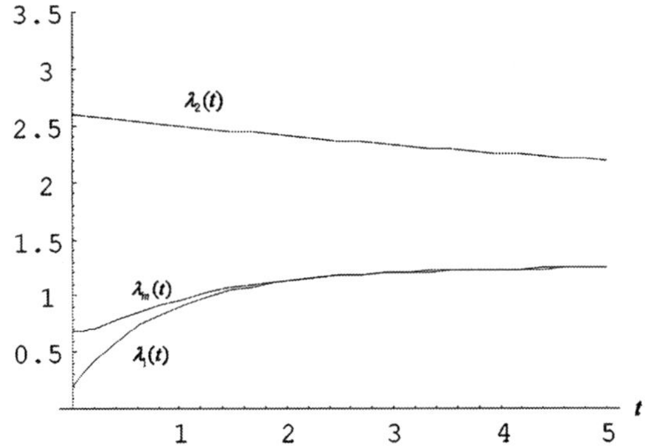

Fig. 9.6 Mixture failure rate

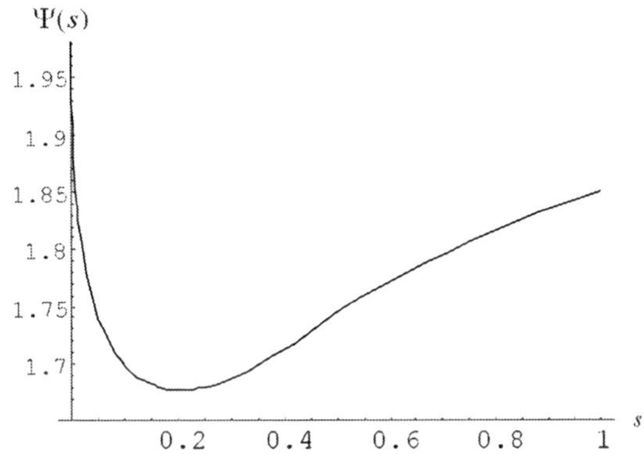

Fig. 9.7 $\Psi(s)$

corresponding strictly increasing mixture failure rate $\lambda_m(t)$ is given in Fig. 9.6. Let the failure rate of $G(s)$ be $r(s) = \exp\{-s\} + 1, \rho(s) = \sqrt{s}, 0 \le s \le 1/2; \rho(s) = s + (1/\sqrt{2} - 1/2)\exp(0.5 - s), s \ge 1/2.$ and $\tau = 3$. Then for $g(x) = x^2$, $\Psi(s)$ is given in Fig. 9.7. It can be numerically shown that there exists s_0 such that $\frac{\rho'(s)r(\rho(s))}{r(s)} > 1$, for $s < s_0$, and $\frac{\rho'(s)r(\rho(s))}{r(s)} < 1$, for $s > s_0$, and, as illustrated by Fig. 9.7, there exists the finite optimal severity level ($s^* \approx 0.20$). Note that, as the failure rates are ordered, minimization of $\Psi(s)$ in (9.48) is equivalent to maximization of the proportion of the strong components, $\pi_s(z_1|0) \equiv \pi_s$. Therefore, the optimal severity in this case does not depend on the value of τ and this is also the optimal severity level for **Criterion 1**.

Note that the proportion of the strong subpopulation after the shock burn-in is $\pi_s \approx 0.86$. (compare with 0.80 before burn-in). In addition, it can be shown graphically that the mixture failure rate in this case has also been decreased for all $t \ge 0$, as in Fig. 9.5.

9.3.2 Continuous Mixtures

As in the previous parts of this chapter, consider now the case of the 'continuous' mixing model for a heterogeneous population, i.e.,

$$F_m(t) = \int_0^\infty F(t,z)\pi(z)\mathrm{d}z,\ f_m(t) = \int_0^\infty f(t,z)\pi(z)\mathrm{d}z, \tag{9.49}$$

where $F(t,z) \equiv F(t|z), f(t,z) \equiv f(t|z)$ are the Cdf and the pdf of subpopulations indexed (conditioned) by the frailty parameter Z and $\pi(z)$ is the pdf of Z with support in $[0,\infty)$ Then the mixture failure rate $\lambda_m(t)$ is defined as in (9.15), (9.16).

As in the discrete case, let our subpopulations be ordered in the sense of the failure (hazard) rate ordering:

$$\lambda(t,z_1) \le \lambda(t,z_2), \quad z_1 < z_2, \forall z_1, z_2 \in [0,\infty),\ t \ge 0. \tag{9.50}$$

We choose an item from a heterogeneous population at random (or alternatively, our item is described by the unobserved frailty parameter Z). Thus, the mixture (population) failure rate of this item is $\lambda_m(t)$. Throughout this subsection, similar to the Basic Assumption 1, the impact of a shock is described by the following *general assumption* [6].

Basic Assumption 2 A shock either 'kills' an item with a given probability or does not change its stochastic properties with the complementary probability. The more 'frail' (e.g., with the larger failure rate during normal operation) an item is, the larger is the probability that a single burn-in shock will 'kill' it.

As we implicitly assume that shocks during normal operation constitute one of the main causes of failure, the above assumption can be justified. Note that, clearly, the burn-in procedure is the same as in the discrete case. The described setting can be defined probabilistically in the following way: Let $\pi_s(z)$ denote the pdf of the frailty Z_s (with support in $[0,\infty)$) after a shock and let $\lambda_{ms}(t)$ be the corresponding mixture failure rate. In accordance with (9.49):

$$\lambda_{ms}(t) = \int_0^\infty \lambda(t,z)\pi_s(z|t)\mathrm{d}z, \tag{9.51}$$

where, similar to (9.50), $\pi_s(z|t)$ is defined by the right-hand side of (9.50) with $\pi(z)$ substituted by $\pi_s(z)$.

Let $q(z)$ be "the survival probability" of an item with frailty z after the shock. Then $\pi_s(z)$ is [10]:

$$\pi_s(z) = \frac{q(z)\pi(z)}{\int_0^\infty q(z)\pi(z)\mathrm{d}z}, \tag{9.52}$$

where, in accordance with Basic Assumption 2, $q(z)$ is a decreasing function of z and therefore, $\pi_s(z)/\pi(z)$ is decreasing [the denominator of (9.52) is just a normalizing constant for the density]. That is, population frailties before $(\pi(z))$ and after $(\pi_s(z))$ a shock are ordered in the sense of the likelihood ratio (Sect. 2.8)

$$Z \geq_{LR} Z_s.$$

Define the functions $p(z)$ and $q(z)$ as functions of the frailty variable z and the severity parameter $s \in [0, \infty)$, $p(z, s)$, and $q(z, s)$. Assume that $q(z, s)$ is decreasing in z for each fixed s and is decreasing in s for each z. Denote the corresponding failure rate and survival functions by $\lambda_{ms}(t; s)$, and $\overline{F}_{ms}(t; s)$, respectively. Similar to (9.52) and (9.50):

$$\pi_s(z, s) = \frac{q(z, s)\pi(z)}{\int_0^{\infty} q(u, s)\pi(u)du}, \pi_s(z, s|t) \equiv \pi_s(z, s)\frac{\bar{F}(t, z)}{\int_0^{\infty} \bar{F}(t, u)\pi_s(u, s)du}. \quad (9.53)$$

For this continuous mixture case, the criteria defined for the discrete case can obviously be generalized as follows:

Criterion 1C Find s^* which minimizes

$$\overline{L}(t|s) = \int_0^{\infty} g(\lambda(t, z))\pi_s(z, s)dz, \text{ for all } t \geq 0.$$

Criterion 2C Find s^* which minimizes

$$\Psi(s) = \int_0^{\infty} \frac{\int_0^{\tau} g(\lambda(t, z))dt}{\tau} \cdot \pi_s(z, s)dz.$$

The following example illustrates the application of **Criterion 2C**.

Example 9.7 Suppose that $\lambda(t, z) = 0.1z\exp\{0.1t\} + 0.02t + 1$, and let Z be exponentially distributed with parameter $\theta = 0.5$. For brevity, we omit the graph showing that the mixture failure rate is strictly increasing in this case. Let $q(z, s) = 0.95e^{-zs} + 0.05$, $\tau = 3.0$, and $g(x) = x^2$. Then $\Psi(s)$ is given in Fig. 9.8.

Thus the optimal shock severity is $s^* \approx 2.03$. As in Example 9.6, the shock burn-in in this case has decreased the mixture failure rate (we omit the corresponding figure for brevity), which obviously cannot be attained by the ordinary time burn-in, as the mixture failure rate of our population is increasing. The frailty distributions before and after burn-in are given in Fig. 9.9.

It can be seen that the frailty density before the shock is much flatter allowing larger proportions of items with higher failure rates (weaker).

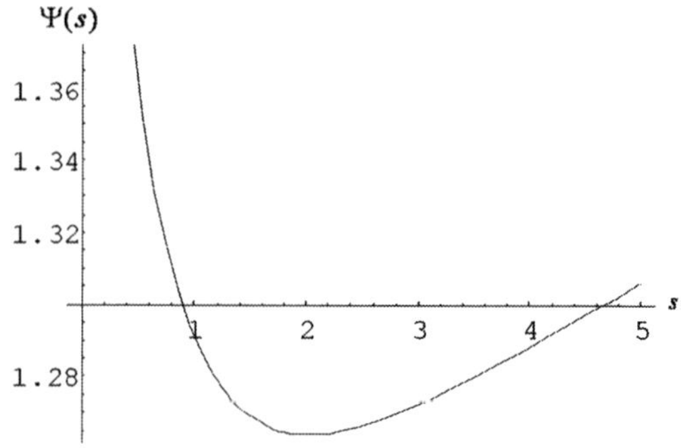

Fig. 9.8 $\Psi(s)$

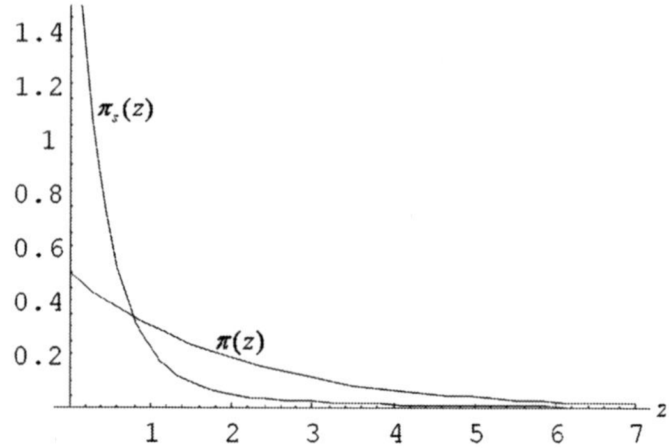

Fig. 9.9 Frailty densities before and after Burn-in

9.3.3 Optimal Shock Burn-in Based on Conservative Measures

Sometimes, failures of items may result in catastrophic or disastrous events. For example, failures in jet engines of aircrafts or those in gas safety valves may cause fatal consequences. Similarly, failures during important missions can cause huge economic loss. In these cases, we need to define some 'marginal quality' of the population that describes in some sense the "worst scenario". That is, if this worst scenario quality is still acceptable then the quality of our population as a whole is considered to be satisfactory. Thus, the marginal quality can be used as a conservative (safe) measure (or bound) for the quality of a population in such cases.

In this subsection, we consider the optimal burn-in procedure which optimizes the conservative measures and modify the approach that was developed in Cha and Finkelstein [7] (see also Sect. 8.3) for the time burn-in with respect to the shock burn-in. Obviously, this refers only to the continuous mixtures case.

Denote by $\Pi_s(z, s)$, the conditional distribution function which corresponds to $\pi_s(z, s)$, defined in (9.53). Define the following measure:

$$\lambda_\alpha(t|s) = \lambda(t, z(\alpha|s)),\ t \geq 0, \tag{9.54}$$

where $z(\alpha|s) \equiv \inf\{z : \Pi_s(z,s) \geq \alpha\}$ and α is usually close to 1 (e.g., 0.9 or 0.95). Thus, $\lambda_\alpha(t|s)$ is the (residual) failure rate of an item after a shock with magnitude s, which corresponds to the αth percentile $z(\alpha|s)$ of the conditional distribution of frailty $\Pi_s(z,s)$. When α is close to 1, this operation describes the αth worst scenario, which is the 'αth worst subpopulation' in the defined way. Based on the above setting, we can define the αth worst mean remaining lifetime (MRL) of the population after the shock burn-in with severity s:

$$M_\alpha(s) \equiv \int_0^\infty \exp\{-\int_0^t \lambda_\alpha(u|s)du\}dt.$$

Therefore, the following criterion can be applied:

Criterion 3 Determine the optimal severity s^* as the minimal severity s such that $M_\alpha(s) \geq m_r$, where m_r is the MRL that corresponds to the αth worst scenario.

Implementation of this approach can be clearly seen while considering the following meaningful example.

Example 9.8 Let the conditional failure rate and the mixing distribution be $\lambda(t,z) = z$ and $\pi(z) = \theta\exp\{-\theta z\}$, respectively. It is well known (see e.g., [2] that the mixture failure rate strictly decreases in this case. Let $q(z,s) = e^{-za(s)}$, where $a(s)$ is nonnegative strictly increasing function with $a(0) = 0$ and $\lim_{s\to\infty} a(s) = \infty$. In accordance with (9.53):

$$\Pi_s(z,s) = 1 - \exp\{-(\theta + a(s))z\}.$$

Then

$$z(\alpha|s) = -\frac{\ln(1-\alpha)}{\theta + a(s)},$$

and [see (9.54)]:

$$\lambda_\alpha(t|s) = -\frac{\ln(1-\alpha)}{\theta + a(s)},\ t \geq 0.$$

The criterion for the shock burn-in is as follows: Find the minimum shock severity such that, after burn-in, the mean (residual) lifetime of the lower $(1-\alpha)\,\%$ quality of items is, at least, m. As the lifetimes are exponential (for the fixed frailty), this MRL is, obviously,

$$M_\alpha(s) = 1/z(\alpha|s) = -(\theta + a(s))/\ln(1-\alpha).$$

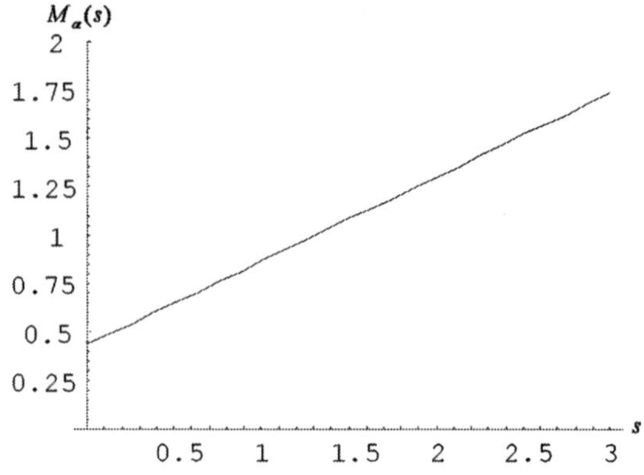

Fig. 9.10 $M_\alpha(s)$ for $\alpha = 9$, $\theta = 1.0$, $a(s) = s$

Let $\alpha = 9, \theta = 1.0$ and $a(s) = s$. Then the corresponding linear function is given in Fig. 9.10.

If, for instance, $m = 1.25$, then the corresponding minimum shock severity: $s^* \approx 1.88$.

The conservative measure (9.54) can be modified (generalized) to account for the *average* of the lower $(1-\alpha)\,\%$ quality of items in the survived population after the shock with severity s. Then, after the shock with severity s, the *initial conditional frailty distribution* [which corresponds to $\pi(z)$ in (9.50)] for the items whose quality is lower than $(1-\alpha)\,\%$ is given by

$$\frac{\pi_s(z,s)}{1-\alpha}, z(\alpha|s) \le z \le \infty,$$

where, as previously, $z(\alpha|s) \equiv \inf\{z : \prod_s(z,s) \ge \alpha\}$. Thus the conditional density after time t (in usage), which corresponds to $\pi(z|t)$ in (9.51) is

$$\pi_\alpha(z,s|t) \equiv \frac{\pi_s(z,s)}{1-\alpha} \frac{\bar{F}(t,z)}{\int_{z(\alpha|s)}^{\infty} \bar{F}(t,z) \frac{\pi_s(z,s)}{1-\alpha} \mathrm{d}z}, z(\alpha|s) \le z \le \infty.$$

Therefore, the mixture failure rate for the items in the survived population whose quality is lower than $(1-\alpha)\,\%$ after the shock with severity s is obtained by

$$\lambda_m(t|s,\alpha) = \int_{z(\alpha|s)}^{\infty} \lambda(t,z)\pi_\alpha(z,s|t)\mathrm{d}z.$$

Example 9.9 (Example 9.8 Continued) As $z(\alpha|s) = -\ln(1-\alpha)/(\theta + a(s))$ and

$$\int_{z(\alpha|s)}^{\infty} \bar{F}(t,z) \frac{\pi_s(z,s)}{1-\alpha} \mathrm{d}z = \frac{1}{(1-\alpha)} \cdot \frac{\theta + a(s)}{\theta + a(s) + t} \cdot (1-\alpha)^{\frac{\theta + a(s) + t}{\theta + a(s)}},$$

we have,

$$\pi_\alpha(z,s|t) \equiv \frac{\pi_s(z,s)}{1-\alpha}\frac{\bar{F}(t,z)}{\int_{z(\alpha|s)}^{\infty}\bar{F}(t,z)\frac{\pi_s(z,s)}{1-\alpha}\mathrm{d}z}$$
$$= (\theta + a(s) + t)\cdot(1-\alpha)^{-\frac{\theta+a(s)+t}{\theta+a(s)}}\cdot\exp\{-(\theta+a(s)+t)z\}.$$

Thus

$$\lambda_m(t|s,\alpha) = \int\limits_{z(\alpha|s)}^{\infty}\lambda(t,z)\pi_\alpha(z,s|t)\mathrm{d}z = -\frac{\ln(1-\alpha)}{\theta+a(s)} + \frac{1}{\theta+a(s)+t}, t\geq 0.$$

The criterion for the shock burn-in is as follows: Find the minimum shock severity such that, after burn-in, the mean (residual) lifetime of the items whose quality is lower than $(1-\alpha)\,\%$ is at least m. Then we have to obtain the MRL of the items whose quality is lower than $(1-\alpha)\,\%$ after the shock burn-in at each severity level s, which is given by

$$\int\limits_0^{\infty}\exp\left\{-\int\limits_0^{x}\lambda_m(t|s,\alpha)\mathrm{d}t\right\}\mathrm{d}x,$$

Let $\alpha = 9, \theta = 1.0$ and $a(s) = s$ and $m = 1.25$. Then it can be easily found numerically that the optimal shock severity is $s^* \approx 2.47$.

9.4 Burn-in for Systems in Environment with Shocks

Burn-in procedures are usually applied to items with large initial failure rate which operate under static operational environment. Similar to previous sections, we consider shocks as a method of burn-in, but in this section we assume that there are two competing risks causes of failure—the 'usual' one (in accordance with aging processes in a system) and environmental shocks. We also suggest a new type of burn-in via the *controlled* (laboratory) test shocks and consider the problem of obtaining the optimal level (severity) of these shocks that minimizes the overall expected cost (burn-in + field use). Furthermore, also to minimize these costs, we combine the conventional burn-in procedure with burn-in via shocks in one unified model. We start with the general description of the basic stress-strength model. In Sect. 4.7 and Sect. 4.10.3 we have already used some specific cases of this model for discussing the operation of thinning of point processes and processes with delay and cure.

9.4.1 Strength–Stress Shock Model

In this subsection, we consider a rather general stress-strength shock model, which will be used as an important supplementary result for considering burn-in problems of the subsequent subsections.

As in Chap. 4, consider a system subject to the nonhomogeneous Poisson process (NHPP) of shocks $N(t), t \geq 0$, with rate $\lambda(t)$ and arrival (waiting) times $T_i, i = 1, 2, \ldots$. Let S_i denote the magnitude (stress) of the ith shock. Assume that $S_i, i = 1, 2, \ldots$ are i.i.d. random variables with the common Cdf $M_f(s) = \Pr(S_i \leq s)$ $(\overline{M}_f(s) \equiv 1 - M_f(s))$ and the corresponding pdf $m_f(s)$. Let U be a random strength of the system with the corresponding Cdf, Sf, pdf, and FR $G_U(u)$, $\overline{G}_U(u)$, $g_U(u)$ and $r_U(u)$, respectively. For each $i = 1, 2, \ldots$, the operable system survives if $S_i \leq U$ and fails if $S_i > U$, '*independently of everything else*'.

Let T be the lifetime of the system described above and $r(t)$ be the corresponding failure rate function, which will be derived in the rest of this subsection. Then the following theorem presents the formal and a more detailed proof of Eq. (4.50):

Theorem 9.10 *The failure rate function of the system lifetime $r(t)$ is given by*

$$r(t) = p(t)\lambda(t), \tag{9.55}$$

where

$$p(t) \equiv \frac{\int_0^\infty \int_0^v \exp\{-\overline{M}_f(r)\int_0^t \lambda(x)\mathrm{d}x\} \cdot g_U(r)\mathrm{d}r\, m_f(v)\mathrm{d}v}{\int_0^\infty \exp\{-\overline{M}_f(r)\int_0^t \lambda(x)\mathrm{d}x\} g_U(r)\mathrm{d}r}.$$

Proof Observe that

$$\begin{aligned} &P(T > t|N(s), 0 \leq s \leq t, S_1, S_2, \ldots, S_{N(t)}) \\ &\quad = P(U > \max\{S_1, S_2, \ldots, S_{N(t)}\}) \\ &\quad = \int_0^\infty (M_f(r))^{N(t)} g_U(r)\mathrm{d}r. \end{aligned}$$

Thus,

$$\begin{aligned} P(T > t) &= \int_0^\infty \left(\sum_{n=0}^\infty (M_f(r))^n \frac{(\Lambda(t))^n}{n!} \exp\{-\Lambda(t)\}\right) g_U(r)\mathrm{d}r \\ &= \int_0^\infty \exp\{-(1 - M_f(r))\Lambda(t)\} g_U(r)\mathrm{d}r \\ &= \int_0^\infty \exp\{-\overline{M}_f(r)\Lambda(t)\} g_U(r)\mathrm{d}r, \end{aligned}$$

where $\Lambda(t) \equiv \int_0^t \lambda(u)\mathrm{d}u$. The corresponding failure rate is

$$\begin{aligned} r(t) &= -\frac{\mathrm{d}}{\mathrm{d}t} \ln P(T > t) \\ &= \frac{\int_0^\infty \overline{M}_f(r) \exp\{-\overline{M}_f(r)\Lambda(t)\} g_U(r)\mathrm{d}r \cdot \lambda(t)}{\int_0^\infty \exp\{-\overline{M}_f(r)\Lambda(t)\} g_U(r)\mathrm{d}r} \\ &= \frac{\int_0^\infty \int_r^\infty m_f(v)\mathrm{d}v \exp\{-\overline{M}_f(r)\Lambda(t)\} g_U(r)\mathrm{d}r \cdot \lambda(t)}{\int_0^\infty \exp\{-\overline{M}_f(r)\Lambda(t)\} g_U(r)\mathrm{d}r} \\ &= \frac{\int_0^\infty \int_0^v \exp\{-\overline{M}_f(r) \int_0^t \lambda(x)\mathrm{d}x\} \cdot g_U(r)\mathrm{d}r\, m_f(v)\mathrm{d}v}{\int_0^\infty \exp\left\{-\overline{M}_f(r) \int_0^t \lambda(x)\mathrm{d}x\right\} g_U(r)\mathrm{d}r} \lambda(t). \end{aligned}$$

□

The expression for $p(t)$ is formally rather cumbersome, but it has a simple and meaningful probabilistic meaning, which is shown in the following remark.

Remark 9.14 Observe that

$$P(T > t | N(t) = n, U = u) = P(u \geq \max\{S_1, S_2, \ldots, S_n\}) = (M_f(u))^n.$$

Thus,

$$\begin{aligned} P(T > t, U > u) &= \int_u^\infty \sum_{n=0}^\infty (M_f(r))^n \frac{\left(\int_0^t \lambda(x)\mathrm{d}x\right)^n}{n!} \exp\left\{-\int_0^t \lambda(x)\mathrm{d}x\right\} \cdot g_U(r)\mathrm{d}r \\ &= \int_u^\infty \exp\left\{-\overline{M}_f(r) \int_0^t \lambda(x)\mathrm{d}x\right\} \cdot g_U(r)\mathrm{d}r, \end{aligned}$$

and

$$P(U > u | T > t) = \frac{\int_u^\infty \exp\{-\overline{M}_f(r) \int_0^t \lambda(x)dx\} \cdot g_U(r)\mathrm{d}r}{\int_0^\infty \exp\{-\overline{M}_f(r) \int_0^t \lambda(x)\mathrm{d}x\} g_U(r)\mathrm{d}r}.$$

Therefore, it can be seen that

$$p(t) = \int_0^\infty P(U < v | T > t) m_f(v) \mathrm{d}v. \tag{9.56}$$

As U is a random strength of our system and $m_f(v)$ is the pdf of the magnitude of any shock, $p(t)$ can be interpreted as the probability of a failure under a shock that had occurred at time t given that it did not occur before. The important feature of (9.56) is conditioning on the event $T > t$, which obviously has the Bayesian interpretation via the updating of the distribution of the system's strength. That is, even though random strength does not actually change, its distribution (on the condition that $T > t$) is updated as t increases, which eventually yields a time-dependent $p(t)$. This conditioning was overlooked in Cha and Finkelstein [7], which resulted in $p = \int_0^\infty P(U < v) m_f(v) \mathrm{d}v$. Relationship (9.56) will be very useful for our further discussion.

9.4.2 Optimal Level of Shock's Severity

We consider a system (a component, an item) that operates in an environment with shocks. Assume that in the absence of shocks, it can fail in accordance with the baseline distribution $F_0(t)$ with the corresponding failure rate function $r_0(t)$. In addition to this type of the 'baseline' failure, the environmental shocks can also cause system's failure. Assume that each shock, with probability $p(t)$ results in immediate system's failure and with probability $q(t) = 1 - p(t)$ it does not cause any change in the system. We use the same notation, as in (9.56), because $p(t)$ in (9.56) as an 'overall characteristic' can be also obviously interpreted in this way. If the shocks follow the NHPP with intensity $\lambda(t)$, then it is well known that the survival function of the system for this setting is given by

$$\begin{aligned} P(T > t) &= \exp\left(-\int_0^t r_0(u) \mathrm{d}u \right) \exp\left(-\int_0^t p(u)\lambda(u) \mathrm{d}u \right) \\ &= \exp\left(-\int_0^t r_0(u) + p(u)\lambda(u) \mathrm{d}u \right), t \geq 0, \end{aligned}$$

and thus the resulting failure rate is

$$r(t) = r_0(t) + p(t)\lambda(t). \tag{9.57}$$

Coming back to the burn-in setting, as in Sect. 9.4.1, we now further assume that the magnitude (stress) of the ith shock $S_i, i = 1, 2, \ldots$ are i.i.d. random variables with the common Cdf $M_f(s) = \Pr(S_i \le s)$ $\left(\overline{M}_f(s) \equiv 1 - M_f(s)\right)$ and the corresponding pdf $m_f(s)$. For each $i = 1, 2, \ldots,$ the operable system survives if $S_i \le U$ and fails if $S_i > U$, *independently of everything else,* where U is the random strength of the system. When we apply the shock of the controlled magnitude s during burn-in, this means that the strength of the component that had passed it is larger than s, and the distribution of the *remaining strength* U_s (given that the strength is larger than s) is

$$G_U(u|s) \equiv \Pr[U \le u | U > s] = 1 - \overline{G}(u)/\overline{G}(s), u > s.$$

Let T_s be the lifetime of the system that has survived the shock burn-in with the controlled magnitude s. Then, in accordance with the discussion in Sect. 9.4.1 and the result given by (9.55), the failure rate in (9.57) should now be modified to

$$r(t,s) = r_0(t) + p(s,t)\lambda(t), \tag{9.58}$$

where

$$\begin{aligned} p(s,t) &= \frac{\int_0^\infty \int_0^v \exp\{-\overline{M}_f(r)\int_0^t \lambda(x)\mathrm{d}x\} \cdot g_U(r|s)\mathrm{d}r\, m_f(v)\mathrm{d}v}{\int_0^\infty \exp\{-\overline{M}_f(r)\int_0^t \lambda(x)\mathrm{d}x\} g_U(r|s)\mathrm{d}r} \\ &= \frac{\int_s^\infty \int_s^v \exp\{-\overline{M}_f(r)\int\limits_0^t \lambda(x)dx\} \cdot g_U(r)dr\, m_f(v)dv}{\int_s^\infty \exp\{-\overline{M}_f(r)\int_0^t \lambda(x)\mathrm{d}x\} g_U(r)\mathrm{d}r}, \end{aligned} \tag{9.59}$$

and $g_U(u|s)$ is the corresponding pdf of $G_U(u|s)$, which is given by

$$g_U(u|s) = \begin{cases} 0, & \text{if } u \le s \\ \frac{g_U(u)}{\overline{G}_U(s)}, & \text{if } u > s \end{cases}.$$

Therefore, similar to (9.56), Eq. (9.59) can be written in a compact and a meaningful way (via the corresponding mixture) as

$$p(s,t) = \int\limits_0^\infty I(v \in [s,\infty)) P(U_s < v | T > t) m_f(v)\mathrm{d}v, \tag{9.60}$$

where

$$P(U_s < v | T > t) = \frac{\int_s^v \exp\{-\overline{M}_f(r)\int_0^t \lambda(x)\mathrm{d}x\} \cdot g_U(r)\mathrm{d}r}{\int_s^\infty \exp\{-\overline{M}_f(r)\int_0^t \lambda(x)\mathrm{d}x\} g_U(r)\mathrm{d}r}.$$

and the indicator $I(v \in [s,\infty))$ accounts for the fact that after the shock burn-in with magnitude s, the system's strength with probability 1 is larger than s.

In order to justify the shock burn-in, we must show that $p(s,t)$ in (9.59) is decreasing in s for each fixed t. Thus, by increasing the magnitude of the burn-in shock, we decrease the corresponding failure rate in (9.58). This property, which is important for our reasoning, is proved by the following simple theorem:

Theorem 9.11 *The function $p(s,t)$ is strictly decreasing in s for each fixed t.*

Proof Observe that

$$\begin{aligned}\frac{\partial}{\partial s}P(U_s<v|T>t) &= \frac{1}{\left(\int_s^\infty \exp\{-\overline{M}_f(r)\int_0^t \lambda(x)\mathrm{d}x\}g_U(r)\mathrm{d}r\right)^2}\\ &\times\left[-\exp\left\{-\overline{M}_f(s)\int_0^t \lambda(x)\mathrm{d}x\right\}g_U(s)\cdot\int_s^\infty \exp\left\{-\overline{M}_f(r)\int_0^t \lambda(x)\mathrm{d}x\right\}g_U(r)\mathrm{d}r\right.\\ &\left.+\exp\left\{-\overline{M}_f(s)\int_0^t \lambda(x)\mathrm{d}x\right\}g_U(s)\cdot\int_s^v \exp\left\{-\overline{M}_f(r)\int_0^t \lambda(x)\mathrm{d}x\right\}g_U(r)\mathrm{d}r\right]<0.\end{aligned}$$

This implies that $P(U_s<v|T>t)$ is strictly decreasing in s for all fixed v and t. Observe that the indicator in (9.60) is also strictly decreasing in s for all fixed v. Therefore, it can be concluded that $p(s,t)$ is strictly decreasing in s for each fixed t.□

Based on the new results obtained above, we now reconsider some of the previous burn-in models.

An item is chosen at random from our population and is exposed to a shock of magnitude s. If it survives, it is considered to be ready for usage, otherwise the failed item is discarded and the new one is chosen from the population, etc. This procedure is repeated until the first survived item is obtained. Let c_{sr} be the shop replacement cost and c_s be the cost for conducting a single shock. Let $c_1(s)$, as a function of s, be the expected cost for eventually obtaining a component which has survived a shock. Then

$$c_1(s)=\frac{c_s+c_{sr}G(s)}{\overline{G}(s)}=-c_{sr}+\frac{c_s+c_{sr}}{\overline{G}(s)},\tag{9.61}$$

where $1/\overline{G}(s)$ is the total number of trials until the first 'success'.

Let K be the gain for the unit of time during the mission time. Then the expected gain during field operation (until failure) is given by

$$c_2(s)=-K\left(\int_0^\infty \exp\left\{-\int_0^t (r_0(u)+p(s,u)\lambda(u))\mathrm{d}u\right\}\mathrm{d}t\right)$$

and the total expected cost $c(s)$ is

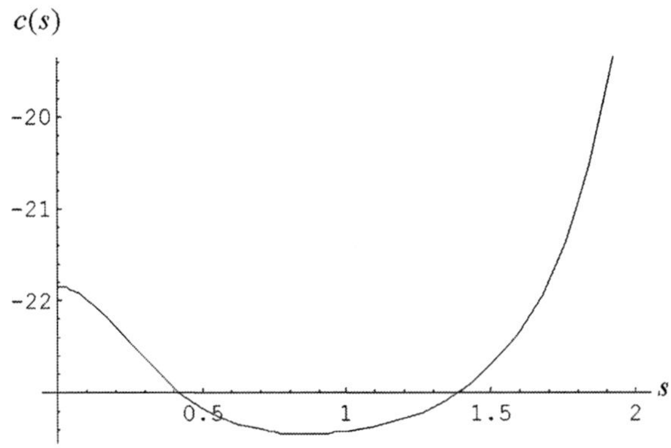

Fig. 9.11 Graph for $c(s)$ (Weibull $G(s)$)

$$\begin{aligned} c(s) &= c_1(s) + c_2(s) \\ &= -c_{sr} + \frac{c_s + c_{sr}}{\overline{G}(s)} - K\left(\int_0^\infty \exp\left\{-\int_0^t (r_0(u) + p(s,u)\lambda(u))\mathrm{d}u\right\}\mathrm{d}t\right), \end{aligned} \tag{9.62}$$

where $p(s, u)$ is given by (9.59). The function $c_1(s)$ is strictly increasing to infinity and $c_2(s)$ is strictly decreasing (Theorem 9.11) to $-K\mu_0$, where μ_0 is the mean time to failure, which corresponds to the distribution with the failure rate $r_0(t)$. Therefore, there should be a *finite* optimal severity. Then, based on (9.62), the optimal severity level s^* that satisfies

$$s^* = \arg \min_{s \in [0,\infty]} c(s)$$

can be obtained.

In the following example, the strength of a system is described by the Weibull distribution.

Example 9.10 Assume that $\overline{G}_U(u) = \exp\{-u^2\}, u \geq 0, \overline{M}_f(s) = \exp\{-6s\}$, $s \geq 0, \lambda(t) = 1, t \geq 0$, and $r_0(t) = 0.06t + 0.2$, $t \geq 0$. Let $c_{sr} = 0.1, c_s = 0.01$, and $K = 8.0$.

Optimal severity in this case is given by $s^* = 0.86$ and the corresponding minimum cost is $c(s^*) = -23.46$ (Fig. 9.11).

Similar reasoning holds when our gain is defined by the success of the mission during the fixed interval of time τ. Let:

- The cost c_m is incurred by the event $\{T_s \leq \tau\}$ (Failure of the Mission);
- The gain g_m results from the event $\{T_s > \tau\}$ (Success of the Mission).

Then the burn-in costs are the same as in (9.61), whereas the expected cost during field operation, $c_2(s)$, is given by

$$c_2(s) = -g_m\left(\exp\left\{-\int\limits_0^{\tau}(r_0(u)+p(s,u)\lambda(u))\mathrm{d}u\right\}\right) + c_m\left(1-\exp\left\{-\int\limits_0^{\tau}(r_0(u)+p(s,u)\lambda(u))\mathrm{d}u\right\}\right)$$

$$= -(g_m + c_m)\left(\exp\left\{-\int\limits_0^{\tau}(r_0(u)+p(s,u)\lambda(u))\mathrm{d}u\right\}\right) + c_m.$$

It is clear that $c_2(s)$ is strictly decreasing to

$$-(g_m + c_m)\left(\exp\left\{-\int\limits_0^{\tau} r_0(u)\mathrm{d}u\right\}\right) + c_m,$$

and all further considerations are similar to those when the gain is proportional to the mean time to failure.

9.4.3 Burn-in Procedure Combining Shock and Conventional Burn-in

In this subsection, we will deal with the combined burn-in procedures considered in Cha and Finkelstein [7] using the results of the previous subsections. We have two possibilities: $B(b,s)$, the strategy when the systems are burned-in for time b (we will call it the 'time burn-in') and then the shock burn-in with severity s is applied to the systems, which survived the burn-in time b, whereas the strategy $B(s,b)$ applies shock first and then the survived systems are burned-in for time b. Unless otherwise specified, we assume that, during the time burn-in, the system is also subject to environmental shocks (as in field usage). In Cha and Finkelstein [7], the simple case of the homogeneous Poisson process of environmental shocks with intensity λ was considered, whereas in the current setting we are able to deal with the general NHPP case. In fact, the shock intensity during time burn-in and that during the field operation can be different. Let $\lambda_b(t)$ be the shock intensity at time t from the starting point of the burn-in and $\lambda_f(t)$ be the shock intensity at time t from the starting point of the field operation. Then the overall intensity function is

$$\lambda(t) = \begin{cases} \lambda_b(t), & \text{if } t \le b \\ \lambda_f(t-b), & \text{if } t > b, \end{cases}$$

where b is the burn-in time.

Let the assumptions and notation for the burn-in strategies under consideration be the same as before. As for the conventional burn-in procedure, assume additionally that the burn-in cost is proportional to the total burn-in time with proportionality constant c_0.

Consider first, the strategy $B(s,b)$. Let $h_1(s,b)$ be the expected burn-in cost for $B(s,b)$ and T_s be the lifetime of the system that has survived the shock burn-in. As our shock is of the fixed magnitude s, the corresponding survival function after the shock, in accordance with (9.58), is

$$\overline{F_s}(t) = \exp\left(-\int_0^t (r_0(u) + p(s,u)\lambda(u))\mathrm{d}u\right),$$

where $p(s,t)$ is defined in (9.59). Then, by similar arguments as those described in Cha and Finkelstein [7], we have:

$$h_1(s,b) = c_0 \frac{\int_0^b \overline{F_s}(t)\mathrm{d}t}{\overline{F_s}(b)} + \frac{c_s + c_{sr}}{\overline{F_s}(b)\overline{G}(s)} - c_{sr}. \tag{9.63}$$

On the other hand, when our system is not exposed to environmental shocks during the time burn-in, (9.63) changes to

$$h_1(s,b) = c_0 \frac{\int_0^b \overline{F_0}(t)\mathrm{d}t}{\overline{F_0}(b)} + \frac{c_s + c_{sr}}{\overline{F_0}(b)\overline{G}(s)} - c_{sr},$$

where $\overline{F_0}(t) = \exp(-\int_0^t r_0(u)\ du)$.

Consider a gain proportional to the mean time to failure in field usage, as in (9.62). Then the total expected cost $c_1(s,b)$ is

$$\begin{aligned} c_1(s,b) = {} & c_0 \frac{\int_0^b \overline{F_s}(t)\mathrm{d}t}{\overline{F_s}(b)} + \frac{c_s + c_{sr}}{\overline{F_s}(b)\overline{G}(s)} - c_{sr} \\ & - K\left(\int_0^\infty \exp\left\{-\int_0^t (r_0(b+u) + p(s,b+u)\lambda(b+u))\mathrm{d}u\right\}\mathrm{d}t\right), \end{aligned} \tag{9.64}$$

whereas the substitution of $\overline{F_s}(t)$ by $\overline{F_0}(t)$ and assuming that $\lambda_b(t) = 0$ corresponds to the case when there are no environmental shocks during the time burn-in.

As Cha and Finkelstein [7] did not take into account the existing dependence of the distribution of strength on time, the failure rate that corresponds to (9.58) was erroneously obtained as $r(t,s) = r_0(t) + p(s)\lambda$ for $\lambda(t) = \lambda$. In accordance with this equation it was stated that "the failures due to shocks during the time burn-in do not contribute to improvement of reliability characteristics in field use, but increase only the cost of burn-in" as time burn-in does not decrease the second term "$p(s)\lambda$". However, the following theorem shows that shocks during time burn-in *do contribute* to improvement of reliability characteristics in field use.

Theorem 9.12 *The function $p(s,t)$ is strictly decreasing in t for each fixed s.*

Proof Observe that

$$\begin{aligned}\frac{\partial}{\partial t}P(U_s<v|T>t) &= \frac{1}{\left(\int_s^\infty \exp\left\{-\overline{M}_f(r)\int_0^t \lambda(x)\mathrm{d}x\right\}g_U(r)\mathrm{d}r\right)^2}\\ &\times\left[-\lambda(t)\int_s^v \overline{M}_f(r)\exp\left\{-\overline{M}_f(r)\int_0^t \lambda(x)\mathrm{d}x\right\}g_U(r)\mathrm{d}r\cdot\int_s^\infty \exp\left\{-\overline{M}_f(r)\int_0^t \lambda(x)\mathrm{d}x\right\}g_U(r)\mathrm{d}r\right.\\ &\left.+\lambda(t)\int_s^\infty \overline{M}_f(r)\exp\left\{-\overline{M}_f(r)\int_0^t \lambda(x)\mathrm{d}x\right\}g_U(r)\mathrm{d}r\cdot\int_s^v \exp\left\{-\overline{M}_f(r)\int_0^t \lambda(x)\mathrm{d}x\right\}g_U(r)\mathrm{d}r\right].\end{aligned}$$

The numerator of the above equation becomes

$$\begin{aligned}&\left[-\lambda(t)\int_s^v \overline{M}_f(r)\exp\left\{-\overline{M}_f(r)\int_0^t \lambda(x)\mathrm{d}x\right\}g_U(r)\mathrm{d}r\cdot\int_v^\infty \exp\left\{-\overline{M}_f(r)\int_0^t \lambda(x)\mathrm{d}x\right\}g_U(r)\mathrm{d}r\right.\\ &\left.+\lambda(t)\int_v^\infty \overline{M}_f(r)\exp\left\{-\overline{M}_f(r)\int_0^t \lambda(x)\mathrm{d}x\right\}g_U(r)\mathrm{d}r\cdot\int_s^v \exp\left\{-\overline{M}_f(r)\int_0^t \lambda(x)\mathrm{d}x\right\}g_U(r)\mathrm{d}r\right]\\ <&\left[-\lambda(t)\int_s^v \overline{M}_f(r)\exp\left\{-\overline{M}_f(r)\int_0^t \lambda(x)\mathrm{d}x\right\}g_U(r)\mathrm{d}r\cdot\int_v^\infty \exp\left\{-\overline{M}_f(r)\int_0^t \lambda(x)\mathrm{d}x\right\}g_U(r)\mathrm{d}r\right.\\ &\left.+\lambda(t)\int_v^\infty \overline{M}_f(v)\exp\left\{-\overline{M}_f(r)\int_0^t \lambda(x)\mathrm{d}x\right\}g_U(r)\mathrm{d}r\cdot\int_s^v \exp\left\{-\overline{M}_f(r)\int_0^t \lambda(x)\mathrm{d}x\right\}g_U(r)\mathrm{d}r\right]\\ =&\left[-\lambda(t)\int_s^v \overline{M}_f(r)\exp\left\{-\overline{M}_f(r)\int_0^t \lambda(x)\mathrm{d}x\right\}g_U(r)\mathrm{d}r\cdot\int_v^\infty \exp\left\{-\overline{M}_f(r)\int_0^t \lambda(x)\mathrm{d}x\right\}g_U(r)\mathrm{d}r\right.\\ &\left.+\lambda(t)\int_v^\infty \exp\left\{-\overline{M}_f(r)\int_0^t \lambda(x)\mathrm{d}x\right\}g_U(r)\mathrm{d}r\cdot\int_s^v \overline{M}_f(v)\exp\left\{-\overline{M}_f(r)\int_0^t \lambda(x)\mathrm{d}x\right\}g_U(r)\mathrm{d}r\right]\\ <&\left[-\lambda(t)\int_s^v \overline{M}_f(r)\exp\left\{-\overline{M}_f(r)\int_0^t \lambda(x)\mathrm{d}x\right\}g_U(r)\mathrm{d}r\cdot\int_v^\infty \exp\left\{-\overline{M}_f(r)\int_0^t \lambda(x)\mathrm{d}x\right\}g_U(r)\mathrm{d}r\right.\\ &\left.+\lambda(t)\int_v^\infty \exp\left\{-\overline{M}_f(r)\int_0^t \lambda(x)\mathrm{d}x\right\}g_U(r)\mathrm{d}r\cdot\int_s^v \overline{M}_f(r)\exp\left\{-\overline{M}_f(r)\int_0^t \lambda(x)\mathrm{d}x\right\}g_U(r)\mathrm{d}r\right]=0\end{aligned}$$

as $\overline{M}_f(r)$ is strictly decreasing in r. Therefore, $P(U_s<v|T>t)$ is decreasing in $t$and, due to the fact that

$$p(s,t)=\int_0^\infty I(v\in[s,\infty))P(U_s<v|T>t)m_f(v)\mathrm{d}v,$$

$p(s,t)$ is strictly decreasing in t for each fixed s. □

Therefore, the second term of the failure rate in (9.58), $p(s,t)\lambda(t)$ is decreasing in t for each fixed s when $\lambda(t)$ is nonincreasing. Or, even if $\lambda(t)$ is increasing, $p(s,t)\lambda(t)$ can be decreasing in t (for each fixed s) in some cases. Therefore, in this sense, shocks during time burn-in *do contribute* to improvement of reliability characteristics in field use.

Similar considerations can be used for describing the strategy $B(b,s)$. Let $h_2(s,b)$ be the expected burn-in cost. Then by similar arguments as those described in Cha and Finkelstein [7]:

$$h_2(s,b) = \frac{1}{\overline{G}(s)}\left(c_0 \frac{\int_0^b \overline{F}(t)\mathrm{d}t}{\overline{F}(b)}\right) + c_s \frac{1}{\overline{G}(s)} + c_{sr}\frac{1}{\overline{F}(b)\overline{G}(s)} - c_{sr}, \tag{9.65}$$

where

$$\overline{F}(t) = \exp\left(-\int_0^t (r_0(u) + p(0,u)\lambda(u))\mathrm{d}u\right).$$

Note that just after time burn-in (before performing the shock burn-in), as follows from Remark 9.14, the initial distribution of U is

$$\overline{G}_U(u;b) = P(U > u | T > b) = \frac{\int_u^\infty \exp\left\{-\overline{M}_f(r)\int_0^b \lambda(x)\mathrm{d}x\right\} \cdot g_U(r)\mathrm{d}r}{\int_0^\infty \exp\left\{-\overline{M}_f(r)\int_0^b \lambda(x)\mathrm{d}x\right\} g_U(r)\mathrm{d}r},$$

and, if we further perform the shock burn-in with the magnitude s, then the resulting pdf for U is

$$\begin{cases} 0, & \text{if } u \le s \\ \frac{g_U(u;b)}{\overline{G}_U(s;b)}, & \text{if } u > s \end{cases},$$

where $g_U(u;b)$ is the pdf which corresponds to $\overline{G}_U(u;b)$:

$$g_U(u;b) = \frac{\exp\left\{-\overline{M}_f(u)\int_0^b \lambda(x)\mathrm{d}x\right\} \cdot g_U(u)}{\int_0^\infty \exp\left\{-\overline{M}_f(r)\int_0^b \lambda(x)\mathrm{d}x\right\} g_U(r)\mathrm{d}r}$$
$$= \frac{\exp\left\{-\overline{M}_f(u)\int_0^b \lambda_b(x)\mathrm{d}x\right\} \cdot g_U(u)}{\int_0^\infty \exp\left\{-\overline{M}_f(r)\int_0^b \lambda_b(x)\mathrm{d}x\right\} g_U(r)\mathrm{d}r}.$$

In accordance with (9.59), the failure probability at the 'field use age' t is

$$p(b,s,t) = \frac{\int_s^\infty \int_s^v \exp\{-\overline{M}_f(r)\int_0^t \lambda_f(x)\mathrm{d}x\} \cdot g_U(r;b)\mathrm{d}r\, m_f(v)\mathrm{d}v}{\int_s^\infty \exp\{-\overline{M}_f(r)\int_0^t \lambda_f(x)\mathrm{d}x\} g_U(r;b)\mathrm{d}r}. \tag{9.66}$$

Finally, from (9.65) and (9.66), the total expected cost $c_2(s,b)$ is

$$c_2(s,b) = \frac{1}{\overline{G}(s)}\left(c_0 \frac{\int_0^b \overline{F}(t)\mathrm{d}t}{\overline{F}(b)}\right) + c_s \frac{1}{\overline{G}(s)} + c_{sr}\frac{1}{\overline{F}(b)\overline{G}(s)} - c_{sr}$$
$$- K\left(\int_0^\infty \exp\left\{-\int_0^t (r_0(b+u) + p(b,s,u)\lambda(b+u))\mathrm{d}u\right\}\mathrm{d}t\right).$$

Note that, $p(b,s,u)$ (not $p(s,b+u)$) should be used in $c_2(s,b)$ above. From Theorems 9.11 and 9.12, it is clear that $p(b,s,t)$ is strictly decreasing in both s and t for each fixed b, respectively. By similar procedure as before (Theorem 9.12), it can also be shown that the function $p(b,s,t)$ is strictly decreasing in b for each fixed s and t.

In Cha and Finkelstein [7], two stage optimization procedures for minimizing the cost functions are discussed. Similar approach can be applied to the modified results of the current paper. For example, for obtaining optimal (s_1^*, b_1^*) which minimizes, $c_1(s,b)$ defined by equation (9.64), we can follow the following procedure:

1. Fix $b \geq 0$, then find optimal $s^*(b)$ which satisfies

$$c_1(s^*(b),b) = \min_{0 \leq s < \infty} c_1(s,b), \text{for fixed } b \geq 0.$$

Note that, as $c(s,b)$ is eventually increasing in s to infinity for each fixed b, such $s^*(b)$ exists for all b.

2. Find optimal b^* which satisfies

$$c_1(s^*(b^*),b^*) = \min_{0 \leq b < \infty} c_1(s^*(b),b).$$

Then, finally, such $(s^*(b^*),b^*)$ is the optimal solution of the problem. However, in this modified model, even if we assume that $r_0(t)$ is the bathtub-shaped failure rate with two change points t_1 and t_2, t_1 is not necessarily the uniform upper bound for the optimal burn-in time. However, if we assume additionally that $r_0(t)$ is *increasing to infinity* after t_2, there obviously should be the uniform upper bound for the optimal burn-in time and the standard numerical procedures can be used for obtaining optimal solutions in this case.

References

1. Bagdonavicius V, Nikulin M (2009) Statistical models to analyze failure, wear, fatigue, and degradation data with explanatory variables. Commun Stat—Theory Methods 38:3031–3047
2. Barlow RE, Proschan F (1975) Statistical theory of reliability and life testing. Holt, Renerhart & Winston, New York
3. Beard RE (1959) Note on some mathematical mortality models. In: Woolstenholme GEW, O'Connor M (eds) The lifespan of animals. Little, Brown and Company, Boston, pp 302–311
4. Block HW, Mi J, Savits TH (1993) Burn-in and mixed populations. J Appl Probab 30:692–702
5. Cha JH (2006) An extended model for optimal burn-in procedures. IEEE Trans Reliab 55:189–198
6. Cha JH, Finkelstein M (2010) Burn-in by environmental shocks for two ordered subpopulations. Eur J Oper Res 206:111–117
7. Cha JH, Finkelstein M (2011) Burn-in for systems operating in a shock environment. IEEE Trans Reliab 60:721–728
8. Cha JH, Finkelstein M (2013). Burn-in for heterogeneous populations: How to avoid large risks. Commun Stat—Theory Methods (to appear)
9. El Karoui N, Gerardi A, Mazliak L (1994) Stochastic control methods in optimal design of life testing. Stoch Process Appl 52:309–328
10. Finkelstein M (2008) Failure rate modelling for reliability and risk. Springer, London
11. Finkelstein M (2009) Understanding the shape of the mixture failure rate (with engineering and demographic applications). Appl Stoch Models Bus Ind 25:643–663
12. Mi J (1996) Minimizing some cost functions related to both burn-in and field use. Oper Res 44:497–500
13. Reddy RK, Dietrich DL (1994) A 2-level environmental-stress-screening (ESS) model: a mixed-distribution approach. IEEE Trans Reliab 43:85–90
14. Vaupel JW, Manton KG, Stallard E (1979) The impact of heterogeneity in individual frailty on the dynamics of mortality. Demography 16:439–454
15. Wu S, Xie M (2007) Classifying weak, and strong components using ROC analysis with application to burn-in. IEEE Trans Reliab 56:552–561
16. Yan L, English JR (1997) Economic cost modeling of environmental-stress-screening and burn-in. IEEE Trans Reliab 46:275–282
17. Yang G (2002) Environmental-stress-screening using degradation measurements. IEEE Trans Reliab 51:288–293

Chapter 10
Stochastic Models for Environmental Stress Screening

There are different ways of improving reliability characteristics of manufactured items. The most common methodology adopted in industry is burn-in, which is a method of 'elimination' of initial failures (infant mortality). As was mentioned previously, the 'sufficient condition' for employing the *traditional* burn-in is the initially decreasing failure rate. For example, when a population of items is heterogeneous, and therefore consists of subpopulations with ordered failure (hazard) rates, it obviously contains weaker (with larger failure rates) subpopulations. As the weakest populations are dying out first, the failure rate of this population is often initially decreasing and burn-in can be effectively applied.

It should be noted that not all populations of engineering items that contain 'weaker' items to be eliminated exhibit this shape of the failure rate. For example, the 'weakness' of some manufactured items can result from the latent defects that can create *additional* failure modes. The failure rate in this case is not necessarily decreasing (see Example 10.1), and therefore traditional burn-in should not be applied. However, by applying the short-time excessive stress, the weaker items in the population with increasing failure rate can be eliminated by the environmental stress screening (ESS), and therefore the reliability characteristics of the population of items that have successfully passed the ESS test can still improve. This is the *crucial distinction of this operation from burn-in.* In fact, the formal difference between the ESS and burn-in has not been clearly defined in the literature. In our discussions, we understand the ESS as the method of elimination of items with additional (nonconventional) failure modes, whereas burn-in targets elimination of weaker items with conventional failure modes and it is effective *only* when the population failure rate is initially decreasing. Another important distinction of the proposed model from burn-in is that the ESS can also create new defects in items that were previously defect-free.

Numerous stochastic models of burn-in have been intensively studied in the literature during the last decades. Although some practical engineering approaches to the ESS modeling were reported (e.g., [2, 4]), to the authors' best knowledge, there has been little research dealing with adequately advanced stochastic modeling and analysis of the ESS.

M. Finkelstein and J. H. Cha, *Stochastic Modeling for Reliability*, Springer Series in Reliability Engineering, DOI: 10.1007/978-1-4471-5028-2_10,

In this chapter, we develop a stochastic model for the ESS, analyze its effect on the population characteristics of the screened items and describe related optimization problems. We assume that, due to substandard materials of faulty manufacturing process, some of the manufactured items are susceptible to additional cause of failure (failure mode), i.e., shocks (such as electrical or mechanical shocks). We define the ESS as a procedure of applying a shock of the controlled magnitude, i.e., a short-time excessive stress. In practice, for example, a shock can be understood as a short-time electric impulse. For the ESS to be effective, the corresponding magnitude should be reasonably larger than the magnitude of shocks that occur in field usage.

Our modeling is within the framework of the general shock models. We will consider two different types of ESS models in this chapter. In the first model, the failure of an item occurs when the magnitude of the stress (shock) exceeds its strength. The larger magnitude of the ESS shock (within 'physical limits') implies the better reliability characteristics of survived items in field usage but at the same time, the larger cost of the ESS as more items with defects are discarded. An important feature of our model is that we assume that the item during field usage is exposed to the point process of environmental shocks of an ordinary, not excessive magnitude. These shocks can obviously destroy only defective items that have passed the ESS or were induced by the ESS. In the second model, an external shock can either destroy an item with a given probability or increase the 'size of the defect' by a random amount. We also analyze the effect of the ESS on the population characteristics of the screened items and discuss related optimization problems. We will extensively use the general stress–strength model described in Sect. 9.4.1.

10.1 Stress–Strength Type ESS Model

10.1.1 Stochastic Model for ESS

The description and assumptions of our model are as follows. During the manufacturing process, the items with the failure rate $r(t)$ and the corresponding lifetime T_N (which is only due to 'normal' failure mode) and also the defective items with the lifetime T_D are produced. Let the proportion of the nondefective items be π and that of the defective items be $1 - \pi$.

The defective items, in addition to the normal failure mode of the nondefective items, are characterized by a new additional failure mode. In this chapter, we assume that this additional failure mode describes susceptibility to external shocks. For example, consider the case when the normal (nondefective) items, in accordance with specifications, should not be susceptible to electrical or mechanical shocks. However, due to substandard materials or a faulty manufacturing process, some of the produced items are susceptible to these shocks [4]. For instance, during the manufacturing process, the items can be exposed to a strong electric shock and this shock may result in some defective items which are even sensitive to electrical shocks of a 'normal' magnitude, whereas the nondefective items are not sensitive to it [3]. Another example is when a small crack in a material of the defective item is sensitive to mechanical impulses (e.g., vibration) in field use, which eventually can result in its failure. Thus, we assume that shocks of a 'normal' magnitude also occur in field operation, and therefore the defective items can fail due to this failure mode. On the other hand, the nondefective items do not fail from external shocks of this type in field operation as they do not have the corresponding failure mode.

In accordance with our description, the survival function of T_N is

$$P(T_N > t) = \exp\{-\int_0^t r(u)\mathrm{d}u\}.$$

Let the two failure modes of the defective items be independent. Then, the corresponding survival function is given by the competing risks model (series system):

$$P(T_D > t) = \exp\{-\int_0^t r(u)\mathrm{d}u\} \cdot P(T_E > t), \tag{10.1}$$

where T_E is the lifetime that accounts only for the external shock failure mode.

Suppose that during the field operation, the external shocks occur in accordance with the NHPP $\{N(t), t \geq 0\}$ with rate $\lambda(t)$. Denote by S_i the magnitude (stress) of the ith shock and assume that $S_i, i = 1, 2, \ldots$ are i.i.d. random variables with the common Cdf $M(s) = \Pr(S_i \leq s)$ $(\bar{M}(s) \equiv 1 - M(s))$ and the corresponding pdf $m(s)$. The defective item is characterized by its random strength U, i.e., the resistance ability to external shocks. Here, the strength is understood as the *'maximum stress level that the defective item can survive'*. The corresponding Cdf, Sf, pdf, and FR of U are denoted by $G_U(u)$, $\bar{G}_U(u)$, $g_U(u)$ and $r_U(u)$, respectively. For each $i = 1, 2, \ldots$, the operable system survives if $S_i \leq U$ and fails if $S_i > U$, *'independently of everything else'*. Then, in accordance with Theorem 9.10, Eq. (10.1) reads now

$$P(T_D > t) = \exp\{-\int_0^t r(u)\mathrm{d}u\} \cdot \exp\{-\int_0^t p(u)\lambda(u)\mathrm{d}u\}, \tag{10.2}$$

where

$$p(t) \equiv \frac{\int_0^\infty \int_0^v \exp\{-\overline{M}(r)\int_0^t \lambda(x)\mathrm{d}x\} \cdot g_U(r)dr\, m(v)\mathrm{d}v}{\int_0^\infty \exp\{-\bar{M}(r)\int_0^t \lambda(x)\mathrm{d}x\} g_U(r)dr}. \tag{10.3}$$

From (10.2), we see that the lifetimes of the nondefective and defective items are obviously stochastically ordered: $T_D <_{fr} T_N$, where "$<_{fr}$"denotes, as usual, the failure (hazard) rate ordering of two random variables.

Denote the population lifetime by T. As it consists of defective and nondefective items with given proportions, the corresponding survival function is the following mixture

$$\begin{aligned}\bar{F}(t) &\equiv P(T > t) \\ &= \pi \exp\{-\int_0^t r(u)du\} + (1-\pi)\exp\{-\int_0^t r(u)du\} \cdot \exp\{-\int_0^t p(u)\lambda(u)du\}.\end{aligned} \tag{10.4}$$

Thus, (10.4) defines the survival function of an item in field usage that is chosen at random from the population of manufactured items.

In what follows, we will describe the impact of the ESS on the population structure and on the corresponding population lifetime distribution. Therefore, we must define first the ESS that we consider in this chapter.

ESS Process

During the ESS, all items are exposed to a single shock with the fixed magnitude s. If the strength of a defective item is larger than s then it survives; otherwise it fails. Depending on the magnitude s, a proportion of nondefective items, $\rho(s)$, $0 \leq \rho(s) < 1$, *becomes defective, where* $\rho(s)$ *is an increasing function of its argument. The items failed during the ESS are discarded and only the survived items are put into the field operation.*

Thus the ESS, in principle, can induce defects. Furthermore, as those with induced defects but not failed are not identifiable, they are also put into the field operation.

Recall that shock's magnitudes in field operation are i.i.d. random variables. We assume that the corresponding mean is substantially smaller than the magnitude of stress allowed for the ESS (otherwise there is no reason to perform the ESS). Therefore, the shocks in field operation can hardly 'produce' defective items out of nondefective ones (or this effect is negligible). On the other hand, these shocks can still destroy the defective item with a given strength.

Denote the population lifetime after the ESS with magnitude s by T_{ESS}.

Theorem 10.1 *Under the given assumptions, the population distribution and the corresponding failure rate (after the ESS) are*

$$\begin{aligned}\bar{F}_E(t,s) = P(T_{ESS} > t) &= \exp\{-\int_0^t r(u)\mathrm{d}u\} \cdot \frac{(1-\rho(s))\pi}{(1-\rho(s))\pi + \rho(s)\pi + \bar{G}_U(s)(1-\pi)} \\ &+ \exp\{-\int_0^t r(u)\mathrm{d}u\} \cdot \exp\{-\int_0^t p(u)\lambda(u)\mathrm{d}u\} \cdot \frac{\rho(s)\pi}{(1-\rho(s))\pi + \rho(s)\pi + \bar{G}_U(s)(1-\pi)} \\ &+ \exp\{-\int_0^t r(u)\mathrm{d}u\} \cdot \exp\{-\int_0^t p(s,u)\lambda(u)\mathrm{d}u\} \cdot \frac{\bar{G}_U(s)(1-\pi)}{(1-\rho(s))\pi + \rho(s)\pi + \bar{G}_U(s)(1-\pi)},\end{aligned} \tag{10.5}$$

and

$$\begin{aligned}\lambda_E(t,s) = r(t) \cdot \frac{\pi(1)\bar{F}_1(t)}{\sum_{i=1}^{3} \pi(i)\bar{F}_i(t)} + [r(t) + p(t)\lambda(t)] \cdot \frac{\pi(2)\bar{F}_2(t)}{\sum_{i=1}^{3} \pi(i)\bar{F}_i(t)} \\ + [r(t) + p(s,t)\lambda(t)] \cdot \frac{\pi(3)\bar{F}_3(t)}{\sum_{i=1}^{3} \pi(i)\bar{F}_i(t)},\end{aligned} \tag{10.6}$$

respectively.

Proof Observe that there are now three subpopulations after the ESS and we can define the corresponding frailty variable Z:

(i) the subpopulation with nondefective items ($Z = 1$); (ii) the subpopulation with defective items which were originally nondefective ($Z = 2$); (iii) the subpopulation with defective items which were originally defective but have survived the ESS ($Z = 3$). Then, in accordance with our notation, the distribution of Z is given by

$$\begin{aligned}\pi(1) \equiv P(Z=1) &= \frac{(1-\rho(s))\pi}{(1-\rho(s))\pi + \rho(s)\pi + \bar{G}_U(s)(1-\pi)}, \\ \pi(2) \equiv P(Z=2) &= \frac{\rho(s)\pi}{(1-\rho(s))\pi + \rho(s)\pi + \bar{G}_U(s)(1-\pi)}, \\ \pi(3) \equiv P(Z=3) &= \frac{\bar{G}_U(s)(1-\pi)}{(1-\rho(s))\pi + \rho(s)\pi + \bar{G}_U(s)(1-\pi)}.\end{aligned}$$

Therefore,

$$\bar{F}_1(t) \equiv P(T_{ESS} > t | Z = 1) = \exp\{-\int_0^t r(u)\mathrm{d}u\}$$

and

$$\bar{F}_2(t) \equiv P(T_{ESS} > t|Z = 2) = \exp\{-\int_0^t r(u)\mathrm{d}u\} \cdot \exp\{-\int_0^t p(u)\lambda(u)\mathrm{d}u\},$$

where $p(t)$ is given by (10.3).

Derivation of $P(T_{ESS} > t|Z = 3)$ is not so straightforward. Indeed, it should be taken into account that when we apply a shock of the controlled magnitude s during the ESS, this means that the strength of the defective item that had passed it is larger than s and, therefore, the distribution of the *remaining strength* U_s (given that the strength is larger than s) is

$$G_U(u|s) \equiv P(U \le u|U > s) = 1 - \bar{G}(u)/\bar{G}(s),\ u > s.$$

Thus, the function $p(t)$ in (10.3) should be modified to

$$\begin{aligned} p(s,t) &= \frac{\int_0^\infty \int_0^v \exp\{-\bar{M}(r)\int_0^t \lambda(x)\mathrm{d}x\} \cdot g_U(r|s)dr\, m(v)\mathrm{d}v}{\int_0^\infty \exp\{-\bar{M}(r)\int_0^t \lambda(x)\mathrm{d}x\} g_U(r|s)dr} \\ &= \frac{\int_s^\infty \int_s^v \exp\{-\bar{M}(r)\int_0^t \lambda(x)\mathrm{d}x\} \cdot g_U(r)dr\, m(v)\mathrm{d}v}{\int_s^\infty \exp\{-\bar{M}(r)\int_0^t \lambda(x)\mathrm{d}x\} g_U(r)dr}, \end{aligned} \tag{10.7}$$

where, $g_U(u|s)$ is the corresponding pdf of $G_U(u|s)$, which is given by

$$g_U(u|s) = \begin{cases} 0, & if\ u \le s \\ \frac{g_U(u)}{\bar{G}_U(s)}, & if\ u > s \end{cases}.$$

Finally,

$$\bar{F}_3(t) \equiv P(T_{ESS} > t|Z = 3) = \exp\{-\int_0^t r(u)\mathrm{d}u\} \cdot \exp\{-\int_0^t p(s,u)\lambda(u)\mathrm{d}u\}.$$

Therefore, Eqs. (10.5) and (10.6) hold.

□

We will now discuss the effect of the ESS on the quality of the population after the screening by comparing $\bar{F}_E(t,s)$ with the survival function without screening, $F(t)$ defined by Eq. (10.4). As the ESS in our model can create defective items, theoretically in some cases this operation may have a negative effect on the population of items.

Definition 10.1 The severity (stress) level s is said to be *inadmissible* under the survival function criterion if

$$\bar{F}(t) \geq \bar{F}_E(t,s), \text{ for all } t > 0.$$

Otherwise, the severity (stress) level s is said to be *admissible.*

Obviously, the inadmissible severity levels should not be considered in the ESS practice as reliability of items in field use is worse than that without the ESS in this case. Note that the condition for the 'admissibility' in Definition 10.1 means that $\bar{F}(t) < \bar{F}_E(t,s)$ for some $t > 0$ and not for all $t > 0$. However, for obvious practical reasons, we are mostly interested in the latter case. The following definition addresses this setting.

Definition 10.2 The severity (stress) level s is said to be *positively admissible* under the survival function criterion if

$$\bar{F}(t) < \bar{F}_E(t,s), \text{ for all } t > 0.$$

Theorem 10.2 (i) *If* $\rho(s) < (1-\pi)G_U(s)$, *then this severity level s is positively admissible under the survival function criterion.*

(ii) *If* $\rho(s)\pi > \pi(1-\pi) + (1-\pi)^2\bar{G}_U(s)$, *then this severity level s is inadmissible under the survival function criterion.*

Proof Denote for convenience, $\lambda_1(t) \equiv r(t)$; $\lambda_2(t) \equiv r(t) + p(t)\lambda(t)$; $\lambda_3(t) \equiv r(t) + p(s,t)\lambda(t)$. Note that Eq. (10.7) can be written in a compact and meaningful way as

$$p(s,t) = \int_0^{\infty} I(v \in [s,\infty))h(s,t,v)m(v)\mathrm{d}v,$$

where

$$h(s,t,v) \equiv \frac{\int_s^v \exp\{-\bar{M}(r)\int_0^t \lambda(x)\mathrm{d}x\} \cdot g_U(r)dr}{\int_s^{\infty} \exp\{-\bar{M}(r)\int_0^t \lambda(x)\mathrm{d}x\} g_U(r)dr}.$$

and $I(\cdot)$ is the corresponding indicator. Observe that, for all fixed t and v,

$$\frac{\partial}{\partial s}h(s,t,v) = \frac{1}{\left(\int_s^{\infty} \exp\{-\bar{M}(r)\int_0^t \lambda(x)\mathrm{d}x\}g_U(r)dr\right)^2}$$
$$\times\left[-\exp\{-\bar{M}(s)\int_0^t \lambda(x)\mathrm{d}x\}g_U(s)\cdot\int_s^{\infty}\exp\{-\bar{M}(r)\int_0^t \lambda(x)\mathrm{d}x\}g_U(r)dr\right.$$
$$\left.+\exp\{-\bar{M}(s)\int_0^t \lambda(x)\mathrm{d}x\}g_U(s)\cdot\int_s^{v}\exp\{-\bar{M}(r)\int_0^t \lambda(x)\mathrm{d}x\}g_U(r)dr\right] < 0,$$

for all $s > 0$. Therefore, the function $p(s,t)$ is strictly decreasing in s for each fixed t. This implies that $p(s,t) < p(t)$, for all $t > 0$ and $s > 0$. Thus we have the following failure rate ordering:

$$\lambda_1(t) < \lambda_3(t) < \lambda_2(t), \quad \text{for all } t > 0, \tag{10.12}$$

and accordingly,

$$\bar{F}_1(t) > \bar{F}_3(t) > \bar{F}_2(t), \quad \text{for all } t > 0,$$

where $\bar{F}_i(t) \equiv \exp\{-\int_0^t \lambda_i(u)\mathrm{d}u\}$, $i = 1,2,3$. Observe that, in accordance with (10.4),

$$\bar{F}(t) = \pi\bar{F}_1(t) + (1-\pi)\bar{F}_2(t),$$

whereas in accordance with (10.5),

$$\bar{F}_E(t,s) = \pi(1)\bar{F}_1(t) + \pi(2)\bar{F}_2(t) + \pi(3)\bar{F}_3(t).$$

Therefore, if $\pi(2) + \pi(3) < 1 - \pi$, or equivalently, $\pi(1) > \pi$, then

$$\begin{aligned}\bar{F}_E(t,s) - \bar{F}(t) &> (\pi(1)-\pi)\bar{F}_1(t) + [\pi(2)\bar{F}_2(t) + \pi(3)\bar{F}_2(t) - (1-\pi)\bar{F}_2(t)] \\ &= (\pi(1)-\pi)\bar{F}_1(t) - (\pi(1)-\pi)\bar{F}_2(t) > 0,\end{aligned}$$

for all $t > 0$. The condition $\pi(2) + \pi(3) < 1 - \pi$ is equivalent to $\rho(s) < (1-\pi)G_U(s)$. This completes the proof of (i).

By a similar reasoning, if $\pi(2) > 1 - \pi$, or equivalently, $\rho(s)\pi > \pi(1-\pi) + (1-\pi)^2\bar{G}_U(s)$, then the severity level s is inadmissible under the survival function criterion.

□

Remark 10.1

(i) The conditions in Theorem 10.2 do not imply the admissibility/inadmissibility of the corresponding severity level under the failure rate criterion. That is, the condition $\pi(2) + \pi(3) < 1 - \pi$ does not imply $\lambda_T(t) > \lambda_E(t, s)$, for all $t > 0$, where $\lambda_T(t)$ is the failure rate which corresponds to $\bar{F}(t)$ defined in (10.4).

(ii) The failure rate ordering (10.8) will be important for our further reasoning. This ordering implies that the quality of defective items improves after the ESS, but they are still obviously 'worse' than the nondefective items.

Remark 10.2 The effect of applying two consecutive shocks with severity s during the ESS can be also considered. After this type of the ESS, we also have three subpopulations with failure rates $\lambda_1(t) = r(t)$, $\lambda_2(t) = r(t) + p(t)\lambda(t)$ and $\lambda_3(t) = r(t) + p(s, t)\lambda(t)$ and the corresponding proportions

$$\pi^{(2)}(1) = \frac{(1 - \rho(s))\pi(1)}{(1 - \rho(s))\pi(1) + \rho(s)\pi(1) + [\pi(3) + \bar{G}_U(s)\pi(2)]},$$
$$\pi^{(2)}(2) = \frac{\rho(s)\pi(1)}{(1 - \rho(s))\pi(1) + \rho(s)\pi(1) + [\pi(3) + \bar{G}_U(s)\pi(2)]},$$
$$\pi^{(2)}(3) = \frac{\pi(3) + \bar{G}_U(s)\pi(2)}{(1 - \rho(s))\pi(1) + \rho(s)\pi(1) + [\pi(3) + \bar{G}_U(s)\pi(2)]}.$$

10.1.2 Optimal Severity

In this subsection, we will consider the problem of determining the optimal severity level (magnitude) of the ESS. Let τ be the mission time of an item in the field operation. If it does not fail during this time, then the mission is considered to be successful. Thus, the probability of the mission success needs to be maximized and we should find the optimal severity level s^* that satisfies

$$\bar{F}_E(\tau, s^*) = \max_{s > 0} \bar{F}_E(\tau, s).$$

Alternatively, let $MRL(s)$ be the mean time to failure of an item in the field operation as a function of s, i.e., $MRL(s) \equiv \int_0^\infty \bar{F}_E(t, s)\mathrm{d}t$. Then, the optimal severity level which maximizes the mean time to failure should be obtained as

$$MRL(s^*) = \max_{s > 0} MRL(s).$$

For defining the optimal severity, we should consider the admissible severity class rather than the positively admissible class as we have to take into account all admissible severity levels. It is often more convenient to describe the dual inadmissible class. The following theorem provides the upper bound for the optimal severity level that maximizes the mission success probability or mean time to failure in field usage.

Theorem 10.3 *Suppose that $\rho(\infty) \equiv \lim_{s\to\infty} \rho(s) > (1-\pi)$ and let*

$$s_0 \equiv \inf_{s\geq 0}\{s : \rho(s)\pi > \pi(1-\pi) + (1-\pi)^2\bar{G}_U(s)\}.$$

Then the severity levels in (s_0, ∞) are inadmissible. Therefore, s_0 is the upper bound for the optimal severity level.

Proof From Theorem 10.2, the condition for inadmissibility is

$$\rho(s)\pi > \pi(1-\pi) + (1-\pi)^2\bar{G}_U(s).$$

Here, the function $\rho(s)\pi$ is increasing from 0 to $\rho(\infty)\pi$, whereas the function $\pi(1-\pi) + (1-\pi)^2\bar{G}_U(s)$ decreases from $(1-\pi)$ to $\pi(1-\pi)$. Thus, if $\rho(\infty)\pi > \pi(1-\pi)$, or equivalently, $\rho(\infty) > (1-\pi)$, then there exists $s_0 \in (0,\infty)$ such that the severity levels in (s_0, ∞) are inadmissible. Therefore, s_0 is the *upper bound* for the optimal severity.

□

Remark 10.3 It is reasonable to assume that in practice, $\lim_{s\to\infty} \rho(s) = 1$ and that the proportion of the defective items $(1-\pi)$ is relatively small. Therefore, the condition $\rho(\infty) > (1-\pi)$ can be satisfied in almost all practical cases.

Example 10.1 Let $r(t) = 0.1t,\ t\geq 0,\ \lambda(t) = 1,\ t\geq 0,\ m(s) = 3\exp\{-3s\}$, $s\geq 0,\ g_U(u) = 4u\exp\{-2u^2\},\ u\geq 0,\ \pi = 0.7,\ \tau = 4.0$ and

$$\rho(s) = \begin{cases} 0, & 0\leq s<1, \\ 1-\exp\{-0.05(s-1)\}, & s\geq 1 \end{cases}.$$

Note that the failure rate of the population distribution before the ESS, which is obtained based on (10.4), is given by Fig. 10.1.

Therefore, as $\lambda_T(t)$ is increasing, the burn-in procedure *should not be applied to this population.* On the other hand, as $\rho(s)$ is strictly increasing for $s\geq 1$, there exists a unique solution of the equation

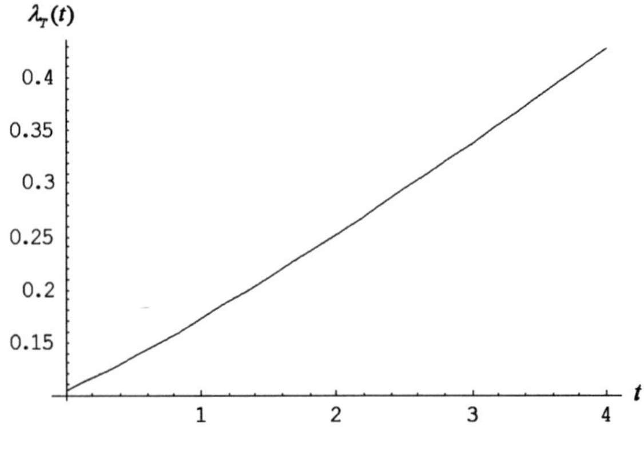

Fig. 10.1 The graph of $\lambda_T(t)$

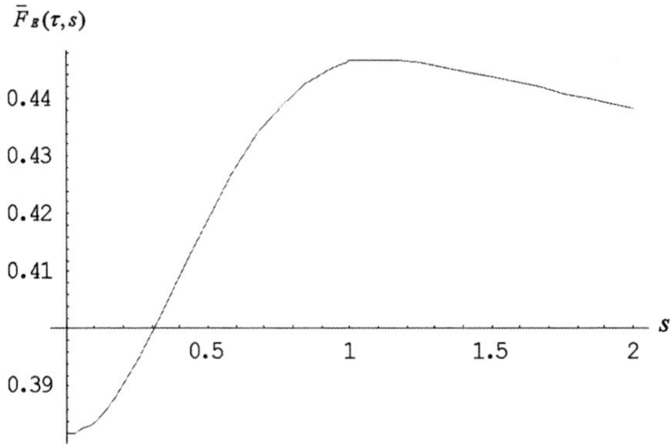

Fig. 10.2 The graph of $\bar{F}_E(\tau, s)$

$$\rho(s)\pi = \pi(1-\pi) + (1-\pi)^2\bar{G}_U(s),$$

which is the upper bound for the optimal severity level. Therefore, the ESS as a method of elimination of defective items is justified in this case. Solving this equation numerically results in $s_0 \approx 8.13$. Therefore, it is now sufficient to search for the optimal severity level in the interval $[0, 8.13]$. The graph of $\bar{F}_E(\tau, s)$ is presented in Fig. 10.2. The optimal severity level in this case is $s^* \approx 1.08$ and the maximum probability of the mission success is $\bar{F}_E(\tau; s^*) \approx 0.447$.

Based on the foregoing results, we can consider now certain cost structures for determining the cost-based optimal severity level. As previously, an item is chosen at random from our initial population and is exposed to a shock of magnitude s during the ESS. If it survives, it is put into the field operation, otherwise the failed item is discarded and the new one is chosen from the population, etc. This procedure is repeated until the first survived item is obtained. Let c_{sr} be the shop replacement cost (actually, it is the cost of a new item) and c_s be the cost for conducting the ESS. Let $c_1(s)$, as a function of s, be the expected cost for eventually obtaining a component which has survived the ESS. Then

$$c_1(s) = \frac{c_s + c_{sr}[1 - \{\pi + (1-\pi)\bar{G}_U(s)\}]}{\pi + (1-\pi)\bar{G}_U(s)},$$

where $1/[\pi + (1-\pi)\bar{G}_U(s)]$ is the total number of trials until the first 'success'.

Assume that if a mission (of length τ) is successful (in field operation), then the gain K is 'earned'; otherwise a penalty C is imposed, where $K > C > 0$. Then the expected gain during the field operation is

$$c_2(s) = -K\bar{F}_E(\tau, s) + CF_E(\tau, s) = -(K + C)\bar{F}_E(\tau, s) + C \qquad (10.9)$$

and the total expected cost $c(s)$ is

$$
\begin{aligned}
c(s) &= c_1(s) + c_2(s) \\
&= \frac{c_s + c_{sr}[1 - \{\pi + (1-\pi)\bar{G}_U(s)\}]}{\pi + (1-\pi)\bar{G}_U(s)} - (K + C)\bar{F}_E(\tau; s) + C.
\end{aligned}
$$

The objective is now to find the optimal severity level s^* that satisfies

$$
s^* = \arg \min_{s \in [0,\infty]} c(s).
$$

Similar to Theorem 10.3, if $\rho(\infty) \equiv \lim_{s\to\infty} \rho(s) > ; (1-\pi)$, then the optimal severity level which minimizes $c_2(s)$ [maximizes $\bar{F}_E(\tau, s)$, as follows from (10.9)] does not exists in the interval (s_0, ∞), where s_0 is also defined by Theorem 10.3. Furthermore, as $c_1(s)$ is strictly increasing to infinity, we can conclude that the optimal severity level s^* should exist in the interval $[0, s_0]$.

Assume now that during field operation, the gain is proportional to the mean time to failure. Therefore, the total average cost function in this case is

$$
c(s) = \frac{c_s + c_{sr}[1 - (1-\pi)\bar{G}_U(s)]}{(1-\pi)\bar{G}_U(s)} - K\int_0^\infty \bar{F}_E(t; s)\mathrm{d}t.
$$

By the similar arguments, the optimal severity level s^* should exist in the interval $[0, s_0]$.

10.2 ESS Model with Wear Increments

10.2.1 Stochastic Model

In this subsection, we develop a stochastic model for the shock and wear based ESS. We assume that, during the manufacturing process due to substandard materials or other faults some defective items with latent defects such as, e.g., a microcrack may be produced. Such defective items are susceptible to failure from mechanical or electrical shocks during field operation. Thus the defective items, in addition to the normal failure mode of the nondefective items, are characterized by a new *additional* failure mode. On the other hand, the nondefective items do not fail from external shocks in field operation as *they do not have the corresponding failure mode.*

Denote the lifetime of the nondefective items by T_N with the corresponding failure rate $r(t)$. In accordance with our description, obviously, the survival function of T_N is defined by

$$P(T_N > t) = \exp\{-\int_0^t r(u)\mathrm{d}u\}.$$

During the field operation, the items are subject to the nonhomogeneous Poisson process (NHPP) of 'ordinary' environmental shocks $\{N(t), t \geq 0\}$ with rate $\lambda(t)$ and arrival times $T_i, i = 1, 2, \ldots$. Let, on the ith shock, the defective item fail with probability $p(T_i)$ (critical shock), whereas with probability $q(T_i)$ it increases the 'defect size' by a random amount W_i (noncritical shock). In the following, for convenience, we will loosely use the term "wear" (or degradation) for the defect size as well. In accordance with this setting, the random accumulated wear of a defective item at time t in the field use is given by

$$W(t) = \sum_{i=0}^{N_q(t)} W_i + W_M,$$

where $N_q(t)$ is the number of noncritical shocks in $[0, t)$ and $W_M > 0$ is the initial wear (defect size of the latent defect). Let R be the random boundary of the item which follows an exponential distribution with parameter θ. The failure due to wear occurs when the accumulated wear $W(t)$ reaches R. Let T_E be the lifetime in the field use that accounts only for the external shock failure mode of defective items (*i.e.*, the lifetime without any other causes of failure). Then, as follows from Eq. (4.4) and the reasoning in Sect. 4.1.2,

$$P(T_E > t) = \exp\left\{-\int_0^t (1 - M_W(-\theta)q(x))\lambda(x)\mathrm{d}x\right\}, t \geq 0,$$

regardless of the distribution of W_M. As there are two independent failure modes for defective items—i.e., the normal failure mode described by $r(t)$ and the additional one due to external shocks, the survival function for the defective items is given by the competing risks model (a series system):

$$\begin{aligned}P(T_D > t) &= \exp\{-\int_0^t r(u)\mathrm{d}u\} \cdot P(T_E > t)\\ &= \exp\{-\int_0^t r(u)\mathrm{d}u\} \cdot \exp\left\{-\int_0^t (1 - M_W(-\theta)q(x))\lambda(x)\mathrm{d}x\right\}, t \geq 0.\end{aligned}$$

Let the proportion of the nondefective items be π and that of the defective items be $1 - \pi$, respectively. Denote the population lifetime by T. Given the structure of our population, the corresponding survival function is the mixture of survival functions for the defective and nondefective items:

$$\bar{F}(t) \equiv P(T > t) = \pi \exp\{-\int_0^t r(u)\mathrm{d}u\}$$
$$+ (1-\pi)\exp\{-\int_0^t r(u)\mathrm{d}u\} \cdot \exp\left\{-\int_0^t (1 - M_W(-\theta)q(x))\lambda(x)\mathrm{d}x\right\}, t \geq 0. \tag{10.10}$$

Thus, (10.10) defines the survival function in field usage of the item that is chosen at random from the population of manufactured items.

In what follows, we will describe the impact of the ESS on the population distribution. Therefore, we must describe first the ESS that we consider in this chapter.

ESS Process

During the ESS, a shock with the fixed magnitude s is applied to all items (e.g., *the mechanical shock). The defective items immediately fail with probability* $\alpha(s)$*, whereas with probability* $1-\alpha(s)$ *an additional wear with magnitude* W_s *is incurred, where* $\alpha(s)$ *is an increasing function and* W_s *is stochastically increasing with s. Furthermore, depending on the magnitude s, a proportion of nondefective items,* $\rho(s)$, $0 \leq \rho(s) < 1$, *becomes defective, where* $\rho(s)$ *is an increasing function of its argument. The failed items are discarded and only the survived items are put into field operation.*

For example, the mechanical shock during the ESS can be executed by the dropping of an item from some height (the "dropping shock"), which can be considered as the magnitude of the shock. Obviously, the assumptions for $\alpha(s)$, W_s and $\rho(s)$ are justified in this case. For instance, the larger height corresponds to the larger wear W_s.

We will now derive the population distribution in field use *after the ESS*. Denote the corresponding lifetime by T_{ESS}. In the following theorem, the distribution of T_{ESS} is obtained.

Theorem 10.4 *The survival function of* T_{ESS} *is given by*

$$P(T_{ESS} > t) \equiv \bar{F}_E(t, s)$$
$$= \exp\{-\int_0^t r(u)\mathrm{d}u\} \cdot \frac{(1-\rho(s))\pi}{(1-\rho(s))\pi + \rho(s)\pi + (1-\alpha(s))P(R > W_s)(1-\pi)}$$
$$+ \exp\{-\int_0^t r(u)\mathrm{d}u\} \cdot \exp\left\{-\int_0^t (1 - M_W(-\theta)q(x))\lambda(x)\mathrm{d}x\right\}$$
$$\times \frac{\rho(s)\pi + (1-\alpha(s))P(R > W_s)(1-\pi)}{(1-\rho(s))\pi + \rho(s)\pi + (1-\alpha(s))P(R > sW_s)(1-\pi)},$$

and the corresponding failure rate is

$$\lambda_E(t,s) = r(t)\cdot\frac{\psi(1)\bar{F}_1(t)}{\sum_{i=1}^{2}\psi(i)\bar{F}_i(t)} + [r(t) + (1 - M_W(-\theta)q(t))\lambda(t)]\cdot\frac{\psi(2)\bar{F}_2(t)}{\sum_{i=1}^{2}\psi(i)\bar{F}_i(t)},$$

where

$$\psi(1) \equiv \frac{(1-\rho(s))\pi}{(1-\rho(s))\pi + \rho(s)\pi + (1-\alpha(s))P(R > W_s)(1-\pi)},$$

and

$$\psi(2) \equiv \frac{\rho(s)\pi + (1-\alpha(s))P(R > W_s)(1-\pi)}{(1-\rho(s))\pi + \rho(s)\pi + (1-\alpha(s))P(R > W_s)(1-\pi)}.$$

Proof Observe that there are formally three subpopulations after the ESS and we can define the corresponding frailty variable Z: (i) the subpopulation with nondefective items ($Z = 1$); (ii) the subpopulation with defective items which were originally nondefective ($Z = 2$); (iii) the subpopulation with defective items which have survived the ESS ($Z = 3$). Then, in accordance with our notation, the distribution of Z is given by

$$\pi(1) \equiv P(Z=1) = \frac{(1-\rho(s))\pi}{(1-\rho(s))\pi + \rho(s)\pi + (1-\alpha(s))P(R > W_s)(1-\pi)},$$

$$\pi(2) \equiv P(Z=2) = \frac{\rho(s)\pi}{(1-\rho(s))\pi + \rho(s)\pi + (1-\alpha(s))P(R > W_s)(1-\pi)},$$

$$\pi(3) \equiv P(Z=3) = \frac{(1-\alpha(s))P(R > W_s)(1-\pi)}{(1-\rho(s))\pi + \rho(s)\pi + (1-\alpha(s))P(R > W_s)(1-\pi)},$$

On the other hand, in field use,

$$\bar{F}_1(t) \equiv P(T_{ESS} > t|Z=1) = \exp\{-\int_0^t r(u)\mathrm{d}u\},$$

$$\bar{F}_2(t) \equiv P(T_{ESS} > t|Z=2) = \exp\{-\int_0^t r(u)\mathrm{d}u\}\cdot\exp\left\{-\int_0^t (1 - M_W(-\theta)q(x))\lambda(x)\mathrm{d}x\right\},$$

$$\bar{F}_3(t) \equiv P(T_{ESS} > t|Z=3) = \exp\{-\int_0^t r(u)\mathrm{d}u\}\cdot\exp\left\{-\int_0^t (1 - M_W(-\theta)q(x))\lambda(x)\mathrm{d}x\right\}.$$

Therefore, although there formally exist three subpopulations after the ESS, due to the exponentially distributed boundary, we actually have two subpopulations. Based on the above results, the population survival function in field use after the ESS with magnitude s is given by the following mixture

$$\begin{aligned}\bar{F}_E(t,s) = P(T_{ESS} > t) &= \sum_{i=1}^{2} \psi(i)\bar{F}_i(t) \\ &= \exp\{-\int_0^t r(u)\mathrm{d}u\} \cdot \frac{(1-\rho(s))\pi}{(1-\rho(s))\pi + \rho(s)\pi + (1-\alpha(s))P(R > W_s)(1-\pi)} \\ &+ \exp\{-\int_0^t r(u)\mathrm{d}u\} \cdot \exp\left\{-\int_0^t (1 - M_W(-\theta)q(x))\lambda(x)\mathrm{d}x\right\} \\ &\times \frac{\rho(s)\pi + (1-\alpha(s))P(R > W_s)(1-\pi)}{(1-\rho(s))\pi + \rho(s)\pi + (1-\alpha(s))P(R > W_s)(1-\pi)},\end{aligned}$$

where

$$\psi(1) \equiv \frac{(1-\rho(s))\pi}{(1-\rho(s))\pi + \rho(s)\pi + (1-\alpha(s))P(R > W_s)(1-\pi)},$$

and

$$\psi(2) \equiv \frac{\rho(s)\pi + (1-\alpha(s))P(R > W_s)(1-\pi)}{(1-\rho(s))\pi + \rho(s)\pi + (1-\alpha(s))P(R > W_s)(1-\pi)}.$$

Then the corresponding failure rate is

$$\begin{aligned}&\lambda_E(t,s) \\ &= \frac{\sum_{i=1}^{2} \psi(i)f_i(t)}{\sum_{i=1}^{2} \pi(i)\bar{F}_i(t)} = \frac{1}{\sum_{i=1}^{2} \pi(i)\bar{F}_i(t)} \left(\psi(1)\bar{F}_1(t) \cdot \frac{f_1(t)}{\bar{F}_1(t)} + \psi(2)\bar{F}_2(t) \cdot \frac{f_2(t)}{\bar{F}_2(t)}\right) \\ &= r(t) \cdot \frac{\psi(1)\bar{F}_1(t)}{\sum_{i=1}^{2} \psi(i)\bar{F}_i(t)} + [r(t) + (1 - M_W(-\theta)q(t))\lambda(t)] \cdot \frac{\psi(2)\bar{F}_2(t)}{\sum_{i=1}^{2} \psi(i)\bar{F}_i(t)}.\end{aligned}$$

□

Therefore, due to the exponential boundary, the ESS in this case does not essentially change subpopulation distributions but only changes the subpopulation proportions.

We will discuss now the effect of the ESS on the quality of the population after the ESS by comparing $\lambda_E(t,s)$ with the failure rate without the ESS, $\lambda_T(t)$, that can be defined by Eq. (10.10). Note that as the ESS in our model can create defective items and theoretically this operation may have a negative effect on the population of items in some cases. Similar to Definitions 10.1 and 10.2:

Definition 10.3 The severity (stress) level s is said to be *inadmissible* under the failure rate function criterion if

$$\lambda_T(t) \le \lambda_E(t,s), \text{ for all } t > 0,$$

where $\lambda_T(t)$ is the failure rate which corresponds to $\bar{F}(t)$. Otherwise, the severity (stress) level s is said to be *admissible.*

Obviously, the inadmissible severity levels should not be considered in the application of the ESS. Note that the condition for 'admissible' is that $\lambda_T(t) > \lambda_E(t,s)$, for "some $t > 0$", not "for all $t > 0$". However, for obvious practical reasons we are mostly interested in the latter case. The following definition addresses this setting.

Definition 10.4 The severity (stress) level s is said to be *positively admissible* under the failure rate function criterion if

$$\lambda_T(t) > \lambda_E(t,s), \text{ for all } t > 0.$$

Theorem 10.5 *If*

$$\frac{1-\rho(s)-\pi}{(1-\pi)(1-\alpha(s))} > P(R > W_s), \tag{10.11}$$

then this severity level s is positively admissible under the failure rate function criterion. Otherwise, this severity level s is inadmissible under the failure rate function criterion.

Proof Denote for convenience, $\lambda_1(t) \equiv r(t); \lambda_2(t) \equiv r(t) + (1 - M_W(-\theta)q(t))\lambda(t)$. Clearly, we have the following failure rate ordering:

$$\lambda_1(t) < \lambda_2(t), \text{ for all } t > 0.$$

Observe that

$$\lambda_T(t) = \lambda_1(t)\cdot\frac{\pi\bar{F}_1(t)}{\pi\bar{F}_1(t)+(1-\pi)\bar{F}_2(t)} + \lambda_2(t)\cdot\frac{(1-\pi)\bar{F}_2(t)}{\pi\bar{F}_1(t)+(1-\pi)\bar{F}_2(t)},$$

and

$$\lambda_E(t,s) = \lambda_1(t)\cdot\frac{\psi(1)\bar{F}_1(t)}{\sum_{i=1}^{2}\psi(i)\bar{F}_i(t)} + \lambda_2(t)\cdot\frac{\psi(2)\bar{F}_2(t)}{\sum_{i=1}^{2}\psi(i)\bar{F}_i(t)}.$$

From this, it can be seen that both $\lambda_T(t)$ and $\lambda_E(t,s)$ are the weighted averages of $\lambda_1(t)$ and $\lambda_2(t)$ with corresponding weights, respectively. Thus, to compare $\lambda_T(t)$ and $\lambda_E(t,s)$, it is sufficient to compare the weights which corresponds to $\lambda_1(t)$, *i.e.*, if the first weight is greater, then the second one is smaller, and vice versa. Note that

$$\frac{\pi \bar{F}_1(t)}{\pi \bar{F}_1(t) + (1-\pi)\bar{F}_2(t)} = \frac{1}{1 + \frac{(1-\pi)}{\pi}\frac{\bar{F}_2(t)}{\bar{F}_1(t)}}$$

and

$$\frac{\psi(1)\bar{F}_1(t)}{\sum_{i=1}^{2}\psi(i)\bar{F}_i(t)} = \frac{1}{1 + \frac{1-\psi(1)}{\psi(1)}\frac{\bar{F}_2(t)}{\bar{F}_1(t)}}.$$

Therefore, if $\psi(1) > \pi$, i.e., if

$$\frac{(1-\rho(s))\pi}{(1-\rho(s))\pi + \rho(s)\pi + (1-\alpha(s))P(R > W_s)(1-\pi)} > \pi, \tag{10.12}$$

then $\lambda_T(t) > \lambda_E(t,s)$, for all $t > 0$. It is easy to show that the condition in (10.12) can be reduced to (10.11).

□

Remark 10.4 (i) In the ESS model considered in this section, a level s can only be positively admissible or inadmissible.

(ii) The condition (10.11) implies the admissibility/inadmissibility of the corresponding severity level under the survival function criterion, i.e., $\bar{F}(t) < \bar{F}_E(t,s)$, for all $t > 0$.

10.2.2 Optimal Severity

For further analysis, we need to describe a model for W_s as a 'function' of the shock's magnitude s. It is reasonable to assume first that if $s_1 < s_2$ then $W_{s_1} \le_{st} W_{s_2}$. Let s_b be some 'baseline severity level' (e.g., $s_b \equiv 1$), with the corresponding 'baseline distribution' of W_{s_b} denoted by $G_0(w)$. Therefore,

$$P(W_{s_b} > w) = \bar{G}_0(w), \; w \ge 0.$$

Then the assumption of the above stochastic ordering for W_s is equivalent to assuming the following accelerated life-type model [1]:

$$P(W_s > w) = \bar{G}_0(\emptyset(w,s)), \; w > 0, \tag{10.13}$$

where $\phi(w, s)$ is a function with the following properties: it is decreasing in s for each fixed w, it is increasing in w for each fixed s, $\phi(w, 0) \equiv \infty$, for all $w > 0$, $\phi(0, s) \equiv 0$, $\phi(\infty, s) \equiv \infty$, for all $s > 0$. Furthermore, clearly, $\phi(w, s_b) = w$, $w \geq 0$. Therefore, (10.13) implies that if $s_1 < s_2$ then $P(W_{s_1} > w) \leq P(W_{s_2} > w)$, for all $w \geq 0$, which is, obviously, the usual stochastic ordering.

We will consider now the problem of determining the optimal severity level (magnitude) of the ESS. Let τ be the mission time of an item in field operation. If it does not fail during this time, then the mission is considered to be successful. Thus, the probability of the mission success needs to be maximized and we should find the optimal severity level s^* that satisfies

$$\bar{F}_E(\tau, s^*) = \max_{s > 0} \bar{F}_E(\tau, s).$$

Alternatively, let $M(s)$ be the mean time to failure of an item in field operation as a function of s, i.e., $M(s) \equiv \int_0^\infty \bar{F}_E(t, s)\mathrm{d}t$. Then, the optimal severity level s^* which maximizes the mean time to failure should be obtained:

$$M(s^*) = \max_{s > 0} M(s).$$

It is clear that, for defining s^*, we can consider only the positively admissible severity class, as the other severity levels are inadmissible. The following theorem provides the upper bound for the optimal severity level that maximizes the mission success probability or mean time to failure in field usage.

Theorem 10.6 *Suppose that* $\rho(\infty) \equiv \lim_{s \to \infty} \rho(s) > (1 - \pi)$ *and let*

$$s_0 \equiv \inf_{s \geq 0} \{s : \rho(s) > (1 - \pi)\}.$$

Then the severities in (s_0, ∞) *are inadmissible. Therefore,* s_0 *is the upper bound for the optimal severity level.*

Proof From Theorem 10.5, the condition for inadmissibility is

$$\frac{1 - \rho(s) - \pi}{(1 - \pi)(1 - \alpha(s))} \leq P(R > W_s),$$

which can now be stated in detail as

$$\frac{\rho(s) - (1 - \pi)\alpha(s)}{(1 - \pi)(1 - \alpha(s))} \geq \int_0^\infty \bar{G}_0(\phi(r, s))\theta \exp\{-\theta r\} dr. \tag{10.14}$$

The inequality in (10.14) can be restated as

$$\int_0^\infty \left(\frac{\rho(s) - (1-\pi)\alpha(s)}{(1-\pi)(1-\alpha(s))} - \bar{G}_0(\phi(r,s)) \right) \cdot \theta \exp\{-\theta r\} dr \geq 0.$$

Observe that for all $r \geq 0$ and for all fixed s,

$$\frac{\rho(s) - (1-\pi)\alpha(s)}{(1-\pi)(1-\alpha(s))} - \bar{G}_0(\phi(r,s)) \geq \frac{\rho(s) - (1-\pi)\alpha(s)}{(1-\pi)(1-\alpha(s))} - 1.$$

Therefore, for a fixed s, if

$$\frac{\rho(s) - (1-\pi)\alpha(s)}{(1-\pi)(1-\alpha(s))} - 1 \geq 0,$$

or equivalently, if $\rho(s) > (1-\pi)$, then for this s the condition (10.14) is satisfied, and accordingly this s is inadmissible. Note that $\rho(s)$ is increasing and, by the assumption in the theorem, $\rho(\infty) \equiv \lim_{s\to\infty} \rho(s) > (1-\pi)$. Hence, there exists $s_0 \in (0,\infty)$ such that $s_0 \equiv \inf_{s\geq 0}\{s : \rho(s) > (1-\pi)\}$ and thus the severities in (s_0,∞) are inadmissible. Therefore, s_0 is the *upper bound* for the optimal severity. □

Remark 10.5 It would be practically reasonable to assume that $\lim_{s\to\infty} \rho(s) = 1$ and the proportion of the defective items $(1-\pi)$ is relatively small. Therefore, the condition $\rho(\infty) > (1-\pi)$ is practically satisfied in almost all cases.

Example 10.2 Suppose that $r(t) = 0.1t,\ t \geq 0,\ \lambda(t) = 1,\ t \geq 0,\ \theta = 1,\ G_0(w) = 1 - \exp\{-w\},\ w \geq 0,\ s_b = 1,\ \phi(w,s) \equiv \frac{w}{s},\ w,s > 0,\ \pi = 0.7,\ \alpha(s) = 1 - \exp\{-s\},\ s \geq 0,\ \tau = 4.0$ and

$$\rho(s) = \begin{cases} 0, & 0 \leq s < 1, \\ 1 - \exp\{-0.05(s-1)\}, & s \geq 1. \end{cases}$$

Furthermore, $p(t) = 0.1,\ t \geq 0$, and the 'failure rate' for W_i's is given by $\lambda_W(w) = 3,\ w \geq 0$. In this case, $M_W(-\theta) = 3/4$ and

$$P(R > W_s) = 1 - \int_0^\infty \exp\{-\frac{r}{s}\} \cdot \exp\{-r\} dr = \frac{1}{1+s}.$$

As $\rho(s)$ is strictly increasing, there exists a unique solution of the equation

$$\rho(s) = (1-\pi),$$

and this solution is the upper bound, which is given by $s_0 = -\{\ln(0.9)/0.05\} + 1 \approx 3.11$. Therefore, it is now sufficient to search for the optimal severity level in the interval $[0, 3.11]$., The graph of $\bar{F}_E(\tau; s)$ is given in Fig. 10.3.

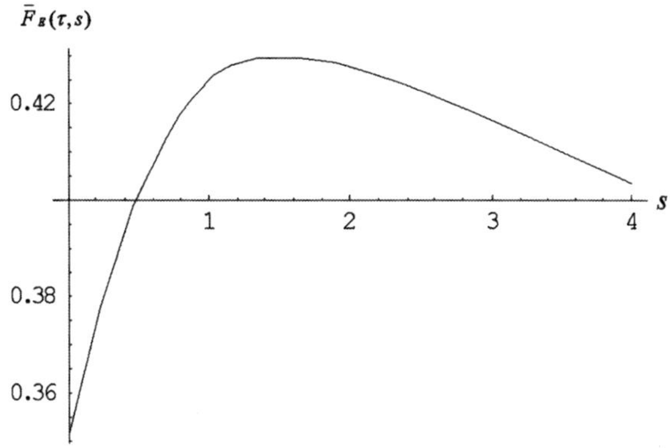

Fig. 10.3 The graph of $\bar{F}_E(\tau, s)$

The optimal severity level in this case is obtained by $s^* = 1.52$ and the maximum probability is $\bar{F}_E(\tau; s^*) \approx 0.43$.

Based on the foregoing results, we can also consider now certain cost structures for determining the optimal severity level. As previously, an item is chosen at random from our initial population and during the ESS it is exposed to a shock of magnitude s. If it survives, it is put into field operation, otherwise the failed item is discarded and a new one is chosen from the population, etc. This procedure is repeated until the first survived item is obtained. Let c_{sr} be the shop replacement cost (actually, it is the cost of a new item) and c_s be the cost for conducting the ESS. Let $c_1(s)$, as a function of s, be the expected cost for eventually obtaining a component which has survived the ESS. Then

$$c_1(s) = \frac{c_s + c_{sr}[1 - \{\pi + (1-\alpha(s))P(R > W_s)(1-\pi)\}]}{\pi + (1-\alpha(s))P(R > W_s)(1-\pi)},$$

where $1/\{\pi + (1-\alpha(s))P(R > W_s)(1-\pi)\}$ is the total number of trials until the first 'success'.

In field operation, assume that if the mission (of length τ) is successful, then a gain K is given; otherwise a penalty C is imposed, where $K > C > 0$. Then the expected gain during field operation (until failure) is given by

$$c_2(s) = -K\bar{F}_E(\tau; s) + CF_E(\tau; s) = -(K + C)\bar{F}_E(\tau; s) + C \tag{10.15}$$

and the total expected cost $c(s)$ is

$$\begin{aligned} c(s) &= c_1(s) + c_2(s) \\ &= \frac{c_s + c_{sr}[1 - \{\pi + (1-\alpha(s))P(R > W_s)(1-\pi)\}]}{\pi + (1-\alpha(s))P(R > W_s)(1-\pi)} - (K + C)\bar{F}_E(\tau; s) + C. \end{aligned}$$

The objective is to find the optimal severity level s^* that satisfies

$$s^* = \arg\min_{s\in[0,\infty]} c(s).$$

Similar to Theorem 10.5, if $\rho(\infty) \equiv \lim_{s \to \infty} \rho(s) > (1 - \pi)$ then the optimal severity level which minimizes $c_2(s)$ (maximizes $\bar{F}_E(\tau; s)$, as follows from (10.15)) does not exist in the interval (s_0, ∞). Furthermore, $c_1(s)$ is strictly increasing to infinity. Therefore, we can conclude that the optimal severity level s^* should exist in the interval $[0, s_0]$.

Assume now that during field operation, the gain is proportional to the mean time to failure. Therefore, the total average cost function in this case is

$$c(s) = \frac{c_s + c_{sr}[1 - \{\pi + (1 - \alpha(s))P(R > W_s)(1 - \pi)\}]}{\pi + (1 - \alpha(s))P(R > W_s)(1 - \pi)} - K \int_0^\infty \bar{F}_E(t; s)\mathrm{d}t.$$

By the similar arguments, the optimal severity level s^* should exist in the interval $[0, s_0]$.

References

1. Finkelstein M (2008) Failure Rate Modelling for Reliability and Risk. Springer, London
2. Fiorentino E, Saari AE (1983) Planning of production reliability stress-screening programs. IEEE Trans Reliab 32:247–252
3. Landers TL, Malstrom RM, Schmitt N, Fant E (1994) Electronics Manufacturing Processes. Prentice Hall, New Jersey
4. Yan L, English JR (1997) Economic cost modeling of environmental-stress-screening and burn-in. IEEE Trans Reliab 46:275–282

Index

A

Accelerated life model (ALM), 31, 39, 151, 155

Accelerated life test, 222

Additive hazards, 221, 230, 232

Admissible, 156, 159, 337, 369, 371, 379, 381

positively, 369, 371, 379, 380

Aging, 15, 16, 37, 51, 72, 81, 144, 152, 192, 194, 349

negative, 5, 144

positive, 5, 144

As good as new, 3, 170, 219, 329

Asymptotic failure rate, 187

Availability, 124, 210- 212

average, 210

limiting, 211, 216

point, 210

steady-state, 210, 212, 216

B

Bathtub (BT), 15, 19, 148, 150, 201, 205, 214, 217, 230, 238, 253, 261, 325, 332, 360

Modified (MBT), 147, 148, 261, 271, 274, 276

Burn-in, 1, 30, 164, 201, 208

accelerated, 6, 202, 222, 225, 231, 235, 237, 255, 256, 259, 313

optimal, 202, 206, 207, 209, 212, 216, 217, 221, 223, 230, 234, 237, 240, 244, 261, 264, 272, 280, 291, 295, 299, 302, 304, 316

time, 202, 206, 207, 211, 216, 223, 225, 230, 232, 234, 240, 242, 250, 253, 264, 269, 272, 275, 279, 285, 290, 294, 298, 309, 316, 325, 327, 360

shock, 339, 340, 342, 346, 349, 353, 356, 359

Calendar time, 38, 80

Change points, 203, 205, 208, 212, 214, 216, 221, 230, 360

Conditional intensity, 80, 134, 135, 137

Conservative measure, 294, 295, 314, 346, 348

Convolution, 23, 24, 86, 209

Copula, 41, 43

D

DFR, 5, 16, 144, 148, 203, 206, 216

Degradation, 5, 31, 51, 52, 55, 153, 189, 192, 375

Distribution

absolutely continuous, 2, 10, 13, 23, 34, 40, 106, 157, 170, 195, 222, 237, 262, 287, 299, 315, 339

baseline, 152, 153, 156, 159, 196, 352, 380

Clayton bivariate, 43

conditional, 34, 137, 172, 178, 179, 295, 346

exponential, 14, 53, 62, 64, 73, 82, 85, 108, 111, 131, 153, 154, 164, 169, 175, 183, 215, 235, 279, 284, 293, 337, 345, 375, 378

'freak', 201, 205

gamma, 57, 144, 147, 148, 150, 152, 156, 159, 162, 183, 188, 193, 196, 203, 261, 337

Gompertz, 72, 73, 151, 152, 157, 183

inverse Gaussian, 16, 54, 160, 183, 191, 194

M. Finkelstein and J. H. Cha, *Stochastic Modeling for Reliability*,
Springer Series in Reliability Engineering, DOI: 10.1007/978-1-4471-5028-2,

Distribution (*cont.*)
main, 201, 205, 291
Makeham, 72
marginal, 41, 43
mixing, 151, 156, 159, 169, 174, 295, 347
stable, 160
DLR, 168
DMRL, 16

E
Environment, 3, 31, 39, 42, 75, 80, 143, 197, 201, 222, 224, 234, 261, 291, 313 , 349, 352, 356
Environmental stress screening (ESS), 7, 313, 366, 368, 374, 380
Exponential family, 161, 162, 281
Exponential representation
bivariate, 43
univariate, 13, 15

F
Failure mode, 8, 364, 365 , 375
Failure rate, 9, 11, 14, 15, 19, 26, 32, 36, 46, 72, 82, 85, 97, 103, 110, 124, 132, 143, 146, 150, 154, 158, 163, 168, 179, 187, 195, 203, 211, 223, 232, 255, 262, 273, 283, 292, 314, 339, 345, 354, 371
bivariate, 43, 45
cumulative, 154, 155, 196, 301, 303, 308, 318
discrete, 12, 13
eventually increasing, 205, 206, 214, 225, 226, 229, 232, 309, 311, 360
mixture, 4, 6, 10, 146, 148, 151, 154, 157, 160, 162, 167, 172, 188, 204, 261, 263, 270, 272, 274, 279, 284, 287, 290, 293, 327, 340, 343
Frailty, 113, 117, 144, 148, 150, 152, 154, 157, 160, 163, 171, 173, 179, 183, 185, 187, 192, 196, 277, 285, 293, 295, 326, 329, 333, 344, 347, 367, 377
bivariate, 187, 188, 192
gamma, 144, 152, 157, 183
multiplicative, 113, 151, 183, 185

G
Gamma process, 54, 109, 112, 184, 192
Geometric process, 5, 79, 126
Glaser's theorem, 17, 18
Gompertz shift model, 76

H
Heterogeneous populations, 143, 165, 170, 176, 183, 184, 194, 196, 262, 312, 327, 333, 337

I
IFR, 5, 16, 144, 148, 195, 203, 206, 261, 273, 309
IMRL, 16
Inadmissible, 369, 370, 379, 382
Intensity process, 3, 21, 24, 25, 28, 36, 88, 95, 134, 170
Inter-arrival time, 73, 84, 88, 108, 111, 112, 115, 122
Inverse problem, 160, 161
Item
defective, 364, 365, 367, 371, 374, 377
non-defective, 364, 365, 367, 371, 374, 377
strong, 6, 164, 261, 263, 287, 289, 298, 338
weak, 164, 261, 263, 287, 288, 308, 313, 326, 339, 341
Jensen's inequality, 60

L
Laplace transform, 24, 60, 62, 64, 69, 96, 160
L'Hopital's rule, 16
Life expectancy, 13
Lifesaving, 75, 76
Loss
function, 217
mean, 289, 290, 342
point, 288, 341, 342

M
Maintenance, 25, 176, 178, 212, 216, 237, 255
policy, 216, 237, 255, 258
preventive, 176, 177, 212, 216
Marginal quality, 295, 296, 346
Mean remaining lifetime (MRL), 13, 14, 16, 18, 204, 206, 250, 261, 262, 269, 272, 275, 279, 285, 290, 347, 349
Mission time, 5, 79, 163, 182, 207, 208, 217, 219, 224, 231, 264, 267, 270, 283, 288, 304, 317, 333, 342, 371, 381
Mixture, 16, 117, 144, 145, 150, 152, 156, 158, 162, 164, 171, 179, 188, 197, 202, 235, 261, 262, 264, 269, 270, 275, 280, 283, 286, 289, 291, 297, 313, 314, 321, 337, 341, 344, 348, 375
continuous, 148, 168, 261, 276, 292, 325, 343, 346

discrete, 262, 314, 339
Mixture model
additive, 155, 160, 161
multiplicative, 155, 156, 159, 161, 197, 329, 337
Mortality plateau, 153, 157, 159, 194
Mortality rate, 11, 27, 72, 74, 76, 153, 159, 195

N
Non-repairable systems, 176, 218

O
Ordering
failure (hazard) rate, 47, 143, 144, 184, 187, 191, 192, 344, 366
likelihood ratio, 47, 168, 327, 333, 345
stochastic, 46, 168, 184, 186, 192, 194, 198, 262, 265, 299, 314, 329, 380

P
Parallel system, 153
Point process, 19, 20, 28, 30, 34, 36, 52, 55, 59, 73, 79, 81, 88, 105, 119, 120, 126, 130, 133, 139, 172, 364
Poisson process
compound, 55
nonhomogeneous (NHPP), 21, 22, 28, 30, 59, 68, 79, 80, 82, 91, 106, 108, 113, 117, 126, 129, 134, 137, 140, 301, 305, 350, 356, 365, 375
spatial, 69
thinning of, 22, 116, 133, 134
Probability of termination, 61, 79, 88
Proportional hazards (PH), 39, 76, 150, 230, 278, 283, 299

R
Random
boundary, 85, 109, 131, 132, 375
environment, 3
failure rate, 3, 164, 165
variable, 3, 9, 12, 13, 16, 17, 22, 28, 38, 45, 47, 52, 54, 58, 70, 81, 87, 90, 93, 99, 104, 106, 113, 117, 121, 124, 131, 136, 143, 163, 187, 192, 194, 244, 251, 264, 299, 317, 340, 350, 366
Regime, 30, 33, 41, 42, 45
Regularly varying function, 158, 160
Renewal density, 23, 25, 34
Renewal equation, 35
Renewal process, 23, 25, 27, 34, 37, 59, 62, 106, 170, 209, 323
g-, 34, 35, 37
terminating, 59
Renewal reward theory, 324
Repair, 3, 9, 23, 26, 30, 34, 36, 61, 65, 68, 103, 114, 124, 170, 176, 201, 210, 220, 224, 224, 231, 237, 241, 244, 299, 304
fast, 65, 66, 68, 125
general, 6, 9, 33, 36
imperfect, 25, 27, 30, 34, 36
instantaneous, 23, 170
minimal, 26, 30, 34, 35, 170, 172, 174, 176, 212, 239, 241, 244, 252, 257, 267, 300, 301, 304
perfect, 3, 27, 30, 34
Repair process, 38, 176
Repairable system, 3, 25, 33, 61, 65, 125, 170, 176, 216, 237
Risk, 2, 11, 39, 40, 45, 56, 58, 163, 165, 187, 262, 286, 288, 296, 298, 314, 320, 338, 342, 349
analysis, 2, 13, 144
weighted, 300, 301, 320

S
Safety, 52, 58, 63, 70
at sea, 51, 69, 70
Scale parameter, 54, 147, 151 , 189, 191, 194, 196, 218
Series system, 39, 41, 44, 84, 375
Severity, 93, 313, 316, 318, 319, 321, 324, 329, 331, 335, 338, 341, 345, 348, 352, 356, 371, 374, 379, 380, 384
optimal, 313, 316, 318, 319, 321, 324, 329, 333, 335, 338, 340, 345, 347, 352, 355, 371, 374, 380, 384
Shape of
failure rate, 205, 207
Shape parameter, 54, 151, 189, 193
Shocks
cumulative, 4, 5, 52, 109, 132
extreme, 4, 5, 59, 79, 119
fatal, 101, 111
non-fatal, 101, 111
Strehler–Mildvan model, 51
Stress-strength model, 117, 349, 364
Subpopulations
strong, 164, 287
weak, 163, 164, 287, 339

Survival function, 12, 13, 25, 26, 39, 41, 45, 59, 82, 86, 89, 97, 101, 106, 110, 112, 117, 130, 145, 148, 161, 164, 206, 209, 211, 238, 264, 278, 287, 294, 316, 328, 330, 339, 345, 352, 357, 365, 368, 375, 380
Spare parts, 36, 37
Spatial survival, 69

U
Upside-down bathtub (UBT), 16, 18, 19, 262, 270, 280, 285

V
Variance, 17, 54, 57, 129, 150, 163, 166, 169, 185, 190, 197
Virtual age, 30
  recalculated, 33
  statistical, 31, 32
Vitality, 73, 74, 134, 183, 184, 187, 189, 190, 192, 194

W
Waiting time, 55, 350
Weakest populations dying out first
  principle, 149, 186, 188, 363
Wear, 16, 31, 51, 54, 80, 84, 91, 95, 108, 111, 132, 203, 205, 225, 309, 374, 375
Wear-out points, 205, 309
Wiener process, 52, 54, 183, 184, 189, 190

CPSIA information can be obtained at www.ICGtesting.com
Printed in the USA
LVOW071443250413

330966LV00001B/1/P

9 781447 150275